Politics and Class in Milan

Politics and
Class in Milan
1881–1901

LOUISE A. TILLY

New York   Oxford
OXFORD UNIVERSITY PRESS
1992

Oxford University Press

Oxford   New York   Toronto
Delhi   Bombay   Calcutta   Madras   Karachi
Petaling Jaya   Singapore   Hong Kong   Tokyo
Nairobi   Dar es Salaam   Cape Town
Melbourne   Auckland

and associated companies in
Berlin   Ibadan

Library of Congress Cataloging-in-Publication Data
Tilly, Louise.
Politics and class in Milan, 1881–1901 / Louise A. Tilly.
p. cm.   Includes bibliographical references and index.
ISBN 0-19-506683-9
1. Working class—Italy—Milan Region—History—19th century.
2. Working class—Italy—Milan Region—Political activity—
History—19th century.
3. Labor movement—Italy—Milan Region—
History—19th century.   I. Title.
HD8490.M52T55   1992   335.5'62'09452109034—dc20
91-15572

1 3 5 7 9 8 6 4 2

Printed in the United States of America
on acid-free paper

Preface

This study of the large-scale structural processes and political struggles that shaped working-class formation in Milan, Lombardy, and Italy at the end of the nineteenth century has been in the making for more than twenty years.

It all began with a term paper written in the spring term of 1966 at Harvard University, when David Landes graciously permitted me to audit his seminar for entering graduate students. The topic that year was "the dark side of industrialization." In Widener Library, I discovered rich collections of the publications of the city of Milan (statistical yearbooks, the minutes of the city council) and of the Società Umanitaria (a Milanese institution which conducted many social surveys and investigations in the first years of this century), and was able to write a well-documented account of workers' conditions in the city. The next fall, I entered the Ph.D. program at the University of Toronto, and began my study of modern European (mostly French and Italian) history. By the fall of 1973 I had completed, under the direction of the late Robert Harney, a doctoral dissertation on Milan's working class which was largely economic and demographic history.

But before the dissertation was even completed, Natalie Zemon Davis, one of my professors at Toronto, had suggested that I write a paper for the American Historical Association conference in December 1972, focusing on women workers in Milan. This paper was presented at one of the first panels on women's history at an American Historical Association convention. After I finished the dissertation, I turned my attention to women's history and began research on women, work, and family in France and England. Although most of my effort went into this exciting new area of research, I continued to visit Milan and explore its Archivio di Stato and libraries for material on working-class politics and collective action. My goal was to combine the economic and demographic history I had already written with an equally thorough examination of local politics.

Teaching and family obligations and the women's history research project kept me from giving much time to writing about Milan for many years, although my shelves and files bulged with notes, books, and journal articles which I continued to accumulate. Two chapters were written in the spring of 1983, but competing projects continued to divert me until fall 1987, when I became a visiting member of the Institute for Advanced Study in

Princeton for the academic year. The opportunity was there, and I managed
to write what I believed was one chapter short of a full draft by August
1988. From then until the fall of 1990, I tinkered and wrote three more
chapters. And here it is.

Despite my long-term commitment to women's history, readers will find
relatively little of it here. My ambition was to integrate questions about
women into a history of working-class formation, but I was severely limited
by lack of good sources. My study deliberately sticks close to structures
and events and the people who shaped them and were in turn shaped by
them. Readers will find no post-structuralist interpretation here but plenty
of local economic, social and political history in which actors have been
given a central place. This history has also been connected to larger questions about working-class formation in Italy and in comparative perspective.

With a gestation period as long as the one just described, I have acquired quite a few debts of gratitude, starting with David Landes, Robert
Harney, and Natalie Zemon Davis, my academic mentors and friends. Clo-
tilde Treves, Luciana Ghizzoni Kinukawa, and Guy Baldwin were efficient
and valued research assistants, and Maria Springer tangled with the manuscript and several computer programs for many hours.

Some of the institutions which have helped me are the Einaudi Foundation, which awarded me a research grant, the Russell Sage Foundation,
which supported me for two months in New York in spring 1983, and the
Institute for Advanced Study, which granted me a fellowship for 1987–88.
Librarians and archivists at the Istituto Giangiacomo Feltrinelli, the Archivio di Stato di Milano, the Biblioteca Nazionale di Brera, Milan's city
library, the New York Public Library, Widener Library, and the libraries
of the University of Toronto, the University of Michigan, Columbia University, Princeton University, and New York University aided me with end-
less bibliographic quests and interlibrary loans.

Colleagues Jan de Vries at Michigan State, Jacob Price and Raymond
Grew at the University of Michigan, Ira Katznelson and Aristide Zolberg
at the New School of Social Research, and members of the Columbia University Seminar on Italian Studies, especially Emiliana Pasca Noether, one
of my undergraduate teachers, offered helpful comments on papers and
chapters presented to them. At the Institute for Advanced Study, Stefano
Fenoaltea, Yasmine Ergas, and Joan Scott (who read the entire 1988 draft)
provided valuable suggestions for revision. I also welcomed advice and
assistance from Franco Andreucci and Patrizia Dogliani. Thanks also to the
anonymous readers for Oxford University Press who gave me important
feedback and recommended what became the directions of the final revision. Suzanne Fujioka drew the figures efficiently and with grace. My sister, Laura Audino, gave me writing advice and moral support as well.

Coming closer to home, my children Chris and Kit, and their friend

Paul Filson did coding for my dissertation in 1972. Chris, Kit, and their sister Laura helped get my graphs into shape to be photocopied in the desperate fall of 1973 as I taught at Michigan State University, involving a 120 mile round-trip commute, while preparing my doctoral dissertation for its defense. In later years it was youngest daughter Sarah who photocopied articles and checked references for the book manuscript. Chuck Tilly encouraged and supported my return to graduate school in 1966, taking on child care responsibilities in the summer of 1968 in Milan, typing the tables for my dissertation, and reading and rereading the evolving book. His sociological imagination—always tempered by his systematic and empirical bent—inspired me to approach the processes and events in this book as social history, or historical sociology. To all Tillys, my thanks and love.

I dedicate this book to the memory of my parents, Piera Roffino and Hector Audino. Born in northern Italy at the beginning of this century, they came as children with their families to the United States by different paths to meet and marry in New York. Their becoming Americans contributed to my addressing the questions about Milan and Italy posed in this book comparatively, based on knowledge of other histories and places. Their encouragement of my curiosity about the past and support for my academic ambitions set me on the path of becoming a historian. The love of learning that they nurtured in me is the root from which this book grew.

New York   L.A.T.
May 1991

Contents

Politics and Class in Milan

1

Italy's Working Class and Its Analysts

In 1886, Andrea Costa, the Romagnol who was the first socialist elected to the Italian Parliament, offered to his fellow deputies an account of working-class formation in Italy. Speaking in defense of the Lombardy-based Partito operaio, whose leaders had recently been arrested and whose newspaper had been muzzled, he told them:

> The rise of the Partito operaio in Italy is not a random occurrence or an individual's caprice. The Partito operaio is a natural product of our society, of our economic and social conditions. These include the economic forces that continue to concentrate the means of production in a few hands, distance the worker more and more from his tools, and separate the class that owns the means of production more and more from the class that puts them to work, putting the two into distinct and opposing classes and rendering class antagonism ever more acute. . . . The Partito operaio is likewise a product of our political conditions that have provided, through national electoral reform, the means by which a good part of the working class, and certainly the more intelligent, can affirm itself as a class apart, as a party that is distinct and separate from the others. . . . Where did this Partito operaio emerge? Not in the Romagna, the southern provinces, or Sicily. It rose in Lombardy, in Milan, where modern industry has penetrated more than elsewhere, where the political and moral level of the working class is certainly higher than in other regions, and where the workers themselves can send their representatives to Parliament without choosing them from other social classes. And when? In 1882, that is, shortly after the passage of electoral reform legislation.[1]

A century ago, Andrea Costa's speech posed historical and theoretical problems that remain unresolved even today. To what extent did (and does) the growth of large-scale industry cause the formation of a connected and self-aware working class—in Lombardy or elsewhere? Under what conditions and how strongly did the variation in origins, living conditions, relations to the organization of production, and ideological dispositions within the working class foster corresponding variations in political position and collective ac-

3

tion? How and how much did the action of the state, of political parties, and of other organizations shape the formation and transformation of the Lombard and Italian working classes? This book addresses such questions through a close examination of changing economic structure and worker experience in a major industrial city and its hinterland, relating those changes to worker organization and collective action in both local and national politics.

By 1886, an irreversible social and economic transformation had begun in Milan, visible in the growth of factory production; the concentration of capital; the massive movement of workers to the city; and the creation of an urban working-class way of life. In 1892, a Socialist party was founded on a national level. Six years later, Milan experienced what has seemed to many contemporaries and historians a classic workers' rebellion—the Fatti di maggio—signaling a high degree of worker militancy in Italy's industrial metropolis. Contemporary observers and historians alike have believed with Costa that Milan, Lombardy, and Italy were undergoing a natural succession from industrial concentration and proletarianization to class consciousness and the political organization of a unified, powerful working class.

Costa's analysis echoes the theoretical process that Marx described: "Economic conditions had first transformed the mass of the people . . . into workers . . . ," he wrote, "capital has created for this mass a common situation and common interests. This mass thus is already a class, as against capital, but not yet for itself." As Costa did, Marx paired his analysis of economic change with a discussion of politics: "In the struggle [between the classes] this mass becomes united and constitutes itself as a class for itself. The interests it defends become class interests. But the struggle of class against class is a political struggle."[2] A working class born of economic change, then, becomes a class for itself when it establishes political institutions through and for class struggle. The critical component of working-class formation, in Marx's account, is the creation of a class party.

Although (or perhaps because) the Partito operaio that Costa discussed was a genuine worker-run party, separate from and opposed to bourgeois parties, it never achieved national status. Further, it never assigned to politics more than a subordinate role, under those of workplace organization and mobilization. Although Costa was wrong in identifying the creation in 1882 of the Partito operaio as a sign of Italian class formation on the national level, his analysis struck a note in common with Marx and with later historians.

Following Marx and Costa, the history of working-class formation in Italy was first written as the history of workers' political institutions (congresses, federations, parties) and their collective action, assuming an unproblematic relationship to economic change as its base.

The Italian Debate on Class Formation

Gastone Manacorda's institutional history (first published in 1953) examines in chronological order the interregional or national congresses of worker associations in order to document his evolutionary scheme, opening with a preunification gathering of workers' fraternal societies and concluding with the 1892 congress that established the Socialist party.[3] It assumes that just as there is a single national state, so also there is a single national working class that, through a single process of economic change and political negotiation, reaches a common awareness of its situation. This common awareness is then assumed to be a prerequisite of effective action at the national scale. It further assumes that the industrialization of textile production in the 1870s and 1880s was the economic change that launched the national process. Ernesto Ragionieri's study of a Florentine suburb and its Socialist municipal government, likewise published in 1953, concluded that the Socialists and those who voted them into office were unable to pursue a vigorous program because they were operating in too-limited an arena. Contrary to the Reformist Socialists' views in the period before World War I, Ragionieri continued, winning local elections was not enough, and in order to understand that period, historians had to examine the development of the workers' movement in the entire national "society." This he sought to do.

In the 1960s, new research on economic development and the occupational structure of the working class challenged Manacorda's assumptions. Economic historians like Alexander Gerschenkron and Luciano Cafagna identified the years immediately before and after 1900 as the beginning of transformative growth in the engineering and metal industries, which moved the entire national economy toward large-scale industrial production. Giuliano Procacci and other social historians noted both the heterogeneity of the Italian occupational structure revealed in the 1901 census and the small number of Italian workers working in large-scale industry. The assumption that a national industrial transformation had preceded the birth of a Socialist party was no longer tenable.[4]

Manacorda and other Italian scholars accordingly moved away from the materialist evolutionary explanation they had posited and toward an idealist one, in which ideas imported from more industrialized European nations where Socialist parties were being established at that time were said to interact with urban corporate and *risorgimento*-linked democratic traditions in Italy. The result, this revised account continued, was a dual set of class institutions: On the one hand, a Socialist party associated with the Second International and active in national-level politics and, on the other hand, the chambers of labor—class-based urban and regional institutions promoting workers' interests on the local level—another "natural" outcome, given the conditions prevailing in Italy.[5]

In 1968 Stefano Merli decisively rejected this view, with its idealist and Gramscian elements, insisting that immiseration was more critical to class formation than were the structural changes—a new mode of production, proletarianized factory workers—that earlier historians had identified as central. "According to Engels," Merli wrote, "misery and the factory lie at the origin of the modern proletariat."[6] Behind immiseration lay the growth of the factory system, the development of factory discipline, and a new division of labor. Factories spawned an industrial proletariat in the two decades before 1900. This factory proletariat—uprooted, immiserated, and atomized—in turn "invented the organization of resistance against capital and created the forms of class power."[7] Merli perceives no anachronism or contradiction in his account. He ignores the historical evidence that had given his colleagues pause and simply asserts that the workers' struggles of the last decades of the nineteenth century were those of a factory proletariat—defined very broadly to be sure, so that it encompasses construction workers and other heterogeneous groups that troubled other historians. In the course of struggle with their bosses, in Merli's account, factory workers organized themselves into resistance leagues, and they established urban-based chambers of labor to help manage the unemployment crisis brought on by the cyclical depression in the years around 1890. Factory proletarians broke earlier patterns of political clientage to bourgeois Democratic Radicals and Republicans and rejected electoral collaboration with them. Thus, Merli concludes, spurred by the spread of the factory and its accompanying misery, a class made itself. But because of their failure to guide the potential revolution, its leaders betrayed it.

Merli's factory proletarianization interpretation did not go unchallenged. His claims were rejected by many: that on a national scale, factory workers were an important component of the labor force and that in the period before 1900 they were the founders of worker institutions and leaders of worker collective action.[8]

In an article taking Merli's study as a foil, Andreina De Clementi offers a "cyclical" alternative to the institutional and factory proletarianization perspectives on Italian working-class formation. Although she uses the evidence that Merli gathered, she severely criticizes his use of theory and shows that neither Engels nor Lenin had made so direct or mechanistic a connection between factory production and class formation as did Merli.[9] De Clementi hypothesizes that the key to the class formation process was the disintegration of the peasantry under the capitalist penetration of agriculture. In spelling out how this worked, she specifies and differentiates "workers," taking into account how the organization of production was changing in many sectors. Once peasant agriculture decayed and rural-to-urban migration brought workers to the cities, they joined one of three groups: (1) "worker-artisans" undergoing proletarianization and "de-skilling" because of mechanization and job competition from rural migrants, (2) urban home workers, or (3) unskilled workers.[10]

The 1870s and 1880s, a period of little industrial strife, were a "20-year truce," according to De Clementi. Uprooted rural migrants in cities had improved their economic condition, often only minimally, over the rural misery that they had left behind. They formed an important section of the urban working class that did not share the urban artisan-workers' interests or press its own interests through collective action. But the sharp cyclical business downturn around 1890 changed this. Worker layoffs, wage cuts, and other attempts by employers to roll back labor costs affected all sections of the labor force. There was a jump in the number of strikes involving previously inactive groups (women and children, rural agricultural workers, textile workers, and rural-to-urban migrants). New workers' institutions were founded. De Clementi also identifies the chambers of labor as significant innovations in the period; their function as employment agencies was an attempt by the workers to control labor markets.[11] The capitalist reaction to the 1888–94 cyclical downturn threatened all Italian workers, and they seized the opportunity to define their common interests and "make" themselves into a class. De Clementi argues that despite later setbacks, unskilled workers had undergone an irreversible process of acquiring a consciousness of common interest with other workers, and that from then on, they shared the other workers' sense of class antagonism.

De Clementi's cyclical interpretation differentiates among workers by industry, rural or urban origin, age and sex, which is a valuable step toward disaggregation. Her interpretation concludes that class formation occurred at a unique historical moment of economic crisis that drove diverse workers to unite.[12] But before we can accept the notion that once acquired, class formation and consciousness are permanent, evidence for a longer period is needed.

The Marxist institutional, factory proletarianization, and cyclical interpretations of Italian working-class formation, despite each one's self-definition in opposition to the others, share several characteristics. First, all four explanations are conceptualized at a national level; they are based on highly geographically aggregated data or use examples from one area, city, or industry in unsystematic ways to make their case for the whole of Italy. Second, all four offer linear explanations. Manacorda's Marxist institutional interpretation is the most mechanistic, echoing Marx and Andrea Costa. The revised structural backwardness approach that Procacci and, later, Manacorda adopted argues that because class formation preceded rapid industrial growth, it cannot be understood in strictly economic terms but must have some organic connection with both urban Italian tradition and ideology imported from abroad. Merli reiterates a reductionist and circular Marxist argument that factory production and immiseration are both necessary and sufficient causes of class formation. Because class formation occurred in the 1890s (i.e., workers [and others] built institutions like resistance leagues, chambers of labor, and the Socialist party and acted collectively on their interests), then the Italian workers who built these class organizations must be factory workers. Although De Clementi sees class formation as a consequence of the changing relationships among sections of the labor force, she attributes to one cyclical downturn a once-and-for-all birth of a class conscious of its interests.

Another by-product of Merli's broadly brushed canvas was a turn by historians to studies of workers' experience rather than institutions; new projects, focused on localities and often inspired by English and American social history, were undertaken and their findings reported.

Giuseppe Berta, for example, took a close look at the artisanal hand weavers of the valleys around Biella (Piedmont) and their struggle to slow down the introduction of factories and power looms. Although the artisans' fight against proletarianization was long and hard, Berta concedes that once the textile factory system was in place, unionization was often "slow and late."[13] As he points out, the transition from household production on hand looms by male weavers to factory production and power looms operated by women and girls resulted in the demobilization of unionized male workers and the substitution of more difficult-to-organize females. Comparing the textile with the engineering industry (located in Turin and Milan), Berta finds that in the latter, workers were less militant in fighting the capitalist reorganization of production in the nineteenth century because they were able to maintain claims to skill and craft. Disputes with capitalist concerns came after the unions based on these claims were well established—in the first decade of the twentieth century—and were better organized and harder fought than were those in textiles.

In 1979, Berta contributed a chapter to a collective history of the workers' movement in Piedmont edited by Aldo Agosti and Gian Mario Bravo.[14] Here he again compares Turin's engineering and Biella's textile industries, but in greater detail. He concludes, echoing E. P. Thompson as well as French and American historians, that class organization often begins as a defense against proletarianization. The Biellese textile workers' organization was horizontal and based on region, whereas auto and engineering workers' organizations were vertical and based on craft or skill. Members of the latter were easier to convert to class-based organizations than were members of the textile workers' unions, linked as they were to a specific region by ties of kin, community, and, sometimes, property holding. Here again, the gender component of some of these differences are noted, but they do not play an integral part in the analysis. Franco Ramella's local study of the Biellese woolen weavers' experience of proletarianization, in contrast, looks at gender relations within households and factories. Berta examines the contrasting capacity for collective action of these groups and discovers that the struggles were cyclical: The regional industries were out of phase with one another.[15] Rich and suggestive as Berta's essays and other local studies are, however, none of them theorizes or studies empirically the relation between the local and national histories of class formation, as Ragionieri contended was needed.

Each position in this greatly simplified version of the scholarly debate over Italian working-class formation has come under fire. The national-level institutional interpretation has been undermined by economic historical analysis of the timing and character of Italian industrialization and social structure. Merli's view that a factory-formed proletariat became a self-conscious revolutionary force through the straightforward modification of the workplace has likewise been challenged by smaller-scale studies that examine the mobilization of specific worker groups and their collective action. The local studies reveal the varying experience of different groups of workers but do not connect this experience to the national process of class formation.

Franco Andreucci and Gabriele Turi, editors of *Passato e presente*, a recently launched journal, asked rhetorically in 1986 whether Italian working-class history had been "ghettoized." They called for the resumption of work on the big questions on this field, new inputs of energy, and ideas building on studies done in other national settings.[16] In response to that call, the rest of this chapter lays out a rationale and theory for a different kind of local study that links changes in that arena to national politics.

Definitions

The simple definition of *class* used in this book is "persons who share a common relationship to the means of production." The *working class,* then, is composed of those who own no capital and must sell their labor power in order to live. The process by means of which more and more persons enter this relationship is *proletarianization.* The next important definition is that of class formation.

A good point of departure for any study of working-class formation is the recently published cross-national comparative study edited by Ira Katznelson and Aristide Zolberg. In his introduction Katznelson examines contemporary theory regarding class formation and the outpouring of histories that followed the publication of E. P. Thompson's *Making of the English Working Class.* In accord with Thompson, Katznelson sees the most "extreme formulation"—class formation as the acquisition of revolutionary consciousness following the emergence of a new class position in the structure of production—as a "political deformation[s] . . . distinguishing as it does between correct and incorrect ways of acting." [17] Although he agrees with Thompson that the elimination of the human agent in history is reductionist, Katznelson nevertheless concludes that the "new working-class history [that Thompson has inspired] has adopted a weak version of the structural 'class in itself—for itself' model of class formation as a hidden and unexamined functioning tool to order the multitude of facts generated by the study of working-class activity and culture." Thompson's richly documented narrative method, with its implicit theory and unformulated questions, successfully illuminates the formation of the English working-class. Nevertheless, Katznelson argues, it also preserves an essentialist perspective in seeing workers as having within themselves, once external material conditions change, the potential of their formation into a class "for itself." [18]

In a meticulous effort to specify their subject, Katznelson and his colleagues first define "class in capitalist societies" as a concept with "four connected layers of theory and history: those of [class] structures, [class-based] ways of life, dispositions, and collective action." This conceptualization undergirds the chapters comparing working-class formation in early industrialization and the mature industrial economy of three nations: France, the United States, and Germany. Three clusters of hypotheses about relationships among and between the "layers" are tested: The first is economy centered and looks internally at capitalist and class development; the second is society centered and seeks the causes for variation in patterns of proletarianization outside the economy, in the links between the layers of class; and the third and "most macrocausal" hypotheses concern the state. Although explored separately, the boundaries between these explanations are not always clear. And Katznelson and Zolberg do not define class formation. [19]

This extraordinarily thoughtful and clear-eyed comparative approach is adopted to varying degrees by Katznelson's coauthors, whose chapters compare national working-class histories and are explicit in posing their problem, comparative in deploying their evidence, and consistently stimulating in their interpretation. Specification of the layers of class and their examination through social and political, as well as economic, lenses introduces greater variation in the process, but it does not offer an alternative to the class in itself—for itself formulation.

The volume's concluding chapter, by Aristide Zolberg, compares the findings of the national studies and lays out some important conclusions. [20] Zolberg notes, first, that the degree of "classness" in either the rank and file or the leadership of the workers' movements is not easily conceptualized or compared. Further, much depends on the historical moment at which the comparison is made. He argues that although the formation of a working class was begun in Britain, France, Germany, and the United States under different political, economic, and ideological conditions and the outcomes seem very different today, there were many signs of convergence in the period immediately before World War I and the Russian Revolution. The economic consequences of the war and the political discontinuities that characterized some countries' experience seem to have led to a resumption of particularistic paths. World historical events like wars have frequently been interpreted as wrenching developments out of their old course and opening up new directions. World War I seems to have had such an effect on the national processes of working-class formation.

Second, Zolberg emphasizes specific economic and political variables as critical to shaping the characteristics of the class formation process and the working class that was its outcome. It is the structure of the economy, he writes, not the timing or the pace of economic growth that matters; the proportion of the labor force in industry, and its capital intensiveness are the key indicators of economic structure that he mentions. The important political variables he identifies are the character of the state and the character of the regime, or "stateness," as he labels it. Again there are two indicators: centralization and the institutional separation of administrative and legal structures. He concludes that "the single most important determinant of variation in the patterns in working-class politics . . . is simply whether, at the time this class was being brought into being by the development of capitalism . . . , it faced an absolutist or a liberal state."[21] Zolberg's analysis offers a useful cross-national comparative framework for the Milanese and Italian cases, and I shall refer to this framework systematically in this study in order to situate it in a broader context.[22]

Looking at Milan, Lombardy, and Italy with the Katznelson–Zolberg formulation in mind reveals several weaknesses in it. First is the curious lack of any definition of class formation. Second is the unexamined relationship of local (urban, regional) and national arenas in terms of economic change, worker interaction with the state, institution building, mobilization, and collective action. Under what conditions do local movements based on regional economic change aspire to and achieve (through coalition or other means) political standing at the national-state level?

The first weakness can be addressed through further definition: *Class formation* is the process(es) by which any class (defined as a group sharing relations of production) increases its capacity for collective action. As E. P. Thompson points out (and Katznelson agrees), collective action based on class interests may precede as well as be a consequence of institution building, or the relationship between them may go both ways.[23] But another concept is needed. *Class transformation* is the ongoing process in which changing economic and political relationships in both region and nation affect a class, altering and even destroying its ability to maintain its institutions, mobilize, and act collectively.

In short, class formation is the process—period and place specific, not "natural" or unmediated—in which some version of a class for itself evolves. (Katznelson might object that his approach is designed to avoid the reductionism remaining in this formulation. Nevertheless, that approach does not afford an alternative to the class in itself–class for itself formulation but, rather, a more complex perspective on the components of the process and a strong reservation about its bipolarity.) Class transformation theorizes working-class formation to be contingent, as it can be shown to be, both empirically and historically. It posits class consequences from continuing economic structural change and political struggle. It avoids a before-and-after linear explanation, but it runs the risk of indeterminacy. I believe that the result—a conceptualization that better fits the historical record—is worth the risk.

In the period up to World War I, the ebb and flow of economic development, proletarianization, and political response continued in Milan, Lombardy, and Italy. Milan became, in some respects, less of an industrial city, resembling Lyons—an even more extreme case—about which Yves Lequin writes that there was a "tendency to crumbling sectors," rather than the classic mode of "the construction of a class with industrialization." He continues; "The industrial proletariat was hardly visible before it began to dissolve into salaried service workers."[24] In the Milanese case, the 1911 census confirms that industries and their workers had moved out of the city, contributing to the growth of nearby suburbs (which also received in-migration from more distant areas); within Milan, the tertiary sector had expanded. The majority of Milan's workers continued to be employed in manufacturing, however, through World War I. At the same time, craft workers became proportionately less important, and factory workers and those in transportation and other services became more important. New groups of urban and rural proletarians organized themselves; some of the older unions amalgamated into national-level federations; new institutional leadership and ideologies emerged; collective action surged and receded; and Socialist party majorities shifted between reformers and radicals.

If we take history seriously, class transformation must be the object of analysis as well as class formation. Neither is, to borrow Philip Abrams's words, "a matter of imposing grand schemes of evolutionary development on the relationship of the past to the present," or simply supplying historical background for present conditions. Both must be understood in terms of the "relationship of personal activity and experience on the one hand and social organisation on the other as something that is continuously constructed in time."[25] Both working-class formation and class transformation are social and political processes that take time to unfold. They are path dependent to the extent that the strategies chosen or outcomes imposed at one time close off alternatives and constrain or facilitate specific outcomes at a second time.[26]

The lack of a specific relationship between local and national arenas of class formation, the second weakness identified in the Katznelson–Zolberg framework, suggests the need for systematic disaggregation, in addition to focus on the national level.

Working-Class Formation: City, Region, and Nation

A process like working-class formation involves changes in relationships that, when conceptualized in linear form at the national level, are overschematized and distorted. Although national indices of industrial production, social structure, or other socioeconomic measures are useful for temporal or cross-national comparative purposes, they are abstractions that miss the texture of reality. Such national data describe precisely and uniformly the characteristics of long trends of economic and social change, but they are not very useful in explaining them. In the case of economic change, in particular, regional disaggregation is critical.

Industrialization does not happen to whole countries. Nor does it happen to individual communities, even if those communities are great cities. Rather, industrialization, properly understood, almost always occurs regionally. In the typical European experience, the expansion of manufacturing in large units of production—a simple working definition of industrialization—came about through an interaction of one or more cities and a contiguous region. In the course of that interaction, not only capital but also labor, entrepreneurs, technology, markets, and sites of particular kinds of production moved back and forth between a dominant city and its hinterland. Most often, the dominant cities served as centers of markets, capital accumulation, entrepreneurship, communication, and consumption. The role of cities in industrialization has historically been problematic, not given. Manchester, for example, became the center of mechanized factory production for the English Industrial Revolution as well as a regional marketing and banking center. Lyons, in contrast, was transformed from a center of French artisanal silk production to a node of banking, entrepreneurship, and commerce, with only a partial and temporary episode of heavy industry. Turin, the other major city of Italian industrialization, developed in a manner more similar to that of Manchester (albeit with a much different industrial base—automobiles instead of textiles) than to that of Lyons or Milan.[27]

What do I mean by region? In principle, I mean a major urban cluster and its tributary area. In practice, I adopt the nineteenth-century state-defined unit of the *regione* or *compartimento,* within which economic, political, and demographic statistics were collected and reported; it provides a reasonable approximation of a functional region. Italian regions were divided into provinces; in Lombardy, these were Milan, Bergamo, Brescia, Como, Cremona, Mantua, Pavia, and Sondrio.[28]

A study of working-class formation and transformation that moves historically beyond the definitions and silences of the national state–based scheme of Katznelson and Zolberg requires (1) establishing the regional pattern of industrialization in the economically advanced regions of a given national state; (2) identifying the major rural–urban industrial and demographic interactions in each region's metropolitan center; (3) examining the processes by which local and national workplace institutions and national political organizations that claim to represent working-class interests are built; and (4) reconstructing relationships between those workers affected differently by the large-scale economic structural change and those workers who build institutions, mobilize, and act collectively with others (in the case of Milan, Lombardy, and Italy, these "others" were both autodidact petty bourgeois and worker [Gramsci's "organic"] intellectuals and professional and intellectual members of international Socialist networks). Household and gender perspectives are incorporated into the analysis as far as possible. Nevertheless, the aggregated nature of available evidence regarding households and the predominance of evidence from formally constituted voluntary associations in the political and economic arenas greatly limit what we can say about the relationship of gender and class formation. Rather than assume that structures of the economy and labor force are homogeneous and distinct nationally, we must examine their evolution at the regional level in order to understand relationships among the forces of change and outcomes.

In this study, I shall consider the changing spatial and temporal patterns of industry, the migration patterns of the labor force, and the place of birth of ordinary workers and activists for the Lombard region. My chief focus is Milan, for this was the pole around which change occurred. This framework has the advantage of a focused perspective, but it neglects close examination of other cities and rural areas in this and other regions.

To analyze and explain the development of worker institutions and collective action, my method is to compare groups of workers with distinctive characteristics (occupation, gender, place of birth, age) and histories in one setting, where the economic forces impinging on each group and its political possibilities can be observed. This is not to suggest that the part (the regional case study) can stand for the whole. Like national indices of economic development or industrialization, national histories of class formation that force regional variation into a single evolutionary process are, by definition, aggregates that obscure diversity. A historically specific case of a city and its region in a limited period cannot explain the process of national class formation; by means of internal comparison, however, it can illuminate relationships that help explain the timing of activism and the groups involved in it.

As a function of the regional structure of the economy and the intense interdependence within regions, then, groups in subnational geographic units (which eventually, of course, formed alliances with one another) developed, at the local level, both institutions and awareness. Analyses of changes in the form of economic activity, structure of the labor force, and conditions of life and work, which are most often regional (as are the economic and demographic forces that shape them) must be separated, however, from analysis of the process of working-class formation, which, although rooted in cities and their regions, is national in its achievement. The emergence of durable national institutions and collective action—the presence of a class as a national political actor—is the outcome of this process.

Milan and Its History

Why Milan? In Italy, two cities dominated the early process of industrialization, Milan and Turin, and I chose Milan as the focus of my study because it was the more important of the two before 1900, both economically and politically. In Milan and Lombardy major changes in the location and structure of industry, and movements of labor from country to city and from city to city, took place in the last decades of the nineteenth century. In the same period, Milanese workers founded the first workers' party, developed resistance leagues, established a chamber of labor, and finally joined with intellectuals from the same city (along with intellectuals and workers from elsewhere in Italy) to establish a Socialist party. The relationship of those changing economic and demographic structures, mediated by political factors, and the actions of workers and their allies in both the local and the national arena, then, form the core of this study.

Volker Hunecke's major synthetic history examines industrialization and the Milanese working class from 1859 to 1892. Hunecke situates himself in the Merli factory proletarianization framework. He claims to be able to say general things about Italian class formation based on this case because Lombardy was one of the most highly industrialized regions of the period, almost unique in the scale of its industry. The years around 1890 are for him (as for the earlier historians of working-class formation just reviewed) the key period of class formation.[29]

To support his interpretation, Hunecke offers, first, a thoughtful and thorough economic history of Milan. The 1880s were a period of economic expansion, based on cheap labor migrating from the Lombard countryside to exploitative conditions and low wages in the city. Although Hunecke sees Milan as a site of change, he does not see it as a differentiated place, thereby missing the significance of the geographic mobility of both capital and labor for patterns of working-class life and organization. Nor does he analyze the powerful evidence that he himself offers, that the workers' experience was not uniform but, instead, highly differentiated. Outcomes varied. Some workers were able to build organizations; others, although they acted collectively in the workplace through strikes, were not. Hunecke's argument, then, is that in Milan, if not in Italy as a whole, industrialization and the creation of the workers' movement occurred together in the prescribed chronological order and that the factory proletariat were the creators of that movement. Like Merli, however, Hunecke does not provide sufficient evidence for the leadership role of proletarianized and immiserated factory workers. He treats such disparate groups as construction workers, former artisans or skilled workers who were losing control over their work, and women and child workers in newly mechanized shops as an indifferentiated mass of factory industrial workers. He also neglects the biographies of worker leaders that he needs in order to make his argument.

Hunecke's account ends in 1892, he explains, because by then industrialization was complete in Milan, and the key worker institutions—local and national— were in place. Yet Milanese industry was then still relatively small scale and not highly capital intensive; the formation of the Socialist party of Italy in that year can hardly be interpreted as Hunecke understands it, as the end point of working-class formation. By concluding his study in 1892, Hunecke avoids major problems for his argument by simply asserting what he believes to be true. Thus he implies that the evolution continued in an already established direction of worker characteristics and class struggle from 1892 to 1900 and later. In fact, in the ensuing period, Milanese worker and Socialist organizations underwent two major government repressions (1894 and 1898–99); elsewhere in Italy, workers' institutions were also repressed (the largest-scale repression was that of the Fasci siciliani in 1894). Workers' and Socialists' collective action, electoral participation, and parliamentary activity were intermittently blocked with impunity by the national authorities. As a consequence, both the workers and the Socialists eventually joined with the Democratic Radicals and Republicans in electoral alliances.

Giuseppe Berta's cogent comment (in his analysis of a regional workers' movement in the Biellese) about the late nineteenth-century Socialist party bears repeating here: "Before the twentieth century," he writes, "the Socialist party was still . . . an instrument for the political defense of social struggles, essentially preoccupied with simply surviving the police repression that precluded for a long time any extended political activity."[30] The class organization and politics that Hunecke sees as the hallmark of working-class formation in 1892 were muted and had all but disappeared by 1899. Because of the repressive state and lack of political rights in Italy, the birth of a national working-class party did not mean that it acquired durable status as a political actor in the national arena. The early formation of a Socialist party at the level of the state, however, is a historical puzzle that deserves systematic examination.

Some regional changes brought migrants to the city; others led industries to leave. The particular conditions of Lombardy's slow and highly regional pattern of industrialization helped some of its workers to build party and other formal institutions. The relatively slow industrialization contributed, moreover, to some skilled workers' maintaining and developing older organizational bases and to their leadership in the Milanese and Italian workers' institution building.

Monographic studies of the changing organization of work and collective action in several groups of urban workers have also been written by Milanese historians, many of them associated with Franco della Peruta, himself a pioneer in studying that city's workers and the archives relevant to them. In Milan and Lombardy, Ada Gigli Marchetti studied printers; Maria Teresa Mereu, masons and associated workers; Luciano Davite and Duccio Bigazzi, metallurgical and machine shop workers; in the area of Como, Luisa Osnaghi Dodi examined textile workers; and for Monza, Andre Cocucci Deretta addressed the case of hatmakers. Documented in detail and exceedingly useful as these articles are, they too are teleological insofar as they accept Merli's factory proletarianization interpretation. Hence they offer similar, rather than differentiated histories of worker experience, linked by the common experience of factory-based proletarianization and driven to collective action by immiseration.[31]

Yet proletarianization did not necessarily involve large-scale factories or immiseration in Milan. Some of those who organized early (e.g., printers) were wage earners with little control over their own labor power or that of others, but they successfully resisted immiseration. Their value to their employers, and their skill, gave them some control of labor markets, and their long-standing organizations placed them in a powerful position as both workers and class leaders. Others, although in large factories (e.g., engineering workers), maintained specialized skills and some craft autonomy in ways that led them to resist class-based organization until after 1900. Still others, such as male skilled textile workers, saw their industry and occupations transformed; although they were skilled in the old ways, their qualifications were no longer vital to their employers. Despite deep organizational roots and strikes, they experienced dislocation as their employers moved their factories out of the city or replaced them with low-waged female workers.

National Politics

Although economic and social factors must be analyzed at the urban and regional levels, the state and politics at the national level have been implicated historically in regional and urban economic and demographic change, as Giorgio Mori points out with examples from English history (sometimes also touching Italian economic development).[32] Tariff policy that favored agriculture, cotton textile production, and the metal industry, but not engineering, greatly influenced the timing and the temporal and geographical patterns of Italian industrialization, as shown in the next chapter. Government contracts for armaments and infrastructure projects (e.g., railroad locomotives and cars, swamp drainage) sometimes favored Italian industries or particular regions but at other times did not. Most important, perhaps, the state's limited resources and the policy choices made in the national arena (e.g., to pursue colonial wars) in regard to spending those resources obviously shaped economic possibilities at the local and regional levels as well.

Any study of class formation must attend to the national level, for two reasons. First, the outcome—the emergence of institutions able to make a class into a political actor—occurs within national politics. Second, the timetable of urban and regional worker militancy and institution building was shaped at the state level by political as well as economic factors. On the one hand were those that facilitated worker action, such as the political mobilization of other groups and the expansion of rights (for suffrage in national elections passed in 1882, and suffrage in local administrative elections passed in 1889). On the other hand were laws that limited strikes and organization and state repression that intermittently dissolved worker institutions and imprisoned their leaders. Although workers differed in their capacity to fight for elementary rights to organize and strike, some of them tentatively entered the local and regional political arena through the action of the Partito operaio. When government repression struck them on the political as well as the economic front, some workers moved to ally with the bourgeois intellectuals in order to create a Socialist party. In Italy's late nineteenth-century illiberal political climate, modest Socialist claims and even more modest successes were met with new repression. Elements in the party allied with bourgeois parties in the fight for political and civil rights and in elections. Their adversary was then the national state, not local government or a regional bourgeoisie.

State repression of the Milanese Fatti di maggio, born of worker resistance to further government infringement on elemental rights and then repression on a hitherto-unknown scale were both an opportunity for the conservative local government to push through some policies opposed by workers and Socialists and at the same time, its own death knell. This government was replaced in 1900 by a new Democratic coalition, supported by the Socialists. The ensuing period saw the emergence on the national level of a liberal coalition dominated by Giovanni Giolitti that granted basic rights and channels for political participation to working-class organizations, including unions and the Socialist party. In terms of economic structural change, shifting political relationships, and evolving working-class capacity for collective action, however, class transformation continued.

Conclusion

The experience of Milan has important lessons to teach about Italian working-class politics and the general conditions under which workers gain and lose the capacity to act together, that is, about working-class formation. First, class formation is a contingent political process in which the actions of workers, intellectuals, industrialists, and politicians are the engine of change, not the unfolding of a logic inherent in economic change. Second, the flow of capital to urban centers, the reorganization of urban and regional economic structures, and the spatial redistribution of workers undergird the process of working-class formation. Third, under some conditions, the regional character of industrialization can lead to concomitant, highly regional periods in class formation. Fourth, when looking at the formation of a working class in other Western countries, it appears that the Italian Socialist party was founded early in relation to national indicators of industrialization, and Italy's state policy was both reactively repressive and difficult to fight, given the lack of political rights and guarantees. Moreover, Italy's industrial capitalism was slow to develop and fragile, and its national state was both centralized and authoritarian yet paradoxically weak because of the country's lack of resources, its highly regional economic structure, and its limits on political participation. Working-class formation in Italy was historically distinctive, but the process was shaped by readily generalizable political and economic factors.

The organization of this book takes its inspiration from Katznelson's scheme of the levels of class. Chapters 2 and 3 look at movements of capital and labor in Italy, Lombardy, and Milan and the resulting changes in economic and social structures. Chapter 4 examines the workers' way of life in Milan and efforts to improve their standard of living. Both the workers' organization in pursuit of their economic interests and the way in which that changed over time are the focus of Chapter 5. Chapters 6 and 7 analyze the relationships between organization and workplace struggle, through the strike statistics for the aggregate level in Italy and in Milan and by means of case studies of specific industries in Milan. The last three substantive chapters look more intensively at the workers' dispositions and the ideologies that inspired their organization and collective action. The regional federation of workers' organization under the patronage of Democratic Radical politicians, the formation of an autonomous workers' party, and its collective action are considered in Chapter 8. Chapters 9 and 10 continue the story of worker and bourgeois organization for collective action in the Socialist party and its political activity, ending with the repression of the Fatti di maggio. Chapter 11 briefly reviews the history of the Socialists' and workers' politics up to the Fascist seizure of power in 1922 and reaches some conclusions about the questions posed here.

2

Economic Change: Italy,
Lombardy, and Milan

In an address celebrating the Milan National Exposition (Esposizione nazionale di Milano) of 1881, Giuseppe Colombo lauded the progress the
Italian engineering industry had made since the Florence Exposition of 1861.
Milanese industry, he noted, was remarkable for its technically advanced
nature, which was evidence of the effectiveness of its entrepreneurs: "[The
engineering industry] has already proved itself strong, energetic, and ready
for the future." One of the major figures of Milanese industry and politics
in this period, Colombo was proud of the machine industry in 1881, and
his report that year for a special volume commemorating the national exposition lingered long on the larger industries of that period, but he was
of two minds. That is, he insisted also on the value of balanced growth
in Milan's economy, with small businesses producing consumer items for
the large urban market and an active commercial sector, in addition to
large industry.[1] Colombo both contributed to and commented on Milan's
industrialization—the expansion of manufacturing in large-scale units of
production—as the material context for the changing social relations of its
workers.

To understand the historical patterns of Milan's industrialization, we
must go back in time and outward spatially: first, to review Italy's economic development in the eighteenth and nineteenth centuries and then to
consider that of Lombardy, the region of which Milan was the capital. The
beginning of Milan's industrialization in the last decades of the nineteenth
century was the outcome of a process of regional economic growth and
capital flows with distant roots in the collapse of its silk manufacturing in
the seventeenth century. What was the relationship of Milan's industrialization to the national and regional economies and their particular characteristics? To what extent did it reproduce long-established patterns or blaze
new paths? When did change accelerate, and when and how did these structural transformations take place? Lombardy's importance in Italian industrialization was out of proportion to its size (population or territorial), but

it, along with Piedmont and Ligury, were the provinces in which that process was concentrated. The task of Chapters 2 and 3 is thus to specify the
material underpinning—industrialization and proletarianization, the twin
processes of structural change that comprise Karznelson's first level of class—
of the process of class formation. This chapter describes the large-scale
structural changes in the Italian, Lombard, and Milanese economies from
the eighteenth to the early twentieth centuries and compares these changes
with one another.

The Italian Economy, 1700–1860

In 1700, the Italian peninsula was politically fragmented and dominated by
foreign powers. The early years of that century saw a turnover of foreign
control from France to the Austrian empire, which acquired Lombardy,
Venice, and the duchy of Tuscany. In the same period, Spanish Bourbons
became both kings of the kingdom of the two Sicilies (Naples and Sicily)
and dukes of Parma. The papacy maintained its control over Rome and
its hinterland, as well as areas in the Romagna and Marches. Only in Piedmont was there an autonomous Italian state, ruled by kings of the house of
Savoy.

Economically, the peninsula was still largely agricultural, but its agriculture varied greatly, from well conducted and relatively flourishing in the
north and center (based on medium-sized units for the most part, with more
capital intensive development in the Po valley), to much less prosperous,
and in some areas quite desolate, in the south. Craft-organized production
of luxury and more mundane consumer products clustered in urban centers;
textiles were produced in a household-based system in the high valleys of
the north and center. By the beginning of the eighteenth century, politically
divided Italy was, as Carlo Cipolla puts it, "a country at once depressed
and overpopulated."[2]

The European industrial expansion of the eighteenth and nineteenth centuries left Italy even further behind England, France, and Prussia. In fact,
the country as a whole did not close the gap in development and prosperity
between itself and the great industrial powers until after World War II.

The national comparison is misleading, however, as it disguises the great
contrast between the enduring economic stagnation south of Rome (which
was exacerbated after unification) and the industrial growth plus extensive
commercialization of agriculture to the north. Milan and Turin were the
foci of wide networks of protoindustry and cash-crop agriculture in the
eighteenth century and developed into major centers of factory production
at the end of the nineteenth century. In one way or another, the entire band

of Italy from Florence north to the Alps shared in the building of nineteenth-century industrial capitalism.

Household manufacturing of textiles and small metalwares in rural households expanded widely in eighteenth-century Europe, and Italy shared in this process. Although textiles were produced in both north and south and factory production was rare in both areas during the early nineteenth century, the organization of production and distribution was quite different. The pre-Alpine valleys of Lombardy led in the growth of textile and small metalwares production, not in industrial cities, but in the countryside; its workers were part peasant and part weaver or nail maker. This form of industry, according to Giuseppe Sacchi in 1847, provided them with "all the benefits possible, avoiding all the economic and moral consequences that, alas, so greatly disturb other populations."[3] By the mid-1850s, some aspects of cotton and silk textile production, particularly cotton spinning and silk reeling and throwing, were concentrated in mills alongside hand-weaving, which continued to be done at home. In the Biella region of Piedmont, the household weaving of woolen cloth similarly flourished in the same period, along with the mechanical spinning of yarn in mills. Indeed, in 1857, Stefano Jacini, then a young patriot, wrote, "Lombardy is destined for an important industrial position in Europe. Its densely settled, intelligent population, its abundant indigenous resources and products, and its location make it an ideal link between Europe and Asia." Although he saw the link as commercial, Jacini was forward-looking and optimistic about industry in Lombardy and was not concerned about possible unfortunate consequences, as was Sacchi.[4]

Protoindustrialization: Continuity and Change

Alain Dewerpe interprets this long-noted phenomenon of rural industry as a case of protoindustrialization. The northern Italian pattern differed in several ways from the specifications laid out in a key article by Franklin Mendels. First, it coexisted, in some areas at least, with a prosperous specialized agriculture (here the raising of mulberry trees for leaves to feed silkworms was most important) and with factories or mills in the same areas. Second, it lasted until the very end of the nineteenth century. It resembled the protoindustrialization model, however, in being rural household-based production (organized by merchant entrepreneurs) of objects for sale in distant markets and in forecasting later industrialization in the same area (the Genoa–Turin–Milan industrial triangle). Dewerpe describes this pre-Alpine protoindustry's origins between 1750 and 1820, its growth and maturity from 1820 to 1880, and its crisis around 1896, followed by decline, as Italy's "industrial revolution" took hold in Lombardy and Piedmont and ended the reign of protoindustry.[5]

The less highly developed iron, metalwares, and machines industries (organized as well in a combination of small shops and household production) were located in the same belt of pre-Alpine Lombardy and Piedmont, but there was a secondary center in the Naples region. In contrast, the household production of textiles in the south, protected by high customs duties, was destined for either autoconsumption or distribution in local markets. Straw plaiting and hat making in Tuscany were other export industries organized in household production.[6]

Dewerpe emphasizes the geographic continuity of protoindustrial regions and the areas that were industrialized by the end of the century. Thus the Alto Milanese continued to be the center of cotton textile production in Lombardy even as it increased in scale and became more concentrated. Silk and wool production lingered and were similarly transformed in the Comasco and the Biellese (Piedmont). Outposts of the metal and machine industry (which clustered closer to Milan as well) in the hilly areas also echoed early maps of protoindustrial production.[7]

Drawing on some of the research discussed here as well as his own investigations of the first half of the nineteenth century, Giorgio Mori recently identified the period as one of "industries without industrialization," referring to manufacturing processes that employed a considerable work force outside the household but did not use complex machines or steam power. In his interpretation, it was these "manufactories," not protoindustrial household production for distant markets, that were the "beginning of industrialization." Building on several of his earlier articles that touched on how to conceptualize the period and, in particular, to what extent protoindustry could be seen as the first stage of industrialization, Mori explicitly criticizes Dewerpe's use of the term and his characterization of growth as gentle and gradual *(croissance douce)*. He objects to the notion of any "beginning" or "precursor" of industrialization that lasted for a century. Besides the sectors already mentioned, Mori notes the contribution of the clothing, tobacco, paper, glass and pottery, printing, food production (grist and sugar mills), coach-building, shipbuilding, and armaments industries to the preunification Italian and international economies. He demonstrates that factories and mechanized mills had begun to appear (here textiles were the most advanced sector, but there were also some large-scale metal-processing and papermaking mills) and insists that a base for industrialization, a beginning of factory-based industry, was present by around 1861, not only in the north, but also around Naples (Campania).[8] In regard to the north, Mori's account seems consistent with Dewerpe's chronology, although it differs in identifying the period in which protoindustry coexisted with manufactories as both a break with the past and the beginning of industrialization. Dewerpe argues instead that protoindustrial household production was not merely a sign of backwardness but, rather, a slow transition whose conditions were contingent, not preordained.[9]

The structure and productivity of agriculture also varied greatly among the states that occupied the Italian peninsula before unification. There were major differences in the organization of agricultural production. Sharecropping and smallholding on medium-sized farms prevailed in the north (with the exception of the Po valley, already characterized by large-scale, capital-intensive farming for the market). The *latifondi* (large estates farmed by virtually landless laborers on long-term contracts—unlike the *braccianti* day laborers of the Po valley—and devoted to wheat, livestock, and market crops such as citrus fruits) were more common in the south. Although based on customary patron–client relations and technologically inefficient, the *latifondi* were at least "quasi-capitalist," as Domenico De Marco terms it, for Sicily (and as Marta Petrusewicz demonstrates for a huge estate in Calabria) until the late nineteenth-century collapse of agricultural exports.[10]

The Economic Consequences of Unification

The first large step of Italian unification—the consolidation of Lombardy, Tuscany, Emilia-Romagna and the Marches, Naples, and Sicily with Piedmont to form the kingdom of Italy—was taken between 1859 and 1861. The process was completed by the addition of Venice in 1866 and the seizure of Rome and Lazio from the papacy in 1870 while its French protectors were elsewhere occupied, fighting the Prussians.

Northern and central agriculture continued in directions set before the wars of unification, but southern agriculture had to adjust to competition from the north and outside Italy, a consequence of the free-trade policy of Piedmont, extended then to the entire kingdom. The rise in imports of cheap grain from North America from the mid-1870s was critical in this regard. Although exports of wine, olive oil, and citrus fruit expanded, changes in the organization of production in southern agriculture led increasingly to the proletarianization of agrarian populations. In Apulia (the heel of the Italian boot) large wheat farms worked by landless gang laborers became important, in tandem with smaller tenant farming, providing incentives for agricultural improvement. In the rest of the mainland south, the *latifondi* declined, not in favor of owner-cultivators, but of rentiers and property managers who drew their incomes from short-term leases and sharecropping contracts. In Sicily both large estates and medium-sized farms were leased or sharecropped in small plots with unstable tenure. There, the estate managers *(gabellotti)* grew increasingly powerful, in alliance with the quasi government of *mafia*. Overall, the proportion of landless agricultural laborers in the deep south grew.

The organization of industrial production changed slowly in the first two decades after 1860, primarily in the north. Most southern industries languished under competition from imports from abroad or more advanced northern industries. Railways (begun before unification in Piedmont, Lombardy, and, to a much lesser extent, Naples) were extended, but there was little feedback to industrial development as track, steel struts for bridges, locomotives, and rolling stock were imported. Railroad building (privately financed and conducted) and public works did improve Italy's infrastructure, however. Most exports were agricultural, often partially transformed products such as oil, cheese, and raw silk. The chief industrial developments in these decades were increased capacity for silk throwing (spinning) and cotton spinning; more iron founding, using scrap iron; increased scale in food processing and paper production; and the birth of the rubber industry, of which more will be said later.

The first protectionist tariff was passed in 1878, followed in 1887 by a tariff that favored the cotton and iron and steel industries even more decisively.[11] Other state interventions, such as privileged access for national industry to bidding on railway equipment and shipbuilding contracts, and subsidies for the Terni steel works, followed in the same decade. The abolition of the incontrovertible currency policy (1880) attracted foreign capital, which flowed into the urban infrastructure and housing. Railroad freight charges were reduced so that coal could be imported more cheaply. A miniboom ensued in the cotton industry, primarily in spinning, but also beginning a changeover from hand to mechanical looms in weaving. This fed the demand for machines, which in turn benefited the engineering industry, as did railroad commissions and orders for municipal streetcar systems. The initiation of steel production in 1887 at the new Terni works suggested new gains in that industry also.

Instead, however, a "tariff war" with France, starting in 1888, contributed to the collapse of agricultural exports and threw that sector into a severe slump. Exports recovered only following more favorable commercial treaties in 1892–93 and 1903–5 with central European countries, and the end of the tariff war with France in 1894. But the industrial depression that hit the world economy between 1889 and 1896 stopped any forward movement, affecting Italian industry especially seriously. The urban construction boom ended, banks failed, and a severe depression affected all branches of the economy.

Italian Industrialization, 1896–1914

Whether the years between 1896 and 1913 comprised Italy's industrial revolution, its "Big Spurt" (the words are Alexander Gerschenkron's), or merely accelerated its development has been vigorously debated.[12] Part of the issue is statistical, part is conceptual, and part is ideological.

Industrialization rather than industrial revolution is the focus here. Clearly, from 1896 to 1914, most of the indicators of growth rose more rapidly than they did in the 1880s. The major statistical series disagree to some extent about the patterns and level of the national rate of change. Production growth estimates suggesting that Italy's industrial product increased two and a half times between the 1880s and 1914 were published by Alexander Gerschenkron in 1955. The official calculations of ISTAT (Italy's statistical institute), however, show a simple doubling of manufacturing production because it calculates no growth during the depression years between 1888 and 1896 (in contrast with Gerschenkron's estimate of slow growth in those years). As part of a collective cross-national comparison of industrial growth sponsored by American foundations, Giorgio Fua revised and corrected the official series, but with no significant change in the estimate of overall growth.[13]

Whether we accept ISTAT's 4.3 percent annual rate (1896 to 1913) of increase or Gerschenkron's 5.4 percent, Italy's annual rate of industrial growth in this period was considerably lower than that of other countries in comparable periods: Sweden, 12 percent from 1888 to 1896; Japan, 8.5 percent from 1907 to 1913; and Russia, 8 percent in the 1890s. Gerschenkron calculates, however, that within the shorter period between 1896 and 1908, Italy experienced 6.7 percent average annual growth rates, compared with 4.6 percent between 1881 and 1888, the earlier period of accelerated growth. Italy thus did enjoy a growth spurt, but one that was not as big a break, or as rapid, as that in other countries.[14]

To what extent was Italy's economy structurally transformed in these years? The proportion of its labor force in manufacturing rose from 23.8 to 26.7 percent between 1901 and 1911 (an increase of 12.3 percent), and those employed in agriculture fell from 58.7 to 54.6 percent (a drop of 5.5 percent). Clearly, there was no massive movement of labor out of agriculture and into manufacturing and services. There was, however, a 102 percent increase of available horsepower in mechanical engines in manufacturing between 1903 and 1911. (Over 80 percent of that energy was concentrated in the north of Italy at both dates.) Coal imports jumped 197 percent from the period between 1881 and 1890 to that between 1906 and 1913. Annual averages of electric production by decades climbed from 3 kilowatt-hours in the 1880s to 752 in the 1900s and 3,192 in the 1910s. There was notable large-scale structural change at the national level, then, but it was unevenly distributed over Italy's territory.[15]

Italian economic historians have paid more attention to analyzing industrial growth patterns within Italy. Luciano Cafagna, for example, shows that Italian industrialization did not mean a large-scale switch in the allocation of resources from consumption to investment. Food prices generally remained below wage increases, partly because greater agricultural productivity accompanied industrialization. The major exports—cotton and silk textiles—were not dependent on large imports of raw materials, and these same industries also satisfied the growing Italian consumer demand. All these factors improved living conditions for ordinary people to some degree. Italy's industrialization was also favored by an "invisible" component in the balance of trade: emigrants' remittances. Although Italy exported its surplus workers, their wages helped pay for consumption at home.[16]

Both old and new industries grew between 1896 and 1914. Textiles, for example, were the major export industry, growing at a slower rate than the new industries were, but on a larger base. This was the period that Dewerpe identifies as that of the dissolution of protoindustrial weaving and the implantation of mechanized electric-powered production of both cotton and silk. Processed food was the other major old export industry.

New industries like hydroelectric power, steel, and engineering also developed, helped by tariff protection. Steel remained only a small proportion of Italian industrial production, however. Tariffs on imported iron and steel made expensive the acquisition of raw materials by the machine industry, hence neither the ship nor the railroad locomotive and car industry succeeded in replacing imports until after World War I. Such industries required skilled workers and vast internal markets, Cafagna argues, and neither of these requisites was present in Italy's early industrialization. Instead, Italian engineering excelled with new products, both consumer oriented (bicycles and automobiles), and producer oriented (electric motors, internal combustion engines, office machines).[17]

Behind Italy's industrialization between 1896 and 1914 lay a favorable international economic climate, state intervention (the protective tariffs that had not been adequate to protect and promote Italian industries in the 1880s but now were more effective), banks established in the mid-1890s (see the following discussion of German capital in Milan), and the development of Italian entrepreneurship. Cafagna concludes that cross-national comparison is not sufficient for understanding Italian economic growth. Although the Italian industrial sector overall remained small and its rate of growth sluggish compared with that of other nations, there was vigorous and rapid development in a territorially restricted area. In 1911, he notes, 58 percent of workers in industrial firms employing more than ten persons were located in the Ligury–Lombardy–Piedmont regions, which contained less than 22 percent of the country's population. Northwest Italy's development was analogous, in Cafagna's view, to the industrialization of an autonomous small country.[18]

It is to Lombardy, a critical component of that "small country," and to Milan, its metropolis, that we now turn.

The Decentralization of Manufacturing in Lombardy

Milan had seen its manufacturing spread into the countryside in the seventeenth century, during a widespread urban crisis. Italian cities lost their leadership position in economic development. As Domenico Sella writes, they became "anachronistic relics of a rapidly fading past."[19] Milan lost its dominance in silk production to Lyons and France. In rural Lombardy, however, the production of raw silk, much of it done in households, continued; linen and wool were spun and woven in cottages; cotton weaving developed in the Gallarate–Busto Arsizio district; rural mills for grinding grain, crushing ores, producing paper, and fulling and dyeing cloth appeared; and small metal products like nails, tools, and arms were also produced on the farms or in tiny rural workshops.

The ruralization of manufacturing continued in the eighteenth century, except for silk braid and ribbon weaving, which remained in Milan. The city enjoyed a certain prosperity as an administrative capital and as the urban residence of wealthy landowners, many of whom sold the agricultural and rural industrial products (especially raw silk) of their estates in urban or export markets. In the second half of the century, larger-scale cotton-spinning and pottery factories, sometimes financed by non-Milanese investors, were built in Milan, Como, Monza, as well as in smaller cities. At the turn of the century, with the French Revolution and Napoleonic Wars, French occupation, and later, a satellite republic in Lombardy, the economy was stagnant or, in some sectors, declined. Once Austrian hegemony was restored after the fall of Napoleon, however, the cultivation of silkworms and the production of raw silk reemerged vigorously, in response to pent-up demand in England and on the Continent. New Austrian regulations also promoted cotton manufacture, still organized primarily on a domestic basis.[20]

A tripartite intraregional specialization emerged. Figure 2-1 shows Lombardy's three clearly demarcated subregions, from north to south. In the Alps, there were small peasant-family holdings that were purely agricultural, raising cattle, chestnuts, and cereals for household subsistence. In the sub-Alpine hill zone lived tenant sharecroppers (mezzadri) with family-run leased smallholdings on which they raised mulberry trees, grapes, and cereals. The mulberry leaves contributed to a cash crop, that is, the cocoons

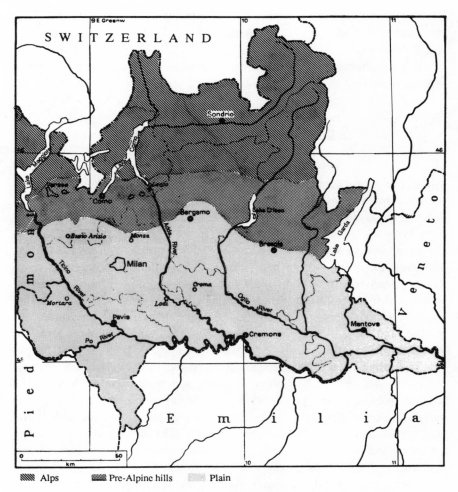

Figure 2-1 Lombardy, with Topographical Subregions

from which silk was reeled. The landlord's share was the cocoons and grapes, and his estate often housed small workshops and reeling mills where the first steps of silk production were performed. Male members of peasant families wove in their cottages in the off-season, and young women worked in the reeling mills when the cocoons were harvested. In the wide plain of the Po River, large irrigated farms, employing wage-earning day labor, produced cereals, primarily rice, and cheese for urban markets.[21]

Capitalist agriculture dominated in southern Lombardy, then, but manufacturing was interspersed with the small-scale farming of the hill area and on the northern edge of the plain. Urban production had changed little since the eighteenth century, except that certain new products were introduced, such as carriages. There was a rapid increase, however, in rural silk reeling, with some increase as well in silk spinning and weaving.[22] Mining and metallurgy, located near the mines in the hinterlands of Brescia, Bergamo, and Como, were in decline. The arms manufacture of Lecco was the only branch of the industry still relatively active.[23] Greenfield calls this intermediate belt a "manufacturing region," although it had few of the urban characteristics of more industrialized manufacturing. "It would generally be necessary to go into the peasant's cottage or to watch the string of horses or wagons on the road to town to detect the presence of industrial activity," he concludes.[24]

By the 1840s, exports were more likely to be in the form of yarn as well as raw silk. Silk weaving, although much less important than in Lyons, had also grown, mostly in Como and Milan.[25] New, mechanized cotton spinning mills had been installed early in the restoration in Busto Arsizio, Gallarate, Monza, Solbiate, all in the Alto Milanese.[26] Linen, wool, and much cotton weaving remained primarily a domestic industry, which Caizzi calls "halfway between agriculture and industry."[27] Larger-scale metallurgy also appeared: The Falck iron and steel mills were established in 1840 at Dongo by an Italian–Alsatian partnership. And in the late 1840s the machine industry began expanding in Milan.[28]

The textile branches had always employed women, but the new large plants crowded young women and children into strictly disciplined shops to work long hours. The factory-industrialized labor of some family members supplemented household-based protoindustry and made it possible for *mezzadri* to remain in agriculture during a period of low prices. In the early days both protoindustry and the mills paid wages below the cost of maintaining even one worker; hence they needed a work force that was partially fed at least by another economic activity, the family farm. Greenfield argues that Lombard manufacturing "grew directly out of its agriculture."[29] Silk mills were established near the source of supply, farms that raised silkworms, rather than in cities. Industrialized cotton spinning likewise was concentrated not in the major cities but in the Alto Milanese where waterpower was available. Both silk and cotton mills were located, then, near available cheap labor, that is, the underemployed "surplus" members of farm households, an aspect that Greenfield does not emphasize but that Dewerpe's more recent study demonstrates. Thus the early textile industrialization of Lombardy took place on the base of rural industry.[30]

The revolution of 1848, in which the five-day uprising of the Milanese expelled the Austrians from their city, interrupted economic growth in Lombardy. Austrian policy became more restrictive of industry and commerce in the second Habsburg restoration. More devastating, however, was pebrine, a silkworm disease, which began to spread in 1854. A twenty-year crisis in the major agricultural cash crop of Lombardy ensued.[31] The last years of Austrian domination and the first years after unification were economically depressed for Lombardy and Milan. The income produced by raw silk exports and the capital accumulation that resulted were drastically reduced, and little new activity replaced silk reeling and spinning.

Industry in Milan and Lombardy After Unification

An immediate consequence of unification for Lombardy was lost access to Austrian markets, including that of the Veneto, its neighboring province in northern Italy, which continued to be part of the Austro-Hungarian Empire. In response to the silk crisis and the new free-trade policy, concentration accelerated in the silk industry: reeling, spinning, and weaving. Fewer but larger mills emerged from the process, mills dependent on silk imported from other Italian regions or overseas. Reeling and spinning mills resumed production in the early 1870s, but the old close links of manufacturing and agriculture had been undermined. Milanese capital played a large role in this process, as the city's financial links to Lombard industry became more institutionalized. Two new banks were founded: the Banca Lombarda di depositi e crediti (1871) and the Banco sete Lombardo (1872). Weaving, however, continued to be depressed by the competition of Lyonnais silk and the loss of Austrian markets.[32]

The cotton industry, which had been recovering in the last years of Austrian domination, also suffered from lost markets—free trade opened Italy to international competition—and, in addition, from the cotton famine of the American Civil War years. The well-established mills of the hills of the Alto Milanese held out and, indeed, contributed to further capital accumulation through new banking institutions such as the Banca di Busto Arsizio, founded in 1872 by the cotton manufacturer Eugenio Cantoni. The 1878 tariff on cotton yarn improved growth prospects for the Lombard cotton industry. Like silk, the cotton industry continued to be located in small provincial cities to the north of Milan, as well as near Como, Brescia, and Bergamo.[33]

There were few changes and little progress in the metal and machine industries in this period. Around 1870, the Lombard iron industry entered a crisis aggravated by the small size and dispersion of productive units, transport difficulties, and financial and technical backwardness. From then until 1886, a slow restructuring of the industry transformed it, according to one analyst, from a "patriarchal regime of small plants" to a "truly industrial regime."[34] The relatively small engineering sector developed especially in Milan but also in smaller cities such as Legnano, where demand for agricultural and textile machines led to new enterprises.[35]

Although Milan's population grew in the 1870s, and the construction industry prospered with "urban renewal" projects, other industrial growth was minimal. Gino Luzzatto noted that with few exceptions, factories in this period were built most often where waterpower and an inexpensive work force were available, that is, in small towns and rural areas. By 1875, however, recovery from the worst of the silk crisis was under way, opening ten years of slowly accelerating development.[36] To what extent and in what ways did this process transform Milan's economic base?

The City of Milan

In the center of that Lombardy in which industries were being established and concentrating, south of the pre-Alpine hills and north of the Po valley, was its capital, Milan. The city had many functions: It was the locus of capital formation and accumulation, the propulsive force of industrial development, the attractive pole for rural migrants in search of work, and the home of economic, political, and intellectual elites.

Centuries of expansion can be traced in Figure 2-2 in the concentric circles, then still separated by canals and walls, of Milan's city plan. In its heart stood the piazza del Duomo, the cathedral square, and the nearby piazza della Scala, in which both the city hall (the Palazzo Marino) and the opera house were located. This nucleus was encircled by a canal, the Naviglio interno (covered over only in 1928), which had begun as the moat for the walls built by the city in its wars against the Emperor Frederick I. The Castello Sforzesco stands at the northwest-quadrant border of this circle.

The next circle out was defined by the sixteenth-century Spanish walls, which formed its circumference. In the 1870s, the commune of Milan was the relatively small city contained within these walls. It was surrounded by an irregularly shaped suburban commune (incorporated in 1781), a consolidation of population agglomerations known as the Corpi santi. The Corpi santi, which had a low population density compared with that of Milan, were nevertheless the second largest commune in Lombardy, second, of course, to Milan.

The population of the Corpi santi grew more swiftly than did Milan's. Migration from the Lombard countryside was the main component of its growth. There was open space in the Corpi santi for housing, and newcomers could live more economically there than in the city. Its inhabitants did not pay the same entry duty *(dazio)* on food and other products as did the Milanese, yet they could participate in the economic activity of the city. Many of these newcomers were involved in the commerce and provisioning of the city, but the outer circle where they lived was also the location of the few large industrial plants in the Milan urban area. In 1873, the Corpi santi were incorporated into the commune, but the large economic and de-

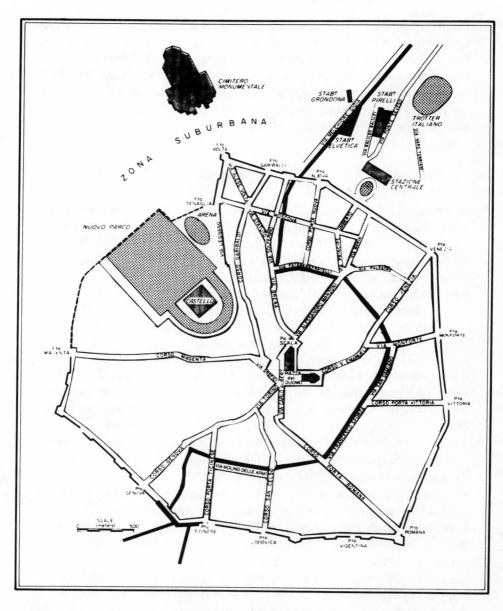

Figure 2-2 Central Milan Around 1910

mographic differences between the internal and external areas of Milan did not disappear. The external *circondario,* as the ex–Corpi were then called, was still not subject to the urban *dazio;* its inhabitants were much more likely than were those of the *interno* to be migrants. This regime favored industrial employers who could (and did) pay their workers less than they did in the internal *circondario* (as the old commune of Milan was now called).

Communication and Transportation
for the City

There were three big navigable canals in nineteenth-century Milan: the Naviglio grande, which went west toward Abbiategrasso and there turned to the northwest to join the Ticino River; the Naviglio di Pavia (a Napoleonic project, officially opened in 1819), which went directly south to Pavia and joined the Ticino there, just north of the confluence of the Ticino and Po; and the Naviglio piccolo, which went to the northeast of Milan to join the Adda River. These canals all joined the Naviglio interno in the heart of Milan.[37]

With the development of the railroads in the 1840s, the city's canals became quiet waterways used for the short-distance movement of goods, along which washerwomen did their laundry. Farther out, the canals were still tapped for irrigation; *trattorie* and country inns were scattered along their banks.

Milan was the point of origin of the second railroad line (the first was in Naples) in Italy, which opened in August 1840 and ran the 13 miles to Monza. In the 1850s, the Austrian regime limited railroad expansion, compared with that of Piedmont. Nevertheless, by 1860 a line to Venice had gone as far as Treviglio, then via Bergamo to Brescia. The line toward Como was two-thirds completed, and that to Gallarate was finished. In 1864, the main railroad station (outside the northeastern section of the walls in what was then still the Corpi santi) was opened. Lombardy's railroad net was completed during the 1870s, and in that same decade, work began on a tunnel through the Saint Gothard pass that would link Lombardy and Milan to central European markets. A second big tunnel through the Simplon pass, a more difficult technical problem, was finally opened in 1906, an event celebrated by an international exposition in Milan. By 1910, Milan had not only its central main railroad station in the external *circondario* but also a commuter rail and a large merchandise station.

Horse-drawn interurban tramways were also built, starting in 1876 with the line from Milan to Monza. In 1879, as part of the preparation for the exposition in 1881, the internal bus system was changed into a tramway, with cars radiating out from Milan's center to the railroad station and the suburban areas outside the walls. Electrification was begun in 1893.

The Industrial Geography of Milan

To what extent was Milan itself an industrial city? Publicity for the 1881 national exposition provides a valuable base line for measuring the development of Milanese industry. Local businessmen had proposed Milan as the site for the fair against the preference of the government that any such event be held in Rome, which had become the nation's capital since the last exposition. The city council, although dominated by a conservative elite with roots in agriculture, supported its business leaders by setting apart a section of the city gardens as a fair site and installing tram service to it. The fair was a huge success: Businesses flocked to the exhibit in such numbers that hundreds had to be turned away. Of the 7,139 exhibitors, 2,872 were Lombard businesses: 55 percent of the engineering, 54 percent of the papermaking, 48 percent of the textile, and 57 percent of the garment-making firms that exhibited their wares.[38]

Satisfaction and pride intermingle in the various declarations and documents produced on this occasion. For example, Giulio Belinzaghi, Milan's mayor from 1868 to 1884, declared: "Our Milan joins itself to a highly significant development—the newly reborn nation of Italy; . . . from these busy exhibition halls a flood of useful economic lessons spreads throughout Italy, a fertile movement of emulation, industriousness, reinvigoration, flowing from Milan."[39] *Milano 1881,* one of several collective volumes extolling the city, contains a chapter entitled "L'Industria," which notes that manufacturing there was the product of "individual owners, that is, workers or other employees who come into some prosperity, and then build a business by the ceaseless work of their own hands, acting at the same time as the directors, engineers, and bookkeepers of their plants."[40]

Giuseppe Colombo's enthusiastic essay in *Mediolanum,* still another celebration of the city published on the occasion of the fair, examines the whole spectrum of Milanese industry. In a great city like Milan, he contends, conditions do not favor large-scale manufacturing. There is no cheap labor, no easily tapped sources of power. Instead, Milan is an important population center that generates demand for consumer products. Clothing, household articles, furniture, and decorative objects are essential to urban life and the "increased prosperity, the refinement produced by the diffusion of culture into all social classes." Colombo is speaking here not only about the production of luxury items but also about the "basic necessities determined by urban taste and style."[41] Colombo understood that the large urban market of a commercial city could support a vigorous consumer products–manufacturing sector: "When we discuss the manufacturing importance of a region . . . we tend to overlook small businesses because they do not employ hundreds of motor horsepower or thousands of workers collected in vast plants."[42] Milan's industrial future, Colombo believed in 1881, lay in a combination of small- and large-scale manufacturing, producer and consumer items, plus the commerce that would move these products far and wide. (Here he echoes the ideas of Stefano Jacini in 1857). Despite his perception of the dual nature of Milan's economy, Colombo's essay spends little time on arts and crafts or consumer industries. His fascination with things technological and entrepreneurial instead led him to focus on large firms, exceptional for their size and their products. Let us retrace his examination of Milan's major industries as an introduction to the workplaces of many of its workers.

Milan's early mechanized plants were located at the sources of hydraulic power, very little of which, however, was available. These plants can be located on Figures 2-2 and 2-3. Ambrogio Binda's paper mill was one of the largest factories in Milan, located in a small valley south of the Ticinese Gate, in the external *circondario;* there his engines harnessed 200 horsepower. At San Cristoforo, also to the south, on the Naviglio grande canal, was Giulio Richard's ceramics plant.[43] These were the only large waterpowered mills, but there were dozens of small water mills: grist mills, rice-polishing mills, saw mills, fulling mills, dyeing mills, and the like. Colombo estimated that the total energy available from hydraulic sources in all of Milan, internal and external *circondarii* (but almost exclusively in the external) was 600 horsepower.

Steam and gas power fed the machine works, the most power-hungry sector of Milan's manufacturing industry. Colombo calculated the total steam motor energy as only 2,000 horsepower, and he believed that as much as the equivalent of 3,000 horsepower of steam was used for heating and industrial processes rather than for turning engines.[44]

The machine industry was the primary user of nonhydraulic motor power when Colombo was writing, as it was in later years also. The big shops were mostly located in the external *circondario* to the north of the city, near the railroad station, in the area between the Porta Garibaldi and the Porta Venezia. Between the vie Bordoni and Viviani were located the engineering works and foundry of the Elvetica, founded in 1846 by a French engineer on the site of a monastery that had formerly housed Swiss monks, hence its popular appellation. In 1881, it employed some 600 workers.[45] Across the canal from the Elvetica was a similar enterprise, Felice Grondona & Company, builders of railroad cars, with 300 workers. The company had started as carriage builders, but as early as 1840, it manufactured the first railroad cars in Lombardy. Grondona also had built cannons for the

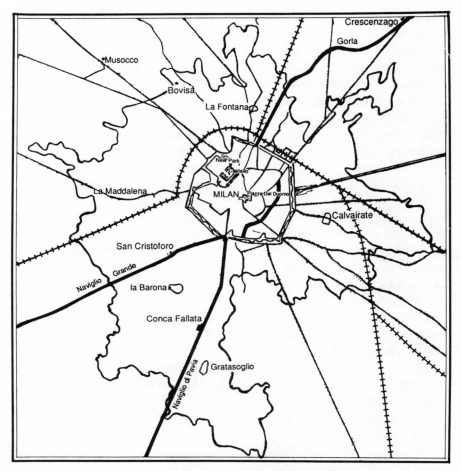

Figure 2-3 Commune of Milan Around 1900

revolutionary provisional government of 1848, to its embarrassment upon the return of the Austrians. After unification, Italian government contracts for military equipment proved safer and more profitable.[46]

The first hydraulic elevator constructed by the Agusto Stigler Company was exhibited at the exposition of 1881. This innovative engineering firm was also located in the northern suburb. Its specialties, in addition to elevators, were gas motors and pumps. Nearby, Suffert and Company, gave work to some 200 workers building steam engines and boilers, railroad equipment, cranes, and also elaborate domes with sliding openings for observatories. And the Invitti Brothers built and installed steel bridges.[47]

On the via di Ponte Seveso in the same northern suburban industrial zone was the Pirelli company, the first Italian manufacturer of rubber products, founded in 1872 by Giovanni Battista Pirelli.[48] His company produced rubberized cloth; insulation for wires, telegraph, and electric cables; and eventually rubber tires for bicycles and automobiles. In 1879, it won a contract for the wires for army field telephones. In 1881, when Pirelli's world renown was still in the future, it had about 250 workers and 150 horsepower in mechanical engines.[49]

Close by, outside Porta Nuova, in quite a different type of industry that also enjoyed national repute, were the Fratelli Branca, producers of vermouth, liquors, and fernet, a medicinal (and alcoholic) digestif. On the via Moscova, inside the walls but adjacent to the northern industrial zone, was the tobacco monopoly factory, with its 1,500 workers, three-quarters of them women, and some small engineering firms. Achille Banfi, another old Milanese company that produced ammonia, boric acid, and Cyprus water was located on the via Solferino, which angled across the via Moscova. Carlo Erba's pharmaceutical plant, which had its origins in the period (1830s) in which its founder practiced in the old Brera pharmacy, was its neighbor on the via Solferino. Finally, on the via Melchiore Gioia, the gas company had one of its big establishments.

To the east and south of the *dazio* walls another band of factories stretched near Porta Genova, Porta Ticinese, and Porta Vigentina. The most notable of these was Miani & Venturi, an engineering firm with 650 workers and 300 horsepower. It too built railroad cars, boilers, and steel bridge parts. In this southern belt also was the Prinetti Stucchi factory, with 400 workers producing sewing machines for home and industry.[50]

On the via Vigevano, running from the Ticinese Gate to Porta Genova station were the headquarters of the Societa telefonica lombarda. On the via Tortona, traversing the same neighborhood, were a series of important shops, manufacturing brass and glass tubing, siphons, and valves. At the eastern end of the southern industrial strip, outside Porta Romana, was Ambrogio Binda's button factory—the first in Italy—which in 1880 employed 600 workers making buttons of horn and metal. It too benefited from military contracts.[51]

In the west and northwest suburbs were two clusters of plants. Directly west of Porta Magenta, at La Maddalena, was the large De Angeli cloth printing and dyeing mill, which consumed large amounts of steam power and employed some 700 workers, many of them women and children. Because of its uncertain profitability, the business, formerly a subsidiary of the Cotonificio Cantoni, had been sold in 1875 to Ernesto De Angeli, its former manager, and a group of investors. De Angeli restructured its production and modernized the plant.[52] At the very northwest edge of the city was Bovisa, Milan's "Little Manchester." Several mills were clustered there, the largest being the Candiani and Sessa Cantu chemical and fertilizer factories, as well as several makers of tallow candles and soaps.[53]

This roughly clockwise geographical tour of the city walls has omitted the textile, garment, and printing industries, as well as the small-scale crafts, all of which were more likely to be located in the heart of Milan rather than on the outskirts on which our description has focused thus far.

The 1881 population census counted 9,233 textile workers in Milan. Giuseppe Colombo wrote in the same year that luxury textiles were the "characteristic industry of Milan . . . now resurgent, full of new energy."[54] There were, he continued, about 100 silk-weaving establishments in Milan, housing 1,000 looms with 1,500 workers. Lower-quality silk was still woven by hand in rural households around Como. In Milan, there were also 10 shops making velvets; 15 producing neckties, ribbons, and fancy articles; and 45 shops weaving braid, fringe, and cord, employing a total of 2,000 persons, many of whom worked at home. A similar system of shop work and homework also produced embroidery and lace. Although Columbo praised the progress that these luxury products had made in recent years, he was wrong about their future: They were the remnants of a once-flourishing industry rather than the harbingers of new development.

Seamstresses who worked in their homes and tailors in small neighborhood shops were concentrated in the central city; the 1881 population census counted 30,000 garment workers. Colombo called Milan's garment-making industry "extremely important," perhaps more than that of any Italian city. The industry included dress and suit making, shoemaking, hat making (nearby Monza supplied felt hat forms to Milan for decoration), white goods, leather goods, and artificial flowers. The industrial survey of 1893, however, indicated only 3,500 employed in large shops.[55]

Homework in the garment industry was not a holdover from earlier household production but a new form, which developed as the industry expanded with increased urban demand. Employers were able by this means to tap the labor of underemployed urban women (many of them married) who might find it difficult to work in a shop, away from home. Their wages were low, and further, the women provided their own sewing machines, light, and heat. In addition to the portion of the garment industry organized as household production, there were several large shops located in the central city. Right off the piazza del Duomo was a workshop for seamstresses and tailors associated with the Unione cooperativa, a consumer cooperative of civil servants. Nearby were the Bocconi brothers' department store and work shops, where clothing, rugs, gloves, furs, and lace were made, and on the via Sempione, to the west, was Savonelli's cloak and suit business.[56]

Milan's vital printing industry, nationally influential, was the outcome of a centralizing process begun in the 1860s and 1870s. In 1873, Milan and its immediate surrounding area (the *provincia,* of Milan) contained 17 percent of the mechanical presses and 13 percent of the workers in the entire Italian printing industry. Of the 70 companies this represented, 60 were in the city itself.[57] From 1875 to 1880, an international business downturn caused small printers to fail and large and medium-sized printers to seek ways to cut costs. Buyouts and consolidation increased the average size of firms, which invested in more mechanized equipment, run by steam. Two publishing houses were dominant: the Fratelli Treves and Edoardo Sonzogno. Both had large mechanized plants and published important newspapers and journals. In 1881, Sonzogno published the influential democratic radical newspaper, *Il Secolo,* which was the largest-circulation paper in Milan.[58] Smaller printers were Reggiani, former publisher of *Il Secolo,* the Tipografia editrice Verri, which published train schedules, and the printing plant of the *Corriere della Sera,* the newspaper second in circulation to that of *Il Secolo.* In 1881, the 216 periodicals printed in Milan made it Italy's largest publishing center.[59]

Three important small-scale industries were located in the Porta Ticinese quarter, to the south of the piazza del Duomo, in the southeastern quarter of the city. First were artisans who fashioned wood, iron, or marble, in the area of the Mill of Arms; second, the silk specialty (braid, ribbon) weaving establishments and silk dye shops already discussed; and finally, the tanneries and slaughterhouses which spread from the old gallows hill, in the piazza della Vetra. Although plagued by overcrowding, this part of Milan maintained the older urban form of specialized artisanal neighborhoods. It also preserved some open space and greenery, although the presence of dye works, tanneries, and slaughterhouses tainted both water and atmosphere.[60]

Despite his evident pride in the large-scale engineering industry, Colombo concluded his 1881 review by returning to his first two points: (1) that Milan had a special industrial character and (2) that the city should promote industry that supplied consumer needs to which taste and style were critical but that did not require motor power or many workers. He argued that it would be impractical and costly to bring the needed energy to Milan and that, furthermore, any increase in large-scale manufacturing would have adverse effects on the city. What most concerned him was the possibility of a business downturn and what he believed would be the consequence—social disorder—in an urban center with "a substantial mass of workers, employed in few but huge factories, in a limited group of industries, carried out on a large scale." Domestic industry was more desirable, for it kept workers apart, "strengthened family ties, and was a guarantee of morality and social peace."[61] Clearly Colombo was, in 1881, a man with both hopes and fears for Milan.

Nevertheless, Colombo had a lot to do with improving the conditions for large-scale industry in the city. Two years after the 1881 exposition, he was one of the Milanese businessmen who established the Società Edison's first electric-generating plant in Italy, at Santa Radegonda. The station came to supply electricity to the city for lighting and power. Perhaps amenities like street lighting and urban transport were what Colombo had in mind, for in a speech to the city council in 1884, he again offered his view of a nonindustrial future for Milan: The city would become "a commercial center of the first rank, of European importance; . . . not a worker city, it will find its place through housing foreign capitalists, with all their demands for urban comfort."[62] Although electricity was exclusively used for lighting until 1891, its ready availability eventually opened up new possibilities for large-scale manufacturing in Milan.[63] In 1911, for example, the local engineering industry got all of its energy from electricity.[64] Whatever Colombo's intent, economical electric power vastly improved the conditions for the engineering and other large-scale industries and set the stage for the development of the electrical supply industry.

Milan's Industry: Structural and Cyclical Change

A brief review of the business cycles in which Milan's industry developed in the last decades of the century starts in 1881, when railroad building and industrial plant expansion were on the upswing. Milan was the center of Europe's commerce in raw and twisted silk. The tariff of 1878 had encouraged cotton-spinning activity with the number of spindles in Lombardy doubling. Although there was relatively little plant expansion in Milan proper, the city's machine industries thrived in the expansive economy. Old companies prospered. One historian argues that the opening of the Saint Gothard pass to rail traffic increased the demand for engineering products and was the critical factor promoting the concentration of the machine industry.[65] New companies were born: Twenty-one businesses were incorporated in Milan between 1882 and 1887. The capital of these new companies was eleven times greater than that of the sixteen companies incorporated between 1874 and 1881. Some of the new companies had familiar names: Pirelli, earlier a partnership but incorporated in 1883, and the Edison system. Others included Vogel, a large chemical fertilizer business, and Larini Nathan, a new addition to the Milanese engineering industry that built pumps, cranes, and boilers.[66] The engineer Ernesto Breda took over the Elvetica works in 1886 and proceeded to expand them. The number of workers employed in Milan's major engineering plants increased by about 60 percent between 1881 and 1888.[67] The good showing of Lombard industry at the 1881 exposition was related to the newly concentrating capital investment in manufacturing industry and to a major cyclical upswing.

Lombardy's agricultural exports also prospered through the 1870s and into the 1880s, untouched thus far by the agricultural depression gathering elsewhere; here agricultural holdings grew in scale. But under these conditions, smallholders found it difficult to compete, so more and more rural people were driven to wage labor in agriculture or protoindustry.[68] The possibility of rural by-employment was declining, however, with the movement of some branches of industry (cotton spinning, especially) to larger cities in the region, and rural-to-urban migration swelled.

Inaugurating his administration in 1885, Mayor Gaetano Negri (Bellinzaghi's successor) summed up the previous mayors' accomplishments in encouraging the city's economic growth: Milan was changing; "the city of luxury and of amusements is turning into an eminently commercial city, becoming a very active center of industry, commerce, and capital."[69]

The city council optimistically adopted an urban plan that called for building a new park and roads around the Castello Sforzesco. Nevertheless, by 1885 there were warning signs of the end of the engineering boom. As the Italian Parliament discussed a plan to transfer the state-owned railroads to private management, orders for rolling stock and equipment dried up. The big Milanese machine works laid off workers and tried to cut labor costs by instituting pay-by-piece work *(cottimo)*.[70] The engineering industry was suffering from lack of internal demand and an unfavorable competitive position in international markets. It depended on government orders to keep factories busy when private-sector demand declined. A cyclical downturn gathered force in the late 1880s and early 1890s.[71]

The worldwide agricultural depression had begun to affect Lombardy even earlier. In 1883, heavy competition from American and Black Sea grain had forced down wheat, rice, and corn prices. The price of raw silk collapsed under pressure from Asian imports into Europe. The only exceptions to the generally poor outlook for Lombard agriculture were the dairy farms of the Po valley, whose products continued to command high prices until the end of the decade. The unilateral withdrawal in 1888 by the Italian government, with the support of Lombardy's businesses from its commercial treaty with France was especially serious for the province, for its reeled silk directed to Lyonnais mills had been a major export. Other agricultural exports declined even further because of the break in trade relations. New plant construction and expansion—and the speculative urban building boom— ended with a crash. A national banking and financial crisis followed.

The banking crisis provided the opportunity for Milan's finance to move to a more central and national role. Following the suggestion of the head of the Bleichroder Bank in Berlin to the Italian ambassador and supported by Prime Minister Francesco Crispi, the Banca commerciale italiana (Milan) was established in 1894, backed by German capital. Although the bank's major role in Italian economic development lay in building up the hydroelectric industry (much of it in Lombardy), it also assisted the expansion of the cotton textile industry of the Alto Milanese in the next decade.[72]

Lombard cotton spinning and silk weaving (around Como) escaped the downturn to some degree because the tariff of 1887 protected them. These industries concentrated their production. Domestic handweaving was phased out, and mechanical looms were installed in the mills. The mechanization and feminization of the work force also permitted some resurgence of mixed-fiber weaving in Milan. As the organizational headquarters of much of this textile industry, the city of Milan shared some of its continuing prosperity, compared with the generally depressed Italian economy.

Machine manufacturing, however, was not protected by the 1887 tariff, and consequently Milan's engineering firms had few contracts during this period. For example, the Breda company, although it was seeking to specialize in locomotive construction, was willing to take a government contract to manufacture shells for the royal navy in 1890, a particularly difficult year.[73] Other firms laid off workers. In a newspaper interview, Breda complained that he and other Milanese entrepreneurs had specialized in locomotives and railroad rolling stock in order to fill national needs, and now there were no orders. Grondona company representatives accused the private management of state-owned railroads of buying railroad cars abroad.[74] By 1890, new housing construction in Milan also had slowed down seriously. The *Corriere della Sera* worriedly called for an investigation into the condition of the city's construction industry.[75]

The nationwide depression lasted until about 1894, with the recovery signaled by the founding of 40 new companies between 1894 and 1897, in the textile, chemical, metal, and utilities industries.[76] As the national and regional economies improved, however, another blow came:[77] Prices increased sharply, peaking in 1898. Wheat prices rocketed because of harvest failures and reduced shipments from the United States. Although the crisis was short-lived, it had significant political consequences, discussed in Chapter 10. Commercial relations with France were restored in 1898, followed by a period of accelerated economic growth, prosperity, and price stability. Gerschenkron's index of industrial output increased, on average, almost 7 percent from 1896 to 1908. Milan's economy, like that of Italy as a whole, showed little effect from the Europe-wide 1900 downturn in the business cycle.[78]

Milan at the End of the Century

A directory of Milan's industries published in 1891 proudly displayed the large scale of some of the city's businesses. In descending order, the top five manufacturing employers in Milan were Ambrogio Binda (paper 1,500 workers); Regia Manifattura Tabachi (tobacco, 1,300); Bocconi Fratelli (garment making, 1,200); Pirelli e C. (rubber, 1,200); and Elvetica/Breda (engineering 1,000). These were followed by Richard Ceramics, De Angeli

Cloth Printing, and three more engineering companies. The average number of workers per firm for these giants is 954.5.[79] The data from the *Annuario* are not classified by sex, but we know from the census that the tobacco, paper, garment, china, and cloth-printing industries were large employers of women. As happened elsewhere in the process of industrialization, the proletarianization of Milan's factories applied to women as well as men. In Milan, however, unlike the English Industrial Revolution, industries employing both men and women developed simultaneously and in the same region.

A more balanced and complete picture can be constructed from a statistical survey conducted in 1893 by Luigi Sabbatini for the Chamber of Commerce of all *opifici* (roughly, "factories") located in the province of Milan. This survey reported 1,564 plants in the city proper, with 50,561 workers, or an average of 32.3 workers per plant. These are certainly some of the larger plants, but it is not certain how they were selected, for in a few subcategories, the number of workers per plant falls below 10.[80] The largest-scale industry was textiles (all processes, including spinning, weaving, dyeing, and bleaching), which employed 9,061 workers in 200 firms (45.3 average). In terms of scale, the machine industry stood below the gas, wood, and brick, as well as the textile, industries. Some 8,815 men worked for 240 companies, an average of 36.7. Although the textile plants were larger on average, than those in engineering, there was no cluster of large plants in the Milanese textile industry like the 5 firms that constructed locomotives that averaged 550 workers each. Textiles and engineering together employed 35 percent of the workers in manufacturing in 1893. Note, however, that Sabbatini's total of over 50,000 industrial workers in the large establishments, accounts for only about 21 percent of the labor force in 1893 (based on interpolation between the population censuses of 1881 and 1901).[81] A very large proportion of the Milanese labor force worked outside industries of the type and scale discussed here.

Giuseppe Colombo, by then a member of the Chamber of Deputies but still an enthusiastic reporter of Italian industrial progress, wrote a commentary on the engineering industry at the Turin National Exposition in 1898.[82] He waxed enthusiastic about the many advances since Milan's 1881 exposition. Gone were his concerns about the possible social problems accompanying industrial growth. He noted that although the important firms had existed in Milan in 1881—Elvetica, Grondona, and Miani e Silvestri—only in 1898 was it possible for them to have metal processed and parts produced in Italy. Indeed, their output was growing so rapidly that they were beginning to look for export opportunities. Colombo credited the development of electrotechnology for this impressive industrial growth. It helped Milan in two ways: First, Milan's companies took a leading role in the manufacture of electricity-run machines and of equipment for generating electricity: Riva

Monneret built turbines; Pirelli, electrical cables; and Belloni Gadda, Brisochi Finzi, and Tecnomasio, other electrical equipment. These machine shops and instrument makers had been established in 1881 and had benefited from the new technology in the intervening years. Second, for a broad range of industries, the newly available electric power provided inexpensive energy.

An observer of the industrial neighborhoods lamented in 1900 that "the Milanese horizon was peopled by smokestacks and factories, blast furnaces, and gas works, rooted in soil in which gardens were dying."[83] (Umberto Boccioni's paintings from the period provide a gritty visual image of this landscape.) Industry had grown as an urban economic implosion brought industries into Milan. Consonni and Tonon observe that at the end of the century,

> Milan contained more than a third of the industrial jobs in the province of Milan: a good 73.3 percent in the metal-machine sector, 74 percent of the chemical and rubber industry, 94.4 percent of the printing industry, and 60.3 percent of the clothing industry; in short, the better part of the total gross product of the province.

Only the textile and wood industries were more important outside the city.[84] Yet there were still many small shops in which clothing and other consumer goods were manufactured. Homework continued to predominate in these industries. Although capital concentration and the accompanying growth of the labor force had come to the printing and publishing, metal and engineering, and rubber industries, these were only part of the picture. In 1900, Milan was an industrial city—indeed, the largest industrial city in Italy—but one in which diversity, not homogeneity, was characteristic.

Equally important, moreover, were Milan's roles as the center of an expanding industrial region, a commercial entrepôt, and a financial capital. The patterns of regional specialization in Lombardy that characterized the mid-nineteenth century still prevailed in 1900.[85] Foundries and steel mills took the place of the old small-scale metal industry in Lecco and the valleys near Brescia. New businesses were founded, combined, and grew after 1898 but even more rapidly proliferated and grew between 1905 and 1907. Silk textile production was still located around Como, whereas cotton was more common in other pre-Alpine areas, especially the Alto Milanese. Silk, the old regional standby, was an exception to Lombardy's business vitality and industrial prosperity. The industry mechanized in an effort to cut prices in response to Asian competition, but in so doing, it exhausted its liquid capital and set the scene for later serious problems. Although silk reeling, in particular, had declined, the combined various branches of silk production were still the largest industry in Lombardy, in terms of capital investment, exports, and number of persons employed. The centrality of the regional silk industry to Milan's prosperity made its problems high-priority concerns for the city's business and financial leaders. Cotton spinning, on the other hand, grew rapidly and vigorously after 1898, in a period of "cotton madness" that ended with a crisis of overproduction in 1907.[86] Export-oriented agriculture was concentrated, as it was earlier, in the plains to the south of the city; cheese and rice continued to be the chief food products. Both manufactured and agricultural products moved on Lombardy's railroad network to and through Milan. Capital markets were also centered in Milan, and the Lombard banks played both a national and a regional role, as did Milan's stock market.

Twentieth-Century Developments

A guidebook for the exposition of 1906 echoes earlier commentary: "Small and middle-sized manufacturing predominates, . . . there are numerous small plants and foundries, shops for nickelplating small metal objects, . . . sacred objects, real and costume jewelry."[87] Nevertheless, when the anonymous author of *Milano nel 1906* takes his readers for the usual guided tour around the giant industries of the city, big changes are evident. Many of the businesses in 1906 were combinations of two or more of the companies already established in the 1890s. Two examples are the incorporated Officine meccaniche, a merger of Miani Silvestri and Grondona machine shops, and the Unione elettrotecnica italiana, a merger of the Gadda and Brioschi Finzi electrical equipment companies. The work forces in these plants numbered in the thousands. Others (such as Breda's Elvetica) had incorporated, and new large firms had been founded or moved into the city. Many more large companies in heavy engineering or small machine making, with hundreds of workers, also are mentioned.[88]

This author hails the exceptional growth of the city's machine industry, working as it did against the disadvantage of having no nearby iron and coal. He attributes that growth "almost exclusively to the generous spirit of initiative, the tenacity of industrialists, and the intelligence and activity of the workers."[89] Although small shops still employed many workers, the large businesses, having substantial capital and taking advantage of new techniques and opportunities, were very evident in the Milan of 1906.

The 1911 industrial census reveals even larger plant sizes, much greater use of electrical energy, and changes in the proportional distribution of workers across industries. This census counted 1,882 firms in Milan with more than 10 workers, employing 124,695 persons (including supervisors and staff), thus 66.2 employees per firm. If we consider production workers only, the average was 59.4 per firm, which contrasts with the 32.3 workers per firm reported in 1893 by Sabbatini, who also considered only the larger

businesses. In 1911, there were another 6,356 firms with 10 or fewer workers, but by now these minibusinesses employed only 13 percent of Milan's
industrial workers. Among the larger firms, 67 percent had mechanical motors: almost 9,000 electric motors, 47 hydraulic, and 263 steam engines.
The electric motors produced almost three times more horsepower (43,948
hp) than did the steam engines (15,378 hp). Only 35 percent of the small
firms used motors at all, and these were most commonly electric, as they
were in the larger businesses.[90] Although the evidence is uneven, it is clear
that units of production were considerably larger in scale than they had been
in 1893 and that mechanical engines of one sort or another were likely to
be used in manufacturing processes. These characteristics of industry indicate that proletarianization—the spread of wage earning and the loss by
workers of control over the means of production—had progressed much
further in Milan's manufacturing industry than in Italy as a whole. And by
1911, industrialization was also well under way in Lombardy.

Over 98 percent of Milan's labor force in 1911 was in manufacturing
and services, with the big gainers being the garment industry, engineering,
utilities, transportation, and communication.

The garment industry had moved to the place formerly held by textiles:
There had been both a large proportionate and an absolute decline in the
number of textile workers and a feminization of its work force as the industry moved out of Milan to rural locations in search of cheap labor. (A
good deal of this downward slide occurred after 1907, when the cotton
branches of the industry entered a crisis of overproduction.)[91] By 1911, in
contrast, the garment industry had more than doubled its share of the labor
force and the number of its workers, as compared with that of 1893. The
scale of the industry, however, remained very small; the "sweating system"—combining shop work and homework—prevailed. Machine making
had grown enormously in both absolute numbers of workers and size of
plant. The number of workers per plant had more than doubled over 1893,
but the machine industry occupied the same rank among industries in terms
of number of workers as it had earlier. By 1911, Breda and other large
firms (including Pirelli) had set up new plants outside the city. Revolution-
ized by the availability of electrical power and new technologies, the utilities, transportation, and communication industries together grew more, proportionately and absolutely, than did any other industry. In 1911, this sector
occupied about 6 percent of all industrial workers, and its labor force had
increased by 700 percent over 1893. The ratio of workers to plants had
increased greatly also. Although directly related to industrial growth, the
greater importance of this sector also reflects the population growth and
urban concentration that were by-products of regional as well as urban economic change.

Conclusion

Industrialization has been conceptualized in this chapter as the expansion of manufacturing (the production of goods, as opposed to service and agriculture) in large units of production. There was no sudden or large break in economic patterns—no industrial revolution—in Italy. Rather, there were slow transformations in patterns of capital investment and organization of production that interacted with one another to produce large-scale structural change. Indices of production show accelerated growth between 1896–1908. By 1911, there had been a decline in the level of the agricultural labor force and an increased proportion of workers in manufacturing and services. High coal imports and much greater use of motors by around 1910 indicate changes in the organization of production and the spread of wage earning. Although these changes were slow and the increments modest, the structure of the economy was being transformed.

Italian industrialization was not historically unique; many of its patterns were similar to those in other European national states that industrialized in the late nineteenth century or before World War I. Protoindustry preceded large-scale concentrated industry in many (but not all) cases of regional development. For example, Tessie Liu and Anne McKernan demonstrate the adaptability and longevity of different types of household production in western France and Ulster, Ireland, respectively. Charles Sabel and Jonathan Zeitlin show that what they call flexible specialization ("high-skill, universal-machine economies") coexisted for decades with mass production in most European countries.[92] Mature industrialism was achieved only slowly.

Industrialization elsewhere was usually uneven, with marked regional differences. The pattern in most European countries was the growth of highly industrialized regions (Lancashire, England; the Nord or the Stéfannois in France; the Rhine or Saar valleys in Germany) near agricultural regions— some capitalist organized and others characterized by peasant agriculture with varying degrees of market involvement—and proindustrial pockets. Industrialization elsewhere was likewise slow to make its impact on national aggregate statistics, and its patterns (occupational structure, industries involved, relation to agriculture, exploitation of mineral resources, connection with urbanization, and cities with different economic functions) varied. In France as well as Italy a high proportion of the labor force remained in peasant agriculture during industrialization.

In northwestern Italy, the pattern was a roughly contemporaneous development of textiles and engineering and metal industries—different from that of England but again not so different from that of France. The chronology of the process varied even among the three cities of the industrial triangle, just as urban economic careers varied in other countries. Milan had a more mixed economy than did Genoa or Turin; it also grew more rapidly in the 1880s and 1890s than did either Genoa or Turin, which later came to specialize in shipbuilding and automobiles, respectively. It was distinctive among the cities of the industrial triangle in its strong regional orientation, which came more and more to outweigh its own importance as a manufacturing center (as space for large new factories disappeared within the city limits).

Milan in some ways resembled Lyons, a regional metropolis whose early industrial developments gave way to the growth of services, and in some ways it resembled Paris. Paris had a vigorous commercial base, alongside a consumer products industry, including souvenirs and trinkets and garments, and also an engineering industry, as did Milan. Further, like Paris, Milan was the country's largest industrial city at the same time that it was the base of intellectuals and shapers of opinion.

As the next chapter will discuss in greater detail, the plurality (47 percent) of Milan's labor force in 1881 worked in manufacturing occupations (46 percent in services, 7 percent in agriculture), according to the population census. Despite some large firms, its manufacturing companies were, overall, small. (Even in 1893, the Sabbatini survey of larger businesses yielded an average of only 32.3 workers per firm.) Most of this manufacturing was organized in old ways, using hand tools and little division of labor. Increasing scale in the ensuing period signaled the onset of industrialization and accompanying proletarianization, which accelerated after 1900 and had spread large-scale industry beyond its borders by 1911. The city's manufacturing workers were themselves a heterogeneous group, and they lived and labored alongside a high proportion of service workers. Milan's labor force was divided as well by the place of birth and gender of its members, by the scale of their workplaces, and by the type of production— consumer or producer—in which they were engaged. Chapter 3 examines the characteristics of Milan's population and its labor force, the raw material of the working class.

3

Population Growth, Occupational Structure, and Migration

For two centuries, Lombardy's capital and industry had traveled to the locations of cheap labor. But toward the end of the nineteenth century, Lombardy's workers began moving massively to the then-concentrating locations of capital and industry, especially Milan. Many of these workers had been wage earners in protoindustry or large-scale agriculture. For them, this move did not involve a new proletarianization. For ex-smallholders, or members of smallholders' households, the move combined migration and proletarianization. The city's population growth and the characteristics of its population were distinctively shaped by this rural-to-urban migration. This chapter looks at the social and demographic components—age, sex, occupation, household composition, distribution in space, and the ways in which these are related to one another—of Ira Katznelson's second level of class. In the next chapter, I shall relate aspects of this level to workers' way of life, and examine the extent to which living conditions changed in the period.

In his autobiographical novel *Alla conquista del pane* (1883), Paolo Valera tells the picaresque tale of a naive village youth who had come to Milan to earn his living. Valera portrays Milanese life in a gritty style, emphasizing the urban ambience (wonderful descriptions of buildings and streets and the crowds of ordinary folk in them) and institutions. He contrasts the miserable and unpredictable life of the poor with the slick comfort and corruption of the bourgeoisie. Like Emile Zola, Valera probes private as well as public vice in all classes; unlike Zola, he tends to account for such vice among the poor in terms of immiseration and, among the rich, in terms of hedonism.

Valera's hero, Giorgio, encounters a rapacious employment bureau, works as a clerk for a lawyer, is sheltered by a prostitute, and becomes a homeless vagabond without work. In desperation, he takes a post as a domestic servant but is soon dismissed for becoming involved with his employer's wife.

He then returns to his rural origins and finds work as a teacher. Dissatisfied, he drifts back to the city and begins a happy love affair with a seamstress. He is still unable to find regular work, however, and ends up on the street with wandering vendors, acrobats, and magicians. *La Plebe,* a socialist newspaper, published a laudatory review of Valera's first novel, hailing him as a *proletarista.* Although the novel concentrates on externals and lacks psychological subtlety, the review concluded, it catches the authentic experience of "that immense swarming anthill of poor creatures that society holds, trapped in its lower depths, trampled upon by the insolent feet of its lordly institutions."[1]

Giorgio's encounters in Milan with fortune and misfortune were echoed in many ordinary men's and women's lives. Valera captures both the insecurity and the shifting prospects of the migrant, who often needed to make her or his way in whatever job came to hand. *La Plebe* emphasized the severe social inequality that migrants experienced in the city and their powerlessness against the impersonal forces shaping their lives. The historical record suggests that even though migrants were at the bottom of urban hierarchies, there were regularities in their paths to the city, in the occupations they entered, and in other aspects of their urban life. Such patterns were linked to the capacities of various groups of workers for political and workplace collective action.

Before discussing those relationships, however, let us examine the demographic and labor force context and patterns of migration. What were the effects on Milan's population of migration from the region and its countryside? Who were the migrants, and what were their characteristics; what was the context of the migration; and once in Milan, what occupations did the new migrants enter? How did the migrants fit into the work force of the industrializing city? To what extent did Valera's solitary but heroic country boy's progress represent the typical migrant's experience?

Population Growth and Patterns in Milan

Between 1881 and 1911, Milan's population grew from 321,839 to 599,200. Movement between country and city was an old pattern, however, as Milan had long grown by means of migration rather than natural increase. During most of this period, migration in Lombardy was intraregional rather than interregional or international. Antonio Golini points out that economic development in the northwest industrial triangle was not strong enough to absorb the migration from the south and northeast, which contributed so heavily to overseas migration. Rural areas in the northwest, however, were also overpopulated in relation to the needs of agriculture, which was intensifying, and of protoindustry, which was collapsing—both consequences of

Table 3-1. Annual Growth Rates, 1901–1911

Geographic Unit	Growth Rates, 1901–1911
Italy	0.7%
Lombardy	1.2
Province of Milan	1.9
Commune of Milan	2.2
Bordering communes	10.7
Northeast suburbs	24.1
Northwest suburbs	10.0
Northern suburbs	4.7
Western suburbs	4.7
Southern suburbs	4.5
Southeast suburbs	7.5

Source: Francesco Coletti, "Zone grigie nella popolazione di Milano," in Carlo Carozzi and Alberto Mione, eds., L'Italia in formazione: Ricerche e saggi sullo sviluppo urbanistico del territorio nazionale (Bari: De Donato, 1970), pp. 107–108.

capitalist expansion. In contrast with the southern agriculturalists, the Lombard peasants forced to find ways to earn their livelihood elsewhere than the countryside were able to find jobs close by in Milan, Turin, and the smaller industrializing cities.[2] In short, northern industrial development did provide work for thousands of rural migrants, most of them from nearby rural areas. The period of this study was one of urbanization, which, defined in ecological terms, is an increase in the proportion of the total population of an area that is living in cities. This occurs when the rate of population growth in the cities is higher than that of non-urban areas.

Overall, Italy's population grew from 28.5 million in 1881 to 34.7 million in 1911. The proportion of the population resident in cities with over 20,000 in population increased from 23.7 to 31.3 percent in the same period. Milan was exceptional in its disproportionate size, compared with other cities in Lombardy (or even compared with the other major industrial cities in the northern industrial triangle, such as Genoa and Turin). From a national perspective, Milan was also distinctive in the temporal simultaneity of its population growth and industrialization; as most of the large and rapidly growing cities in other parts of Italy (e.g., Rome and the Sicilian regional capitals) were bureaucratic and administrative in function. Comparing Milan's growth rate with the growth rates of Italy and Lombardy, and cities over 20,000 in the province shows that in the last two decades of the nineteenth century, Milan grew at an annual rate that was three times higher than Italy and Lombardy and one and a half times higher than the other large cities (Table 3-1). This was a period of great urban implosion, in which Milan itself attracted many rural migrants. More rapid growth of the province, and especially its other large cities, came in the decade after

1901, as Milan's rate of growth decreased with the disappearance of available space for new manufacturing plants or worker housing within the city limits. Giancarlo Consoni and Graziella Tonon write that in the first decade of the new century, Milan "grew by selection of the migration and expulsion of its labor force," and workers left the city for the surrounding communes, which also attracted migrants directly from the rural areas.[3]

Urban historians have debated the relationship of industrialization and urbanization: Does the first cause the second, or does the second create conditions favorable for the first? Paul Bairoch argues provocatively that early industrialization occurred in most western European countries outside traditional urban networks. He speculates that a minimal level of urbanization was necessary, for the cities were the source of technical innovation and markets, but not so much urbanization that capital for industrial investment was scarce, urban consumption demands were high, urban productivity was low, and labor therefore was relatively immobile.[4]

Alberto Caracciolo, reviewing regional and local studies in northern Italy, concludes that in this area commercial, administrative, and culturally important cities provided conditions favorable to industrialization. This indeed seems to fit the case of Milan, but as we have seen, the pre-Alpine belt of Lombardy and Piedmont that was central to both proindustrial development and later industrialization had other advantages as well, particularly cheap and relatively mobile labor, combined with the commercial networks that Caracciolo labels as critical. And the rapid growth of other Italian cities like Rome fits Bairoch's model, as their growth was quite separate from capitalist development.[5] As is often the case, it is not an either/or situation, but one in which the timing and relationship of urbanization and industrialization may differ, depending on historical and contextual factors. Milan's population growth spurt preceded its period of rapid industrial growth, but urbanization and industrialization took place relatively simultaneously in provincial Lombardy. Both patterns were present in Lombardy's urbanization, the first primarily in the 1880s and the second especially after 1896.

A first approximation of the impact of immigration is clearly visible in the demographic structure of Milan at census time, as illustrated by the population pyramids in Figure 3-1. These graphs depict the simplest aspect of population structure: its age and sex distribution. The pyramids plot these characteristics by arraying the number of persons by sex along the horizontal axis, with five-year age groups on the vertical axis. The population structure of Milan is contrasted in this figure with that of Italy as a whole.[6]

The Milanese pyramids for both census years, 1881 and 1901, show a bulge in the numbers of persons in age cohorts 15 to 25, a result of the recent immigration of persons of those ages, both male and female. (The male side includes the military contingent stationed in Milan, which ac-

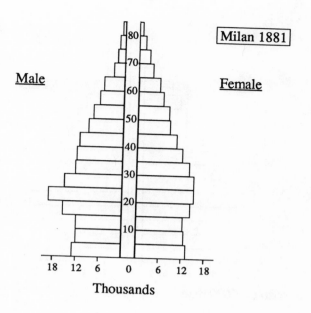

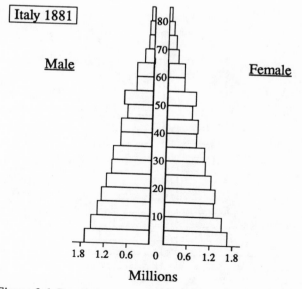

Figure 3-1 Population Pyramids, Italy and Milan, 1881–1911

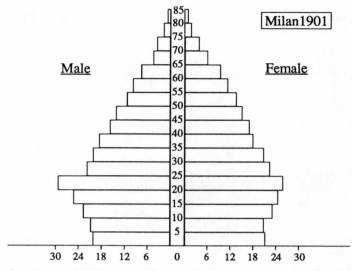

Garrison = 6690 men   Thousands

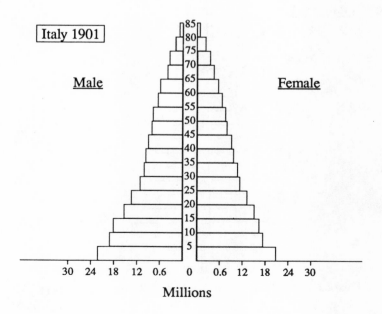

Millions

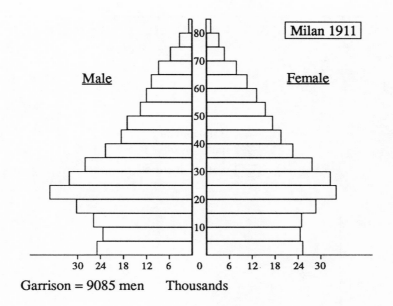

Milan 1911

Male   Female

Garrison = 9085 men   Thousands

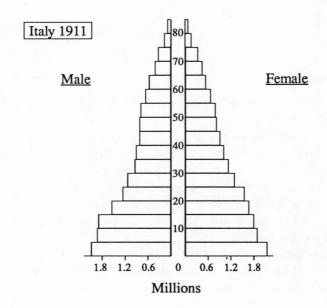

Italy 1911

Male   Female

Millions

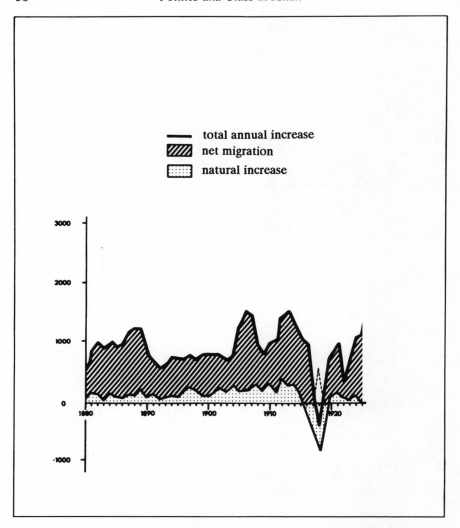

Figure 3-2 Components of Population Growth in Milan, 1880–1925

counts for its higher numbers.) There is a sharp contrast between Milan's population pyramids and those of Italy as a whole. The latter show the typical age distribution of population growing by a high rate of natural increase: a broad base of children and a stepwise reduction from one age group to the next. Those of Milan have a smaller base, reflecting low urban fertility due partly to the celibacy of large occupational groups like domestic servants and the military.

Another graphic portrayal of the importance of migration can be found in Figure 3-2, which displays Milan's annual population growth; Net change is divided into two components, net migration and net natural increase (the difference between births and deaths). Natural increase was much the smaller part of population growth. Over 70 percent of the annual population increase in every year between 1881 and 1911 was due to a very large positive migration balance. In all but three years of the period, net in-migration contributed well over 80 percent of the population growth, with the highest proportion, 96 percent, in 1883. Population growth fueled by migration was typical of contemporary European cities. In the same period, for example, the population growth of French cities such as Marseilles, Lyons, and Bordeaux was even larger, owing to migration.[7]

The outer ring of the city always grew more rapidly, just as it had earlier, before the Corpi santi had been annexed. Over the nineteen years between the census of 1881 and the next one, most immigrants were settling outside the walls in the external *circondario,* or the suburban and rural zones, as the area was called after 1898. Between 1874 and 1878, 59.8 percent of in-migrants moved into the area outside the walls; between 1891 and 1898, the last period for which these data are available, 71 percent of all migrants were settling in the outer zones.[8] According to the 1901 census, 46.8 percent of the population of Milan lived in the suburban zone, and 5.6 percent remained in the rural zone; hence the majority of the urban population lived outside the city center.

People tended to live close to their place of employment. Commerce and consumer production were concentrated in the center zone, and newer, larger-scale industry was located in the suburban zone, outside the Spanish walls. In 1901 the percentage of persons 16 years old and older who were working in industry increased as a function of distance from the center of the city, with the greatest number of manufacturing workers living in the suburban zone. In the center, 30 out of 100 were employed in manufacturing, and in the suburban zone, 45 percent were so employed.

Pottery workers lived in the southwestern rural zone of the city, where the Richard Ginori factory was located. Similarly, glassworkers lived in the south and southwest, where there were glassworks. The most important and visible concentration of factory workers was in the northern suburban zone, outside the Porta Garibaldi and Porta Nuova, where several of the metal and engineering plants, the rubber factory, and the chemical companies were located. In the central core of the city lived skilled craftsmen such as printers, lithographers, precision instrument makers, and garment workers, as well as domestic servants, personal service workers, and those engaged in the production and sale of food. By 1911, the many workers of large-scale industries, living in the outer ring of the city, outnumbered the craft or small-shop workers of the old Milan. Neither silk nor tanning workers (both listed in the census of 1881 as inhabitants and workers in south central Milan) are mentioned in the 1901 census report on residential patterns of workers.[9]

This process continued after 1901. Francesco Coletti, a contemporary geographer, demonstrated the transfer of rapid population growth outside the commune altogether, into the northern belt of suburbs: "The borders of the commune," he explained, "tend to disappear under the overwhelming flood of new arrivals." [10] It was particularly the decade of the 1880s, then, in which unprecedented numbers of migrants flowed into the commune of Milan. This influx resumed early in the twentieth century, only to be displaced outside the commune after about 1905.

Milan's Working Population and Occupational Patterns

The population censuses provide a snapshot view of the working population, revealing extraordinary levels of labor force participation in Milan: Everyone worked—young, old, and those in their middle years. Figure 3-3 shows the resident population of Milan at census time, separated into the labor force and its dependents. (It divides the total population between members of the labor force who listed occupations in the census enumeration, and students, housewives, children, and the nonworking elderly, disabled, or ill—the dependent population.) The dependency index for Milan was very low in 1881; as an enormous proportion of persons declared an occupation. Each working individual supported only about two-thirds of a "person" (not counting himself or herself). The proportion of the population with an occupation decreased between 1881 and 1901, and accordingly the dependency index increased from 669 to 786. This still means less than one dependent person per worker, a much lower ratio than that for Italy as a whole, evidence once again that Milan was the receiver of considerable adult worker rural–urban migration. Intensive labor force participation went along with the in-migration of young working men and women, many of them single and without dependents. The upward trend of the dependency index, which continued until 1911, is linked, as we shall see, to the reduction of labor force participation among women, the aged, and the very young.

Occupational patterns can also be traced through the population censuses. (Occupations are classified into industries, based on the 1881, 1901, and 1911 census categories in Table 3-2, which shows the proportionate importance of industries in the occupational structure.) Not surprisingly, only a small number of workers were employed in the agricultural sector in the city, and its share fell steadily from 1881 on. This typically urban characteristic offers the chief contrast with the labor force of Italy as a whole, where well over 50 percent was employed in agriculture from 1881 to 1911 and afterward.

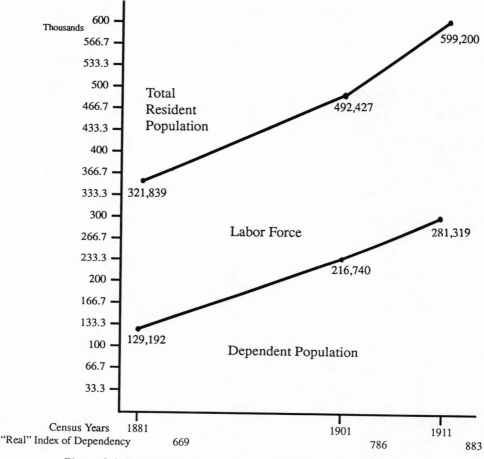

Figure 3-3 Total Population and Dependent Population: Milan, 1881–1911

Even the proportions in Milan's manufacturing and commerce and ser-vice sectors were different from those in Italy as a whole. In Italy, manu-facturing workers averaged about 25 percent of the work force at the times of the censuses; with the commerce and service sector, about 15 to 16 percent. In Milan, the secondary sector (manufacturing), which in 1881 employed a plurality of workers, became the employer of the majority of workers by 1901. At the same time, a very large share of the work force was employed in the tertiary sector, commerce and services. These patterns reflect the city's mixed urban functions; that is, it was both a commercial and service and a manufacturing center. The mix of jobs represented by the occupations that people claimed made Milan the destination of a vast rural–urban migration.

From 1881 to 1911, the pattern of occupations in Milan changed deci-

Table 3-2(a). Sex and Nativity of Working Population in Milan, 1881, 1901, and 1911

Labor Force Category	1881			M & F (%) Native Born
	M	F	F (%)	
Agriculture (total)	6,846	3,949	36.6%	32.2%
Manufacturing (total)	54,301*	37,786*	41.1	47.9
Food	6,263	864	12.1	29.7
Wood, straw	8,620	57	0.6	44.1
Leather, fur	6,758	1,230	15.4	41.2
Textiles	3,354	5,879	63.7	49.0
Garments	4,581	25,173	84.6	53.5
Other agric. products	729	1,303	64.1	57.2
Glass, ceramic	636	163	20.4	42.8
Metal, engineering	10,101	158	1.5	48.9
Precious metals	2,245	380	14.4	67.7
Mixed metals, wood	788	998	55.9	46.3
Chemicals	534	353	39.8	34.8
Rubber	69	65	50.0	26.1
Printing, lithography	3,542	214	5.7	64.2
Construction	5,217	6	0.1	32.4
Paper, cardboard	864	939	52.1	46.0
Professional commerce and service (total)	59,028†	29,821	33.6	31.4
Banking, insurance	6,722	238	3.4	45.4
Professions, arts	7,183	3,185	30.6	44.6
Government	9,398	58	0.6	9.5
Utilities, transport., communic.	6,005	19	0.3	27.7
Domestic service	16,528	22,860	58.0	27.3
Public accommodations	5,053	1,139	18.4	30.4
Retail merchandise	8,159	2,322	22.1	46.3
Nonspecified (total)	808	108	11.7	49.5
Total	120,983	71,664	37.2	39.4

*Includes not-otherwise classified industry.

†Excludes the military regiment stationed in the city, which is the greater part of the "government" category.

Source: Census of 1881.

Table 3-2(b). Sex and Nativity of Working Population in Milan, 1881, 1901, and 1911

Labor Force Category	1901 M	F	F (%)	M & F (%) Native Born
Agriculture (total)	6,253	2,286	26.8%	25.3%
Manufacturing (total)	88,172*	53,421*	37.7	41.0
Food	8,678	1,292	12.9	25.8
Wood, straw	10,108	351	3.3	34.1
Leather, fur	9,342	3,102	24.9	34.8
Textiles	4,173	9,962	70.5	41.9
Garments	5,096	29,903	85.4	48.9
Other agric. products	1,226	1,611	56.8	46.2
Glass, ceramic	1,802	451	20.0	35.9
Metal, engineering	22,369	1,174	4.9	41.5
Precious metals	3,257	394	10.7	62.9
Mixed metals, wood	521	943	64.4	25.0
Chemicals	1,575	648	29.1	30.6
Rubber	855	986	53.5	22.3
Printing, lithography	6,551	732	10.0	55.5
Construction	9,951	14	0.1	27.6
Paper, cardboard	898	1,858	67.4	44.8
Professional commerce and service (total)	82,707†	42,848	34.2	26.3
Banking, insurance	6,388	933	12.7	36.3
Professions, arts	9,407	5,038	34.9	36.2
Government	10,604	68	0.6	10.5
Utilities, transport., communic.	13,558	283	2.0	25.4
Domestic service	13,344	30,202	69.4	23.0
Public accommodations	14,089	2,379	14.4	20.5
Retail merchandise	15,317	3,945	20.4	37.2
Nonspecified (total)	—	—	—	—
Total	177,132	98,555	35.7	33.8

*Includes not-otherwise classified industry.

†Excludes the military regiment stationed in the city, which is the greater part of the "government" category.

Source: Census of 1901.

Table 3-2(c). Sex and Nativity of Working Population in Milan, 1881, 1901, and 1911

Industrial Category	1911 M	F	F (%)
Agriculture (total)	3,557	1,567	30.6%
Manufacturing (total)	110,760*	65,111*	37.0
Food	7,496	1,761	19.0
Wood, straw	10,407	612	5.5
Leather, fur	8,036	2,903	26.5
Textiles	2,661	9,006	77.2
Garments	6,717	35,994	84.2
Other agric. products	953	3,030	76.1
Glass, ceramic	2,784	171	5.7
Metal, engineering	38,904	4,285	9.9
Precious metals	2,723	621	18.5
Mixed metals, wood	277	497	64.2
Chemicals	3,356	714	17.5
Rubber	639	660	50.8
Printing, lithography	5,362	1,342	20.0
Construction	17,899	735	3.9
Paper, cardboard	2,780	522	18.4
Professional commerce and service (total)	97,405†	39,489	29.1
Banking, insurance	5,703	220	3.7
Professions, arts	12,589	5,332	29.7
Government	12,098	176	1.4
Utilities, transport., communic.	26,780	519	1.9
Domestic service	11,652	26,984	69.8
Public accommodations	7,587	1,793	19.1
Retail merchandise	20,996	4,465	17.5
Nonspecified (total)	—	—	—
Total	211,722	106,167	33.4

*Includes not-otherwise classified industry.

†Excludes the military regiment stationed in the city, which is the greater part of the "government" category.

Source: Census of 1911.

sively. Both the number of persons in personal and domestic services and those employed in small-scale industry decreased.

In 1881, 46 percent of the total labor force was employed in commerce, professions, and services, and 47 percent was in manufacturing. The large numbers employed in the service sector were not due, however, to the type of tertiary activity—such as government services, banking, insurance, and office work—which, in highly differentiated economies, grows in importance proportionate to the decline in importance of manufacturing. The service sector in Milan in 1881 was nonindustrial; many of its workers were employed in domestic and personal services and other similar activities. (The other chief service industry was public accommodations, which included hotel and restaurant workers and also street and piazza services, such as newspaper and flower sellers.) A city statistical report in the 1880s remarked on the agriculturalists who migrated to the city to fill these jobs:

> Peasant families, with four or even up to ten children, with the grandpa and the grandma [come]. . . . attracted to Milan by its many resources. Here, they foolishly hope, they will be able to provide for their families, or failing that, seek charity. . . . The problem is that these people do not fit the need for labor in the city because they have nothing to offer but their brute strength; thus they simply increase the competition for manual service jobs.[11]

Domestic and personal service was the largest single industrial category in the city in 1881, employing more than 20 percent of the labor force.

The manufacturing activities that gave jobs to the largest numbers in 1881 were small in scale. Workers reporting occupations in the garment industry, primarily in small shops or their homes, were the largest industrial occupational group in that year. The 1893 Sabbatini survey of industrial plants in Milan listed fewer than 3,000 garment workers in 84 shops, thus an average of 35 per shop; the population census of 1881 listed almost 30,000 and that of 1901, almost 35,000 garment workers resident in the city.[12] Most of the large difference in these figures was due to the homeworkers. Wood, straw, leather, and fur working and food processing, all small scale, accounted for another 16 percent of the urban labor force. The work experience of the majority of Milanese workers in 1881, then, was outside the large-scale factory-based industries. The two major exceptions were the metal and machine, and textile industries, one the industry of the future and the other, that of the past, with each representing around 5 percent of the urban labor force.

The 1901 census industrial patterns indicate change. By then 51 percent of the Milanese labor force was employed in manufacturing, 45 percent in services. The largest single industrial category was still domestic and personal service, but the proportion of the labor force here had declined by almost five points since 1881. The garment industry still accounted for the second largest number of workers, but its share of the urban labor force had also declined, from 15.4 percent to 12.6 percent. The number of persons employed in metallurgy and machine making, chemicals, rubber, and glass in the manufacturing sector, and utilities, transporation, and communications and public accommodations in the service sector had more than doubled in the almost twenty years between the censuses. The number of people working in textiles, meanwhile, had grown at a much slower rate,

53 percent. The largest numerical increase of Milanese workers in manufacturing occupations was in metallurgy and engineering, reflecting the great
expansion of this industry in Milan starting in the last years of the nineteenth century. There also was a very large absolute and proportionate increase in the number of rubber workers over the small base of 1881.

The recovery and burst of economic development that began in the late
1890s and was only partially visible in the 1901 industrial distribution of
the urban labor force had greatly transformed it by 1911. Fifty-five percent
of the workers were then employed in manufacturing. Although the number
of Milanese garment-makers grew by 22 percent, that industry was now
only second among manufacturing employers. The largest category of Milanese workers in manufacturing in 1911 was metallurgy and engineering,
with 43,181 workers, a rise of 83 percent since 1901. The other industrial
group that grew especially rapidly between 1901 and 1911 was construction, which experienced an 87 percent increase in its work force. In contrast, older, craft-organized or small-scale industries, like wood and straw,
leather and fur, precision instruments and precious metals, and printing,
had shrunk proportionately. The number and proportion of rubber workers,
on the other hand, had suffered almost total eclipse, because Pirelli had
built its new plant outside the city limits. Despite such expansion of the
suburban manufacturing zone, however, the 1911 urban labor force distribution by industry illustrates far-reaching changes from that of the earlier
censuses.

Workers in the tertiary sector, which employed 43 percent of the labor
force in 1911, were more likely to be found in services for industry, (utilities, transportation, communications) rather than for persons. Domestic and
personal service workers were still the largest group in the tertiary sector,
but their number had dropped by 11 percent since 1901; their proportionate
share of the labor force had also declined. In the same years, employment
in the industrial services had almost doubled. The larger number of Milanese workers in these services was, of course, linked to the greater importance of manufacturing. Utilities, transportation, and communication workers were the fourth largest category in the labor force in 1911.

Because it was an urban center populated by prosperous potential consumers, Milan continued to be the home of large numbers of service and
commercial workers. Because this city continued to be an enormous market
for consumer goods, it continued to be a center for producing them. For
these reasons and because space for the expansion of manufacturing within
the city limits ran out, Milan's labor force was never wholly dominated by
large-scale industry in the period before World War I. By 1911, the changes
in the city's resident labor force mirrored deep changes in its economic
base. Between 1881 and 1901, however, the city's working population was
still concentrated in industries such as domestic service and garment mak-

ing. Although large-scale industry began to grow rapidly at the end of this period, in 1901 its workers were still outnumbered by service and consumer production workers.

Nativity of the Urban Labor Force

Adna Ferrin Weber, a late nineteenth-century urbanist, wrote: "The effect of the migration of persons in the active period of life to the cities is to wrest away from the city-born the real work of the city." [13] It is true (as shown in Table 3-2) that the proportion of the Milanese labor force born in the city fell from 39.4 percent in 1881 to 33.8 percent in 1901. However, the in-migrants were not equally distributed across occupations and industries. Working-class migrants were employed mostly in the primary and tertiary sectors. The proportion of urban born for the two census dates, 1881 and 1901, in agriculture was 32.2 percent and 25.3 percent, and in services and commerce, 31.4 percent and 26.3 percent. Workers in the manufacturing sector were much more likely to be urban born than were those in the other two sectors: 47.9 percent and 41.0 percent. Former rurals were underrepresented in both the older artisan trades and the newer engineering sector.

Conversely, migrants to Milan were overrepresented in the construction, food, chemical,and rubber industries; a pattern consistent with that found elsewhere. Studies of urban migration have shown that the building trades were a major employer of unskilled male migrants. Well-established migration chains, often leading at first to temporary, seasonal jobs, led masons, stonecutters, ditch diggers, and simple laborers into construction gangs organized by middlemen, the subcontractors *(capimastri)* for large contractors in the city. The urban labor market for these workers was often on a daily or weekly hire basis, with a "shape up" on a designated city street. Such was the *ponte* on the via Tivoli in Milan. The population census of 1881 indicates that more than 5,000 construction workers and 83 *capimastri* lived in Milan but that only 32 percent of them had been born in the city. Some construction workers were doubtless not recorded by the census (which was taken on December 31, 1881) because they did seasonal work and lived in Milan only part time. At the end of the 1880s, when a series of public works projects were in full swing, the industry employed about 10,000. Many of the migrants were unskilled peasants driven out of rural areas by the agrarian crisis that took place after 1884 and by the repression of rural labor organization and strikes in the later years of the decade. Construction workers came from Lombardy and nearby Piedmont to the west. A 1903 survey of workers noted that masons were likely to come from regions such as the Varesoto and Comasco and that ditch diggers were often peasants from villages in the province of Milan, especially from the Abbiategrasso area.[14] Many of these migrants lived around the Porta Tenaglia, inside or outside the gate in the northwest of the city. The growth of Milan as a manufacturing and commercial city and a center of capital markets promoted the concentration of the construction industry and rural-to-urban migration to fill the jobs needed for its activities.

The proportion of Milan-born workers in the food industry resembled those in commerce and services more than those in other manufacturing sectors. This likely reflects the imprecision of occupational titles, as the census categories make it impossible to determine whether a given "butcher" or "baker" was a worker or a small-shop owner. The imprecise occupation title accurately reflects, however, the mixed activity of the industry, which included both the production and the sale of food. Migrants may have moved disproportionately into this kind of occupation because of the few specific skills required or because they migrated in networks that had developed along the lines of marketing farm-grown products in the city. Farmers' markets could offer a link for agriculturalists with potential urban patrons and employment. A similar link may have led rural-to-urban migrants into the chemical industry. The factories in this industry were the fertilizer plants, which offered heavy unskilled labor for unqualified rurals. Thus it is possible that these factories also recruited labor through their connections to the agricultural communities.

Rubber was the manufacturing industry with the lowest percentage of urban-born workers. Two factors probably contributed. Commander Pirelli himself claimed that it was his policy to employ recent migrants, in the hope that their rustic innocence would cut down labor militancy in his factories.[15] Further, rubber was a new industry, and so it was seeking cheap labor to build up its work force. The jobs were unskilled, and native-born Milanese thus had no advantage over migrants. Perhaps the migrants were more willing to try a totally new kind of work, which Milanese with more knowledge of the urban labor market, and more choices, might find smelly, crude, and poorly paid, or perhaps the migrants were directed there by word that there were jobs for new arrivals at Pirelli. Whatever the case, Pirelli found willing ex-rural workers and fulfilled his professed hiring policy.

At the other extreme were the manufacturing industries, in which the urban born were overrepresented: textiles, garment making, metallurgy and engineering, and printing. Workers in all four cases shared the relative youth that went along with urban birth. In a population growing by migration, the younger age groups are more likely to be urban born, the children of older migrant cohorts; young adults are most often recent migrants; and older adults are frequently long-resident migrants. Patterns in garment making and textile work, both heavily female industries, reflected the young age structure of the female workers as well as the youth of the urban-born cohorts. Workers in the garment industry were especially likely to have been born in Milan, although textile workers were also more likely to have been born in Milan than was the labor force as a whole (49 percent as opposed to 39 percent in 1881). As the textile industry feminized, with increasingly unskilled jobs replacing the skilled ones held by craftsmen until the 1880s and 1890s, the proportion of urban-born workers declined. In 1901, the industry was 42 percent Milan born, still well above the labor force average, which had fallen to 34 percent. Note also that the largely female tobacco industry workers (these are classified in Table 3-2 under "Other agricultural products") were also mainly urban born. This statistic most likely was related to the desirability of these jobs and the formal recruitment procedures of the government-owned factory, both of which tended to shut out migrants.

In the predominantly male metal and mechanical and printing industries, youth also reflected urban birth. The printing industry was peopled by highly skilled workers. They retained some craft traditions, in which workers themselves sought to control the labor market through limits on the hiring of their coworkers and successors (apprentices). Long training was necessary to do many of the tasks. The skill requirements and the collective restriction of access to jobs meant that newly arrived rural migrants could be excluded. Indeed, printers were 64 percent Milan born in 1881, 56 percent in 1901. The case was not so extreme in the engineering industry (in which a majority were born outside Milan), but similar skills and collective efforts to exclude have been documented.[16] Rural migrants from agriculture usually lacked the connections and skills that could lead them into the better jobs in the machine-shops. At the same time, some skilled migrants who had done similar work in rural industry or smaller towns followed the occupational migration networks into Milan. Hence in no industry were the skilled workers wholly native born.

Professionals and commerce and service workers were much less likely to be native born than were manufacturing workers. This difference holds even if the resident army unit, which almost completely had been born elsewhere than Milan, is excluded from consideration. Public accommodations and domestic and personal service workers were especially likely to be migrants, who often found their unskilled, unwanted (by city dwellers) jobs in the urban economy through personal connections. Like the food industry jobs, these service jobs were linked to the movement of agricultural products and services (rural wet nurses; sisters, daughters, and sons of carters and haulers; or farmers bringing their products into the city). Sometimes, for servants especially, the connection was a professional or wealthy landowner who had brought a servant from the small town or rural area where he had his estate. That servant might in turn help fellow villagers or relatives find posts in the city.[17]

Transportation, communications, and utilities workers were transferred or circulated from city to city in career migration patterns like those of the army. In some cases, such as the railroads, they were part of a superregional or even national labor market. Among the transportation workers, however, there was also a very different kind of experience: that of the carters, hack drivers, and the like. They found their jobs in the rural-to-urban network that also included construction workers, food sellers, servants, and waiters. Both these factors contributed to the low level of urban birth in the transportation industry, lower, indeed, than the average in the labor force as a whole.

Age of the Labor Force

Age is closely connected with migration patterns of both past and present. In addition to its stratification by place of birth, the labor force was stratified by age. Older workers were clustered disproportionately in agriculture and in the professions, commerce, and services. These sectors shared, then, the older age cohorts of migrants, whereas certain industries classified in the manufacturing sector (construction, rubber, chemicals) welcomed younger migrants. In the professions and arts, education and training requirements kept the minimum age of practitioners quite high, thus affecting the whole age distribution. Agriculture, domestic, and public accommodations workers were quite another case, for they were unskilled and much more likely to be migrants than were other workers. These latter workers migrated to the city and jobs in these industries, and they aged in their jobs. It was the urban-born children of such migrants who were able to move into less servile posts in manufacturing.

The age pattern of the labor force also offers details about the extraordinarily high rates of labor force participation characteristic of Milan in this period. In 1881, 88 percent of the men and 73 percent of the women aged 15 to 20 worked, proportions that were almost identical in 1901. Ninety-five percent of the men between the ages of 21 and 60 worked in both census years; 56 percent of the women in 1881 and 48 percent in 1901. (Unfortunately, these very crude age categories, which include vastly different proportions of the labor force, are the only ones available for comparing the two censuses.) There was also very high labor force participation for men and women over the age of 60: 63.5 percent (1901, 64.2 percent) and 31.2 percent (22.2 percent). The low dependency ratio already observed is due, as speculated, to the high labor force participation of the young, the old, and women. Age patterns also confirm the existence of a typical life cycle–linked pattern of women's employment. Although comparatively large proportions of women declared occupations at all ages, there was a steeper and earlier decline of employment among women than among men, which reflected women's increasing family responsibilities, beginning with marriage in their mid-twenties.

Gender and Jobs

Female workers were concentrated in relatively few industries. A high proportion of workers in textiles, garment making, tobacco, chemicals, rubber, and paper were female. In the tertiary sector about 30 percent of the professional workers were women, primarily teachers, midwives, or performers. These were the only middle-class working women in this period. In no census from 1881 to 1911 was there appreciable employment of women in business or commerce, although women tended disproportionately to be the *padrone* of small food or public accommodations businesses (boarding houses and the like). There were also large numbers of women in tertiary-sector occupations, particularly domestic and personal service. Along with garment making and textiles, these were the quintessential female industries.[18] Note here once again that garment making included a large proportion of home workers, whereas the textile industry in Milan was likely to be confined to factories, often small and specialized, except for the giant De Angeli printing and dyeing works. The textile industry had further feminized by 1901 (and even more so by 1911), as compared with 1881. The term *female industries* is used for those with a higher proportion of female workers than the average 30 to 40 percent. The tobacco, rubber, button, and paper industries also disproportionately hired females. These industries (as well as the printing and dyeing branch of textiles) were represented by few (often only one) very large factories, with (except for tobacco, in which women held skilled jobs) repetitive jobs in mechanized production, not unlike those in the early spinning mills. At the other end of the gender segregation continuum from these female industries were the male ones: most of the crafts, metal and engineering, construction, public accommodations, business, government, transportation, and utilities.

These patterns did not change during this period. In no census is there any significant female presence in industries other than those just listed. This pattern also holds for all age groups of women workers. If we look at the proportion of females in industry by age groups in the census years, females under 15, women aged 15 to 20 and older women all were equally concentrated in the female industries. Industrial segregation by sex was a persistent characteristic of Milan's labor force.

Of these female industries, the only ones with large numbers of workers were textiles, garment making, and domestic service. In fact, in 1881, three-fourths of all Milanese working women were crowded into these three industries: almost 6,000 in textiles, 25,000 in garment making, and 23,000 in domestic and personal service. Although the proportions of working women in these three categories dropped over time, during the entire period under consideration at least three-fourths of employed women worked in female industries, as broadly construed. Furthermore, of those industries that were top employers of women (i.e., textiles, garment making, and domestic service), both textiles and domestic service became even more female over time.

The age distribution of women in the labor force suggests that many women withdrew from regular wage work when they married and had children.[19] There were two major variants of employment, therefore, among Milanese women in the period: work in manufacturing, especially garment making and textiles, starting at a very young age and leaving wage work in the mid-twenties; and work in domestic and personal service, with some older women remaining in live-in service, unmarried, whereas others, more likely to be married, did day work on a less regular basis. These two patterns of wage work made it possible for young women to contribute to the family's income until they married (perhaps as that event approached, they were permitted to save their wages to help set up their new household) and for older women to support themselves or contribute to the family's livelihood.

These patterns of women's wage labor were associated with different birthplaces, as well as with different life-cycle stages. Garment workers were significantly more often born in Milan than were the adult or working populations as a whole; domestic servants were much less often born in the city. Young adult recent female migrants went into domestic service. Some of these women eventually left service and married, and their daughters became the young workers in the garment and textile industries.

Women's employment levels went down from 1881 to 1901 and dropped still further by 1911. Why? Women worked in only a few industries, linked to the consumption patterns of the service and commercial city, not the industrial city, which coexisted throughout the period. As industries like machine manufacturing, transportation, and communication became more important after 1901, and industrial segregation patterns remained unchanged, women's labor force participation declined.

Migration

This structural evidence provides a static picture of the effects of migration on Milan's population. A more dynamic perspective can be derived from the annual reports of the *Anagrafe,* the population register.[20] The usefulness of these figures depends on their completeness, and it is likely that short-term and single migrants were undercounted. Assuming that the recording bias was relatively constant over time, however, two kinds of comparative analysis are safe to do, based on the *Anagrafe* registration: a longitudinal comparison of aggregate levels on in-, out-, and net migration; and a comparison within registered in-migration of the migrants' occupations and status.

The fluctuation of net in-migration roughly corresponds to the timing of the business cycle. Migration was up in the 1880s, down in the 1890s, and recovering and again moving up at the end of the century. Demographic historian Aldo De Maddalena interprets migration patterns wholly in terms of the business cycle: "It is easy to see the correlation between the migratory flows and economic cycles. . . . [I]n-migration movements lagged temporally behind the economic."[21] The steadily increasing migration from 1887 to 1889, a period of urban economic crisis, counters his interpretation, however, unless we assume a three-year lag.

This phenomenon deserves a more satisfactory explanation, which can be found if the industries in which the migrants found work are taken into account. Jobs in the building trades (both the city-sponsored refurbishing of the Castello and the new park around it, and private housing construction) continued to be plentiful, even in the early years of the industrial depression. In the same period, the rural movement for improved wages and conditions for farm laborers in Lombardy was definitively defeated. From 1882 to 1885, the Mantovano; in 1885, the Bassa (the plain); from 1886 to 1887, the Vercellese and the Pavese; and in 1888 and 1889, the Alto Milanese and areas to the west were involved in labor strife in a wave-like fashion. (The last strike wave was both agricultural and industrial, for it included workers in the cotton mills near Milan; the earlier strikes were purely agricultural.)[22] Depressed prices in agriculture gave landowners strong incentives to resist their workers' demands.

Did the defeat of their strikes persuade landless or land-poor workers to leave the countryside? Contemporaries thought so. The *Bollettino dell'agricoltura* remarked in 1888: "The tendency to resort to Milan to seek better luck has become almost general, among workers, among day laborers, and among peasants who must rent land." Another comment from the period: "They come not because things are so good here but because things are so bad there."[23] Conditions in the countryside also contributed to the high levels of in-migration to Milan, by pushing workers out of jobs with deteriorating conditions and pay.

The proportion of migrants who called themselves peasants and general laborers rose by about 10 percent between 1887 and 1889, and these unskilled migrations account for part of the general increase of these years. The other group of workers whose representation in the registered migration expanded in these years were masons and construction workers. The combined increase of unskilled and building-trades workers caused the 1887-to-1889 peak of in-migration to Milan. These particular migrants affected working-class formation and behavior (discussed later). The rate of in-migration in the late 1880s was as high as in any subsequent years in the period covered here. It was quite a specific migration, however, in terms of where it came from, why, and in what sectors its members found work in the city.

Characteristics of the Migration

In-migration to Milan, gross and net, was primarily male until 1900. There was a surplus of women at census times, however. Male mortality was considerably higher than female, a factor that offset the migratory surplus of males and produced an excess of females in Milan's population. The age reports of in-migrants reveal a majority aged 16 to 50. This preponderance of adults in their working years is evident also in the urban age structure illustrated in the population pyramids in Figure 3-1. But there is a substantial proportion—almost 30 percent in 1894—of migrants aged under 16, which appears to be linked to the overrepresentation of households and the concomitant underrepresentation of solitary migrants in the population register. From 1894 to 1900, from 10 percent to 14 percent of registered migrants were students. An additional 11 percent to 14 percent had no occupation. (These were largely children under 6; housewives and pensioners were classified elsewhere.) It appears then that about two-thirds of the migrants under 16 were students or young children. Students, in particular, were likely to be members of higher-status families, which were perhaps overrepresented among the registrants. One-third of the migrants under 16 in the late 1890s who were neither students nor young children were apparently individuals or household members who came to find work. Labor force participation at young ages was characteristic of the migrants as well as of the population at large.

The crude categories of birthplace for registered migrants classify them among those born in (1) the commune of Milan, (2) the province of Milan, and (3) all other provinces of Italy. There are no surprises here. The number of in-migrants who were born in Milan Province (outside the city) was over 50 percent of those registered in 1888, 1896, and 1897 and 40 percent or above in the rest of the 1890s. It fell below 30 percent only in 1906 and later. Birthplace for the population as a whole by smaller geographical categories is available in the 1901 census, which classified the residents' birthplaces among the individual *provincie* of Lombardy and the *compartimenti* for the rest of Italy. According to that census, 31.8 percent of Milan's adult (16 and over) population was born in the city. (For the total population, the respective proportion was 43.4 percent.) The proportion of that total population born in other communes of Milan Province was 22.0 percent. Most of the other inhabitants of Milan in 1901 had been born elsewhere in Lombardy, contributing 17.6 percent to the total. Lombardy, including Milan and its province, was the birthplace of 83 percent of Milan's inhabitants. After 1905, migrants from outside Lombardy became proportionately more numerous. Nevertheless, the 1911 population census shows that most residents of the city who came from outside Lombardy came from the adjoining Piedmont, Veneto, and Emilia regions. Hence, although migration was a common experience for the Milanese, it did not involve long distances.

The temporal patterns of the occupations and status of male migrants to Milan show that skilled workers and white-collar migration increased until 1900. Nevertheless, skilled and unskilled workers were the largest group of male migrants throughout this period. Among women, the proportion of skilled workers changed little, although the proportion of unskilled did decline from decade to decade. Because many of these unskilled migrant women were coming in order to be servants, they were unmarried. As the proportion of migrants into service fell, the proportion of housewives increased. The higher proportion of migrant housewives reflects not only the smaller number of young women becoming servants but also the greater number of skilled worker and white-collar migrants with wives who did not do wage work.

There was about a 25 percent decrease between 1884 and 1900 in the proportion of migrants who called themselves peasants; by 1911, the proportion had declined by 75 percent over that of 1884. The chief reason for this decrease appears to have been the geographic expansion of Milan's urban zone: More and more poor, unskilled, rural migrants came to the surrounding areas rather than to the city itself. In 1884, the first annual volume of city statistics remarked disparagingly that peasant migrants "were unable to take advantage of the opportunities of the city or contribute to it, either."[24] A 1903 survey of Milan's working class concluded that nearly half of the migrants in its sample were peasants or unskilled workers ready to take any job at any wage.[25] In 1917, the rural migrants, by then settling in the outskirts of Milan rather than in the city itself, were described as "recently arrived . . . mostly Lombardese, with a primitive education and a lack of qualifications or skills."[26] The rural-to-urban migration of ex-agriculturalists with few skills thus continued, though much of it was displaced outside the borders of the commune.

Workers' Migration Patterns

Analysis of the censuses has demonstrated the strong links between age and nativity. The 1903 Società Umanitaria survey, which endeavored to reach the entire working-class population, found similar patterns. (This organization, established by a legacy from Prospero Moise Loria in 1892, became an active advocate of reform and, starting at the turn of the century, conducted many surveys to gather data to guide its activities.) This survey's definition of working class was "those who work with their hands." According to the Società's findings, almost 46 percent of the working-class population born in Milan was 13 years old and under; only 8 percent of those born outside Milan were in this age range. (This confirms that most of the children and students among the migrants belonged to higher-status households.) The young in the working-class migration were children 13 to 16 who came as members of households, old enough to earn wages and expected to do so.

Because of the primarily working-class nature of the migration and because of the primarily young adult age of the working-class migrants, the effect of migration on the age structure was large cohorts of very young native-born persons, the children of the migrants. Logic would lead us to expect that this group of young native-born persons should have been proportionately larger in the working-class population than in the population at large (because of the young age structure of the working-class population, leaving aside possible differentials in the groups' birthrates).

A higher proportion of workers and their families were born in the nearby province of Milan than that of the population at large. This difference is almost completely accounted for by the concomitant low (practically nonexistent) proportion of the working-class population born in distant parts of Italy and overseas. The survey concluded, as have most scholars, that migration frequency varies inversely with distance: "The kilometers to cover, whether on foot or by train, are an obstacle to migration that increases with distance."[27]

Information about length of residence is available for the working-class sample but is not cross-tabulated by age. Of the 280,519 persons in the Società Umanitaria survey, some 45 percent were born in Milan, and an additional 29.4 percent had lived in Milan for more than ten years. Three-quarters, then, of the working class had been born in the city or had lived there for ten or more years. This 75 percent was composed of children and old or middle-aged persons, the migrants of earlier years and their children. The other 25 percent of the surveyed working-class population was recent migrants in their young working years.

Conclusion

The late nineteenth–early twentieth-century demographic patterns revealed in this chapter reflect Milan's varied economy, combining consumer products and larger-scale manufacturing industries, service work catering to the bourgeoisie and workers alike (women did laundry and prepared food for workers who lived in rooms without running water or cooking facilities) and commerce. Overall, there was very high labor force participation. Proletarianization was the common experience of workers, but it occurred under different conditions.

Milan and its region were characterized in this period not only by flows of capital from country to city but also by movements of labor in the same direction. The city's population and its labor force were made up of men and women from the region, abandoning the small-scale agrarian and protoindustrial economies of the countryside and moving in order to find work. Most of the migrants came from nearby Lombardy, but over time, their origins were likely to be farther away.

Although many migrants, especially up to 1900, were rural dwellers lacking urban job skills, the migration process was not one of uprooting persons from familiar surroundings and tossing them into the anonymous city, like so many potatoes in a stew, as Valera portrayed the life of his feckless hero Giorgio in the discussion of the story that opened this chapter. Although there is little evidence about individual trajectories or the networks through which these people moved to the city (most often, if we generalize from better-documented accounts, via stops in smaller cities, temporary stays in Milan, and only sometimes settling for long periods or permanently), the patterns observed in the distribution of occupations are consistent with a migration process that was at least informally organized.[28] In many cities, migration and job finding are functions of the same networks. The systematic stratification of migrants and nonmigrants, old and young, men and women, skilled and unskilled, and in different industries and occupations suggests structured processes as well.

The interaction of changes in economic activity and demographic factors like migration and age and/or sex of workers affected a given industry's workers' strategies vis-à-vis their employers, their capacity to protest, and their chances for success in a workplace struggle. This relationship was seldom simply one of uprooting leading to protest but, rather, of connections and structure affecting solidarity and—under some conditions—the likelihood for collective action, as the ensuing chapters will show. The next chapter looks at the conditions of life and work of Milan's diverse but systematically connected working population.

4

Conditions of Life and Work

For working-class Milanese, wage work was instrumental, as it brought in the wherewithal to buy the necessities of life. In 1908, Alessandro Schiavi, a Socialist and journalist, emphasized the simple needs that workers as consumers sought to satisfy: "Today, what is the overwhelmingly salient character of the working class . . . speaking of it as a group of citizens?" he asked rhetorically. And he answered his own question: "That of being consumers. Consumers, above all, of housing, of bread, of milk, meat, vegetables."[1] In this chapter, we take a closer look at workers' way of life, the experiential component of Katznelson's second level of class. The sources for the evidence we discuss are the studies of Milanese reformers and groups interested both in workers' conditions and in improving them. Despite efforts such as these, Antonio Gramsci expressed his discontent with the state of knowledge in 1920, excoriating the elders of reformist Socialism in these words: "Where is your research on the economic conditions of the Italian nation? . . . Have you studied the way of life of the Italian proletariat?"[2] Franco Della Peruta went a long way toward answering Gramsci's questions, but our knowledge is still far from complete.[3]

In this chapter I have assembled exemplary evidence to answer four questions: How badly off were most workers? To what extent were poor conditions systematically linked to particular occupations or statuses, such as that of migrant? Did the conditions of workers and their families, in the workplace or out of it, deteriorate or improve between 1880 and about 1910? (Most of the surveys of working-class conditions were conducted after 1900, so more detail is available for that period. Those aspects of everyday life and work discussed in this chapter—employment, wages and hours, food prices and housing—come from available evidence that was relatively systematic and specific and covered a long span of time.) And finally, what part did workers play in improving their conditions (or in trying to do so)?

Employment and Unemployment

The high proportion of persons working for wages in the Milanese population, as revealed by indices of dependency calculated on census data, was examined in the last chapter. The high labor force participation meant that old people continued working as long as they could and that children entered the job market very young. (It also signified an active labor market, with some kind of job available for all who sought wage work.) The rising dependency indices show that this behavior was more salient in 1880 and 1901 than in 1911, suggesting possible improvements in pay so that the wage labor of the old and very young was not as essential to household well-being. The displacement of the incoming poorer population outside the city limits reminds us, however, that no such facile assumption is justified.

One correlate of the necessity for finding work and its ready availability was fairly low "background" unemployment (i.e., that occurring outside the years of cyclical downturn). No systematic data on recurring unemployment were collected for the period, but two estimates were gathered by self-report for large populations. One was based on the 1901 census, which asked questions about current employment status in addition to occupation. At census time (February 1901), 2.7 percent of men workers and .8 percent of women workers reported that they were unemployed, for a very low overall average of 1.9 percent unemployment. In the Società Umanitaria survey, taken on July 1, 1903, 4.3 percent of men and 3 percent of women workers were unemployed, for an average unemployment rate of 3.9 percent. The larger incidence of unemployment at that date was attributed to the season, summer being the "dead season" for some industries.[4] Seasonality of employment was typical, especially in consumer products and service industries. Both garment making and public accommodations work were very slow in the summer, it is true. Construction, however, had its slow season in the winter, so workers in that industry could count on only 200 to 250 days' work annually. Seasonality worked in both directions, therefore; the different levels of unemployment that the two surveys show probably cannot be attributed completely to the summer season in which the second survey was conducted. All that we can reasonably conclude is that both surveys reveal modest levels of unemployment in relatively prosperous years of the business cycle.

In addition to ascertaining the rate of unemployment as of July 3, 1903, the Umanitaria survey queried its respondents about episodes of unemployment during the previous year. Some 18,720 respondents (11.4 percent of the total number of workers in the sample) reported unemployment during that period. Unfortunately, only 27,365—including all those unemployed at the time—answered this question, hence a rate of 68 percent among respondents. It is reasonable to assume that the 11.4 percent is a minimum and that the true proportion was much higher. It appears, then, that outside the cyclical downturns (some of which were long and deep, such as the winters of 1889–91 discussed in Chapter 5), unstable employment, high turnover, and layoffs or periods of unemployment between jobs are probably better descriptions of the typical situation than unemployment alone.[5]

Not all workers were subject to equally instable jobs: Better-qualified workers were more sought after. In 1895, according to a Chamber of Labor report, there were more jobs for mechanics and wood craftsmen than there were takers, as production returned to higher levels after the long depression. Instead, employers were finding "too few able workers." Ten years later, the chamber reported in a similar vein that there were "plenty of unskilled workers—porters, laborers, and so on"—but a shortage of skilled workers.[6]

Paradoxically, with unstable employment, unskilled workers probably spent less time between jobs than skilled workers did. They had fewer resources and were so dependent on some sort of regular pay that they were obliged to find another job, any job, quickly. It was not easy to piece together wages to support a family in the city if the household head had unstable employment. Such unskilled workers had to spend much time and energy simply to survive in the city. According to the Società Umanitaria's working-class survey of 1903, the relatively well paid metal and machine shop workers had an unemployment rate more than twice that of domestic servants or unqualified general workers.[7] Fifteen out of the seventeen machine workers' leagues had unemployment subsidy programs in 1905.[8] Workers in this sector could be assisted by their unions to ride out layoffs and temporary unemployment; thus they could afford to wait, at least for a while, for a job for which they were qualified.

Unskilled workers were more often between jobs or seeking jobs, but they could not afford to spend much time unemployed. Such was the case for bakers, as they were described in 1905: "All, or mostly all, people from the Milanese countryside and from the Lombard rural areas . . . often unemployed or working at bestial labor, poorly fed, poorly lodged, poorly dressed, ignorant, they cannot adapt themselves to civil, regulated life."[9] Similarly, cooks, waiters, and barbers also were said to be constantly changing jobs, and these all were typical migrant occupations.

If they could not find another job, workers simply left the city. Their migration had covered only a short distance, and assistance from relatives, perhaps a bit of land, and wages of some sort or live-in service might be available in the country for returning unemployed rural-to-urban migrants. In-migration was quite responsive to the demand for unskilled labor, but when such jobs were harder to find, as in the early 1890s or from 1908 to 1910, out-migration was the result. Between 1893 and 1903, unskilled male workers represented 42.6 percent of the in-migrants registered by the *Anagrafe,* or population register, but 54.4 percent of the out-migration. In contrast, skilled workers and artisans made up 43.5 percent of the in-migration but only 32.5 percent of the out-migration.[10] During a long building-trades strike in 1910, the national *Bolletino dell' Ufficio del lavoro* reported that many workers either sought out other jobs or left the city altogether.[11]

Workers who had families in Milan could also get help from family members who were earning wages. This was the case for many of the respondents to the 1903 Società umanitaria survey. The average number of days of unemployment in the past year for respondents to these questions was thirteen. Periods of unemployment were very long for some of the unemployed: three months or longer among one-third of the currently unemployed.[12] Those who were out of a job for so long needed to have other members of their families working in order to survive.

The multiple employment of family members is shown by the household employment data in the same survey. In over two-thirds of the households (excluding single-person households), more than one-half of the members were working. In fact, 70 percent of the males and 40 percent of the females, of any age, in these working-class families had jobs. The results were enormously high ratios of workers to consumers: half of the families with 2 or more members had 1.6 or fewer consumers (including themselves) per worker. According to the index of dependency, this would be .65 dependents per worker. (The index calculated in Chapter 2, is based on the entire population, thus it is higher than the ratio here, which is based on the working-class population only.) Women workers were especially prevalent in families with up to 5 members. In families that were bigger, the rate of labor input for females (and males) started to decline.[13]

Job instability, job changing, and frequent periods of unemployment were said in a Chamber of Labor report to be "chronic in some occupations but present in all" during the period.[14] Cyclical unemployment was much higher and more threatening to public order. In the winters of 1889–91, thousands of construction and engineering workers were out of work, a situation that was different from that in previous times, wrote journalist Ferdinando Fontana in 1893:

> By unemployed I don't mean the bands of workers that until now . . . have been for long or short periods "out of bread"; *unemployed* today refers to the enormous masses of workers laid off from one day to the next; in short, not little clouds on the horizon, but huge clouds threatening storms.[15]

Both workers and city authorities sought ways to stabilize the labor market in the wake of this cyclical downturn.

Finding Work

Charitable institutions, labor organizations, and the city government, helped people find work. Both the Camera del lavoro (Chamber of Labor), founded in 1891, and the Società Umanitaria, established in 1892, assisted workers in finding jobs. The city council, concerned with unemployment as a threat to public order, provided a subsidy to the chamber in support of its *uffici di collocamento* (employment offices).[16]

In its first year, occupational groups ran these employment offices for bakers; masons and construction laborers; metal and wood workers; and cooks, waiters, and associated workers. The general secretariat, headed by glovemaker Giuseppe Croce, handled other types of workers. By May 1892, after eight months of operation, the chamber's general secretariat had found jobs for 1,379 workers, including 354 in construction.[17]

The Società Umanitaria asked about 20,000 workers in its 1903 survey who reported getting jobs in the previous year how they had found their new jobs. Only 18.3 percent declared that they had used the Camera del lavoro employment service; the others had applied directly to employers (30 percent), relied on private employment agencies (22 percent), or asked relatives or friends for help (22.7 percent).[18] A few years later, another report by the Umanitaria remarked how durable the traditional subcontracting or other mediated employment systems were in many industries. It would take a long struggle to eliminate them and the inequities they caused.[19]

The Società Umanitaria and the chamber combined their employment offices in 1906, but some unions, such as the printers, continued to run their own agencies, now sanctioned by contract with their employers but still based in the quarters of the Chamber of Labor. A Catholic workers' labor league began to function at about this time, and although its results were uneven because of the lack of resources, it successfully placed 1,188 workers in 1908.[20]

A sampling of the reports of the Chamber of Labor–Società Umanitaria employment office reveals that in those years, about half the applicants were placed. Table 4-1 shows the record of the agency for the years 1906 and 1907. The jobs in which the registrants were placed were, by a great majority (85 percent), in the city of Milan, but jobs in nearby communes were also listed.

In the same years, the major groups seeking work through the office were unskilled workers and skilled ones thrown out of work by the crisis of 1907, particularly automobile workers.[21] In regard to requests for workers from employers, there was some fit with the registered unemployed in the case of painters, finishers, carpenters, and tailors, for example, but there were also many requests for shoemakers and helpers and workers in paper

Table 4-1. Società Umanitaria–Chamber of Labor Employment Office Data

Year	Number of Workers Registered	Number of Jobs Filled	Registered Workers Employed (%)
1906	8,692	5,089	53.2%
1907	11,331	5,906	48.1

Source: *Bollettino dell' ufficio del lavoro* 9 (1908): 881.

and glass. These often were jobs that unskilled migrants could fill. The best rates of job finding were among the most skilled and the least skilled workers: on the one hand, mechanics and steam engine operators and, on the other, café workers, porters, and messengers. In 1908, the rate of placement for registered unemployed workers was generally lower across the board, as could be expected for a depression year.

The employment service of the well-organized printers did not outperform the Camera del lavoro employment office. Over 99 percent of the registrants at the printers' employment office were union members. Nevertheless, their placement rate (50 percent for compositors and 75 percent for pressmen) was not as high as that of machine shop workers or unskilled workers through the Chamber of Labor–Società Umanitaria employment office.[22]

Despite the good intentions and efforts of reformers and workers' success in building institutions, both classwide, cross-occupational institutions like the Camera del lavoro and single-craft ones like the printers' union, they enjoyed only modest success in their intervention in the labor market. The choice of how many and which workers to hire remained that of the employers, despite union leader Ludovico D'Aragona's extravagant claim in 1907 that the Milanese Camera del lavoro employment service worked "marvelously well."[23]

Wages and Hours

A very full list of actual wages received by the thousands of workers surveyed by the Società Umanitaria in 1903 reveals the hierarchy of workers' wages that applied throughout this period.[24] Table 4-2 summarizes these findings: the daily wages actually received by 112,171 workers with 139 occupations, grouped in the industrial categories used for the labor force analysis in Chapter 3. Because both the actual wages received and the number of workers in each occupation who reported a given daily wage were provided, the averages were weighted by the proportion of workers in each category reporting a wage.

The highest average wages were in the tertiary sector, earned by work-

Table 4-2. Average Daily Wages of a Sample of Milan Workers, 1903

Category	Number of Workers	Avg. Daily Wages Lire/Day*	Rank
I. Agriculture	206	1.3	10
II. Industry			
Food	3,469	1.5	7
Wood, straw	8,364	1.5	9
Leather, fur	4,400	1.3	12
Textiles	8,139	.9	18
Garment making	21,386	.7	20
Other agric. products	738	.9	17
Ceramic, glass	3,264	1.3	10
Metal, machine	17,200	1.6	3
Precision, precious metal	1,236	1.5	8
Chemicals	1,272	1.1	15
Rubber	1,511	1.3	13
Printing, lithography	5,804	1.9	2
Construction	7,878	1.6	5
Paper	2,236	.9	16
III. Commerce, Services, and Professions			
Unspecified workers in commerce and business	8,840	1.5	6
Banking	—	—	—
Professions, arts	—	—	—
Government	—	—	—
Utilities	7,428	1.9	1
Domestic, personal service	8,100	.7	19
Public accommodations	2,259	1.6	4
Retail merchandising	1,404	1.2	14

*Wages weighted by numbers of workers in industry with each wage given.

Source: Società Umanitaria, *Condizioni generali della classe operaia in Milano: Salari, giornate di lavoro, reddito, ecc.* (Milan: Società Umanitaria, 1905), pp. 250–53.

ers in transportation and communications. (The highest single daily wage for an occupation was the 3.28 lire of the tramway mechanics.) The survey estimated that wages lost because of unpaid holidays, short time, and periods of unemployment in most industries meant 10 percent less money in the pocket over a year than would be expected with full-time, year-round employment. In industries like construction, however, there was a much shorter annual employment period than the average. This meant an in-pocket wage of only two-thirds of what construction workers would have earned if the daily rate had pertained to a full fifty-two weeks of employment (six-day week).

The tertiary sector also included domestic servants whose wages placed them among the most poorly paid. The top manufacturing wage earners were printers at 1.9 lire daily, followed by metal and engineering and construction workers. Workers in consumer industries like wood and straw, leather, and food production fell into a middle position. The bottom of the wage hierarchy of manufacturing workers were categories with large numbers of female workers: textiles and garment making, other products of agriculture (tobacco), and paper. (The garment industry, to make matters worse than the very low daily wage suggests, shared with construction a very high proportion of time in dead seasons.)

Evidence from the Pirelli rubber factory illustrates the effect on a category of large proportions of female workers. There, according to an 1896 report, qualified male workers could make an average daily wage of 2.8 lire, but at the same time, women in a similar category were making only 1.5 lire.[25] Because just about half of the Pirelli workers were women, the average wage was low. Similarly, in the 1903 data, among largely male tanning and leather workers, the average wage was 1.4 lire. Adding the glove makers (70 percent of whom were female and were paid .9 lira daily) and the fur workers (75 percent of whom were female and were paid .98 lira daily) dropped the average pay for the combined leather category to 1.3 lire.[26] The message of this cross-sectional comparison of wages is that the largest wage differential was a gender differential. Workers in the "female" industries in which women were concentrated by occupational segregation were usually wretchedly paid.

This gender differential was compounded by the relative youth of female workers. Young workers were the other category that was especially poorly paid. Almost all workers 11 years old and younger received a daily wage of less than a lira, and about three-quarters of the boys aged 12 to 15, but 90 percent of the girls in the same age group, received no more than a lira. Further, over 50 percent of young women 16 to 19 years old also received the same low wage. The highest female wages were clustered around 1.5 lire a day, for women 20 to 49 years old; older women again earned a lira or less.[27]

Migrant-hiring industries do not appear to be particularly low paid in this listing, except for chemicals and rubber. In the rubber industry, average wages were low as well because of its numerous women workers. However, the daily wage measure does not take into account the phenomenon of seasonal employment, which was particularly important to construction, an industry in which many migrants were employed, or the lack of stability in many migrants' jobs. Quite apart from industrial differences in daily wages, the survey found that persons who had lived in Milan for three years or less were much more poorly paid than were those who had lived there longer.[28]

The instability of jobs in all industries is also underestimated, as the respondents were asked to report their daily wage in their current job. Franco Della Peruta labels the survey wage information "probably too pessimistic" because people were likely to lie about wages on the low side. He does not explain why this might be so. Actual daily wages are bound to be lower than either contractual wage rates, which may not apply to all workers in an industry, or wages reported by employers, who are likely to exaggerate on the high side to make themselves look good in the government statistical reports for which most other wage data were collected. Unlike Della Peruta, I accord considerable weight to these data. He himself displays ambivalence when he argues next that the Società Umanitaria survey's wage information is nevertheless "valid evidence of a general situation of hardship" among Milanese workers.[29]

This wage hierarchy provides a different insight into family wage pooling as a counterbalance to unstable employment. Much unskilled male employment was seasonal (construction work especially); so also was much female work (garment making and, to a lesser degree, textiles); women's work and children's work was miserably paid. Although the family could, by throwing everyone into the labor market, make ends meet somehow, life was especially precarious at times when women's and children's wages were the primary sources of the family wage. Indeed, their wages were so low partly because they were perceived by employers as supplemental wage earners, in contrast with the male primary wage earners. Thus they were frequently not paid enough to ensure their own reproduction. But there were other reasons for the low wages of women and children. Their crowding into a relatively small number of occupations meant that all women and children workers were in direct competition with one another. There was also an almost unlimited supply of labor in the 1880s and 1890s. These aspects of women's and children's wages were not noted in the Società Umanitaria report. Although reformers were critical of the low wages, they did not recognize their effect on working-class families. The survey also missed the widespread incidence of such marginal work. As one review of the survey pointed out, although it intended to count all workers, it missed many, compared with the census of 1901. These missed workers were largely the "invisible" women workers already discussed: 11,000 in homework in the garment industry and about 15,000 in domestic service.[30]

When wages are averaged and weighted by the number of workers in an industry receiving them, the high wage rates of some occupations, especially in printing, metal and machines, and utilities and transportation, are obviously exceptional. Overall, however, if only men's wages are considered, the spread of wages in Table 4-2 is not really very wide, partly because it represents an averaged daily wage, not an hourly wage. Long hours of work in industries like food and public accommodations brought their daily wage close to manufacturing wages. Men in these industries could make close to a living wage by working longer, by quickly finding any alternative employment if discharged, and by being uncomfortably dependent on the contributions of poorly paid women and children family members in times of job instability. Although the wages of Milan's workers as revealed in the 1903 survey by the Società Umanitaria were overall quite modest, there were some consistent differentials. Most important were the systematically lower wages for recent migrants and women. These were also the segments of the labor force with the least stability. Because migrants and women were disproportionately concentrated in a few industries, they were in competition with one another, and the result was a surplus supply of labor in these categories. Their ability to act collectively in the workplace with any degree of success was limited by this situation, as shown in Chapters 6 and 7.

Temporal Change in Wages

To what extent did wages improve over the period studied?

In response to this question in 1907, social researchers at the Società Umanitaria warned:

> Increases in salary and reductions in hours cannot be calculated, because reliable documentation is missing and because we cannot be sure that increased wages were generally accepted and actually paid, except for three categories: compositors and pressmen, masons, and bakers. Furthermore, we cannot assume that these increases are an index of general improvement in the working class, especially in those areas that are unskilled and unorganized.[31]

Keeping this warning in mind, hourly wage rates for selected occupations (those just noted, plus textile, foundry and puddling, rubber, machine shop, tobacco, and railroad workers) and for the years between 1885 and 1911 were calculated in terms of the hours of work needed to buy a kilo of bread at current prices (for the internal *circondario,* where the high *dazio* affected bread prices).

Although workers' diet was more varied than just bread, bread prices are a good indication of workers' food budgets as a whole because bread constituted such a large part of their diet. Giovanni Montemartini, a sociologist who had conducted studies for the Società Umanitaria, objected to the varied food market basket constructed by Geisser and Magrini in their ambitious longitudinal study of real wages in Italy in the second half of the nineteenth century. "When dealing with industrial workers," he wrote, "it is essential to take into account the price of bread."[32] Expenses for food amounted to 65 percent of the workers' total budgets in a 1914 study of fifty-one families, and the major item was bread.[33]

The actual wages reported in the Società Umanitaria 1903 survey were in almost all cases much lower than the published wage rates, on which this longitudinal comparison is based, would have us believe. The reason is, of course, that some of the published rates were rates paid in certain factories to particular workers, others were contractual rates established by unions in agreement with some employers. Many workers, even in the same industry, worked outside these settings or unprotected by union contracts. For example, in 1904, 8,000 building-trades workers were reported to have joined leagues.[34] In the 1901 census, however, more than 9,500 men reported occupations in the construction industry, and these were only those who lived in Milan. The union membership figure surely included men who commuted to work in the city and perhaps seasonal workers as well. Union membership sank even lower later in the decade; in 1907, a recession year, only 2,800 construction workers were reported to be members of the Chamber of Labor through their league membership.[35] There were plenty of non-unionized workers who probably were not getting the union wage rate. All of the workers discussed here are to some extent exceptional, as they either worked in large plants chosen as typical for the purposes of data collection, or they belonged to unions that over the years negotiated public wage rates, or *tariffe*.

The wage hierarchy evident in Table 4-2 is apparent in these wage rates as well.[36] Railroad men, printers, some skilled mechanics, and the men working in the tobacco factory were the best-paid workers. Masons followed very close behind. The next highest cluster of wages included those of rubber, chemical, and gas workers. Male bakers and spinners had the lowest wages. Females were lower paid than males in every case, but the tobacco monopoly workers were the best-paid women.

In factory work throughout this period, the 10-hour day was most common. (Masons and associated workers also had an agreement with the *capimastri*—often honored in the breach—for a 10-hour day, starting in 1887.) Textile workers and garment makers usually worked longer hours. The smaller shops in the crafts, such as wood and straw, leather and fur, also had longer hours in their busy seasons. Workers in food production, such as bakers; in public accommodations, such as waiters, cooks, elevator operators, and porters; in personal service, such as barbers, hairdressers, and laundresses; in domestic service; and in some highly seasonal occupations (dependent on good weather), such as sand quarrying and brick making, worked exceptionally long hours, which changed little during this period.

Even in the more bureaucratized factory industries, the heralded 8-hour day was slow to come. Although workers' organizations demanded the 8-hour day as early as May Day 1890, they were unable to achieve this goal before World War I. In 1913, an automobile workers' strike succeeded only in reducing the Saturday workday from the usual 10 hours to 8.

There were two outstanding exceptions. One was the printers. As early as 1902, unionized printers worked only 9 hours. Some who did automatic typesetting worked only 8 hours. The tobacco monopoly plant was the other exception. Tobacco workers enjoyed an 8-hour workday from the 1890s to 1904, when it was reduced to only 7 hours (to be increased again to 8 in 1910).

Real wages in all the occupations that I considered improved at least until 1908. As Alexander Gerschenkron pointed out in his studies of Italian industrialization, its benefits were passed on to workers during the period of growth that started in the last years of the nineteenth century.[37] This growth improved wages not only for industrial workers but for service workers also, at least in the period of consumer price stability up to 1905. As the price of bread then started to edge up, not all wages kept ahead of inflation, and many workers surely felt the pinch. Schiavi quoted the common opinion that "everything is getting dearer, life is getting dear, one cannot live."[38] Some workers like the cotton textile workers (Lombard textiles were in crisis from 1907 on) and the puddling furnace workers suffered losses in their real wages. These were unusual cases, however; it seems likely that real wages in the occupations examined here were stationary or improved somewhat up to 1911. More light can be shed on this trend by looking at indices of food consumption.

Food Consumption

Public health controls regarding the city's slaughterhouses and the records of the *dazio* entry duty, which was levied on products brought into the city, provide evidence concerning the quantities of food products consumed in the city. The statistical yearbook *(Dati statistici)* published annual figures for the weights of cattle slaughtered in the city and fresh meat imported through the *dazio*. Guglielmo Tagliacarne used these figures to calculate (taking account of bone in the dead weight of carcasses reported) an index of annual beef consumption per capita from 1881 to 1932. For the period up to 1911, these amounts oscillated between a minimum of 31 kilos (1904) and a maximum of 38 kilos (1888), with an average of 32 to 33 kilos. The 1911 figure was exactly 33 kilos per capita.

The appropriately named Tagliacarne chose to calculate the consumption of beef alone, not the total of all animals slaughtered, presumably because the latter varied in the types of animals listed over the period he considered (1881–1932). "The index of meat consumption is among those usually held to be most significant as evidence of the improvement of the conditions of life of a population," he wrote. "The calculation is very complex and uncertain; because of this the results cannot be assumed to be correct. Nevertheless, they [are] . . . adequate at least to suggest the tendency."[39] Equally tentative is my calculation of all types of meat slaughtered in Milan, plus freshly slaughtered meat (both minus 20 percent for bone) brought into the city from 1897 to 1911. By this measure there was an increase in available meat per capita of well over one-third from 1897 to 1911.[40] By the end of this period, the poorer population—particularly the new migrants—was more and more likely to live outside Milan proper. Hence, the greater amount of meat per capita in the city may have been an artifact of the declining proportion of low-income persons in the city.

The meaning of these figures is ambiguous also because it is not known how meat consumption was actually distributed in the population. This uncertainty is illustrated in two contemporary comments. The editors of the city's *Dati statistici, 1897* proudly pointed to increased meat consumption over recent years, which they attributed to "the social well-being that is so distinctive in Milan, the fruit of the industrial activity that reigns there." They nevertheless concluded that either

> a large part of the population does not partake of this food at all, or, if they do partake, it is of such a small quantity as to be incalculable. To what causes should we attribute this disagreeable fact? There is no doubt that, first, there are many families that lack the means to buy meat and instead buy foods of lesser value and that, second, Milan's little people prefer vegetables.[41]

Fifteen years later, in 1912, the Milanese Chamber of Commerce pointed out with satisfaction that increased meat consumption over the years was a sign of the population's higher standard of living and that there was need for a bigger meat supply to meet the rising demand. The Chamber of Labor, in the same year, joined the call for municipal action to enlarge the city's supply of fresh meat; however, at the same time, it did not miss the chance to complain about prices.[42]

Family or individual food budgets are a better indicator of food consumption habits by class. But the few available workers' food budgets before Pugliese's detailed 1913 survey of the diet of 51 working-class families in Milan are not very reliable. His sample included 210 people, all from the better-off section of the working class, he claims. The families kept diaries of their purchases and consumption of food, for which he then calculated caloric content and proportions of fat, sugar, and protein. Pugliese set the male requirement at 3,800 calories per day (for a man doing physical labor), which is high. Post–World War II Italian industrial workers consumed about 3,500 calories daily. The ration of an Italian soldier in 1904 provided 3,500 calories, not counting those from wine.[43] How many calories were needed is an academic problem, anyway, as only one-fifth of the families studied surpassed 3,000 per capita.[44] (Weights were assigned for men, women, and children of various ages, so the *caput* mentioned here is a sliding scale in which adult males equal 1, women 0.8, and so on down.) Further, Pugliese showed, the Milanese not only lacked calories; their diets also were deficient in all the necessary components. He concludes that poor diet was an important factor in what he perceived as the progressive immiseration of the working class during industrialization and urbanization.[45]

Curiously, Pugliese found that there was a good supply of protein in most of the diets, as opposed to fat and sugar, which supply more calories per gram. To explain this he observed that urban diets are heavy in animal products rather than plant products,[46] and because animal products cost more per calorie, this contradicts his immiseration argument. (Note also that this conclusion contradicts that of the 1897 *Dati statistici*.) In any case, about 16 percent of Pugliese's families had a per-capita caloric intake that was low by any criterion, though none fell below 2,000 calories per capita. His study reveals what an important part in the families' total expenses that food played. In no case were food expenses less than 50 percent of the entire family budget, with the highest proportion being 97 percent. There were great contrasts among family situations connected with the occupations of the heads of households and, importantly, the number of persons in the household. The worst diets were eaten by large families whose fathers worked as, respectively, a tram driver, a garbage collector, and a gas company worker. One such family, with nine children, ate little else but vegetables, bread (4.1 kilos a day!), and wine, along with 1 liter of milk a day. The whole family's total caloric intake and other food value intake were actually smaller in many aspects than the intake values of a fat, childless couple, who lived on a very rich diet, indeed.[47]

Male heads of households often ate better, relative to the other family members, because they could obtain good dinners near their workplace. In fact, that midday dinner provided an important part of the family protein intake, even though it was eaten by a single family member. Pugliese also attributed the many thin women to malnutrition, arising from mothers' readiness to give up food for their children and their husbands.[48] This suggests that married women probably consumed less than their estimated share—80 percent of that of adult males—and that gender inequality in food consumption was common. (It is also possible, of course, that the women ate less because they were small and needed fewer calories. Most likely there was an interaction in this case between cause and effect.)

Pugliese, like many reformers of the period, was also concerned about alcohol consumption. There was great variation among the families in wine use, but he regretted that some families consumed up to 600 calories daily in this form. Nevertheless, he was limited in his ability to calculate total alcohol consumption because men often drank not only at meals in the family or work setting but also in cafés after work and on holidays.[49]

Like the evidence regarding wages, that on food consumption suggests inequality among occupations and within families, but certainly no deterioration and likely some improvement in living standards across the period. Indeed, the printers' union newspaper could state with satisfaction in 1912 that "modern workers no longer resemble those of the past: They enjoy their comforts; they do not live in indecent hovels; their clothing is clean; they own bicycles; and they buy newspapers."[50] Workers had succeeded, through workplace collective action, in keeping pressure on wages and resisting immiseration. Their situation at the end of the period was not wonderful, but it was not any worse. Any ambitious version of a working-class agenda regarding conditions of life and work, however, was still largely unachieved at the eve of World War I.

Housing

Badly built, poorly maintained, and crowded housing was a problem of long standing in Milan. The debate about the need for "popular housing," as it was called in contemporary discussions, led as early as the 1860s to intervention by charitable and workers' associations. It came to the fore as a political issue in the first decade of this century when the concern with extraordinarily crowded and unsanitary housing and the high mortality related to such conditions finally prompted the city to intervene. Despite much discussion, much investigation, and some building, the problem was only partially relieved by community action, and its immediacy was reduced by private action—workers', especially migrants', choice to live outside the city's boundaries.

During the civic-building boom of the 1860s in Milan, a group of public-spirited citizens formed the first privately funded association to build sanitary and safe homes for workers: the Società edificatrice di case operaie, bagni, e lavatoi pubblici. The city council granted land to the project. Twenty-two large houses containing 822 rooms were built, and apartments were to be rented or sold on long-term contracts. By current standards the houses were comfortable—each unit had running water, for example. In later years, however, one commentator admitted that from the point of view of modern conveniences, the houses left a lot to be desired.[51]

The Consolato operaio, a federation of workers' societies led by Democratic Radicals, launched the Società edificatrice di case operaie in 1879. Borrowing from the Cassa di risparmio (the famous Lombard savings bank), the Consolato association built smaller two-story houses, with good terms for mortgages to finance purchases and make workers property owners. Over the long run, however, most tenants found it impossible to maintain the payments, and the Società itself eventually became the houses' owner.[52]

A project proposed in 1885 by a group of private investors, including Mayor Guilio Belinzaghi, called for the renewal and development of the area around the Castello. Grandiose plans were drawn up by architect Cesare Beruto. Stefano Allochio, a notary active in city affairs, wrote *La Nuova Milano* (New Milan) as his contribution to the debate. Allochio's book, a vigorous polemic, reviewed the housing situation in the city as of the census of 1881.[53] The volume was not intended to promote the building of more inexpensive housing. Far from it. Allochio and the proponents of the plan reasoned that there was a shortage of housing and that it was expensive housing that was needed!

Allochio's evidence regarding the workers' housing comes from the 1881 census and is not broken down by class (see Table 4-3). Nevertheless, it captures the crowding characteristic of Milan's housing. The majority of Milanese lived in apartments of one to three rooms. In the external *circondario*, where the population was largely working class, 86 percent had such apartments. A breakdown by family household size (households of 2 or more related persons, not counting nonrelated groups living together), shows that families of all sizes were jammed into tiny apartments; 162,046 persons, in households ranging from 2 to 16 members, lived in one- or two-room apartments.[54]

The municipal council rejected the Nuova Milano plan, partly on the grounds that inexpensive housing rather than a *quartiere signorile* (aristocratic neighborhood) was needed. A committee headed by Giovanni Battista Pirelli was appointed to study the problem. Its report, in the form of a statement of principles, concluded that any new plan ought to "consider the possibility that in some part of the city there be land allocated at low cost for building healthy housing for the poor."[55] Although the city council accepted these principles, in the end, no low-cost housing was included in the plan. Construction of new streets and private housing for the wealthy did, however, provide jobs for Milanese building-trades workers in a difficult period when agriculture contracted and migrants poured into the city.

Little was achieved on housing issues in the 1890s because of the economic crisis and the conservative city administration's lack of interest. A suggestion by socialist city council member Osvaldo Gnocchi-Viani in January 1891 that the city build workers' housing (part of a larger statement proposing solutions to a wide range of workers' problems) was rejected by the council committee appointed to consider it. The committee declared that housing should be left to the private sector but that construction of dormitories for homeless workers was desirable. (An earlier dormitory project, funded by private donations and built in 1884, had provided beds for 18,000 person nights per year.[56]) After the turn of the century, however, the question reappeared. The Società Umanitaria summed up the alarming situation: Housing for workers was the most urgent public question in Milan.[57]

Table 4-3. Details of Milanese' Housing, 1881

Apartment Size	Available Apts. (%)		Population Living in Apartments (%)		Avg. No. Persons/Room	
	Int. Circ.	Ext. Circ.	Int. Circ.	Ext. Circ.	Int. Circ.	Ext. Circ.
1–3 rooms	69%	90%	60%	86%	1.7	2.5
4–6 rooms	19	8	22	11	.8	1.1
7+ rooms	12	2	18	3	.5	.8

Source: Stefano Allochio, *La Nuova Milano* (Milan: Hoepli, 1884), p. 46.

The shortage of housing was acute and was caused by the resumed heavy in-migration and more strictly enforced housing regulations, which meant that some particularly decrepit buildings were condemned and torn down. The city council's Democratic majority, elected in December 1899 (including Democratic Radicals, Socialists, and Republicans), was slow to respond to the problem. The Chamber of Labor began a campaign in 1901 to study the problem and propose ways of improving it. During 1901 and 1902, committees of workers and other citizens recommended "municipalization"—city-financed public housing. Despite the split between themselves and revolutionaries, the reformist Socialists again joined the democratic coalition in the city council after the 1902 election.[58] When the council took up the recommendation for municipal sponsorship of low-cost housing, the majority felt, in principle, that some kind of city action was desirable, either on its own account or in conjunction with other public or private institutions, such as banks, charitable foundations, or the Società Umanitaria. In March 1903 it approved a plan for a modest number of inexpensive housing units. At the same time, it commissioned the Società Umanitaria to conduct both a statistical study tracing the supply and demand of housing over the previous years and a survey to investigate current conditions.

Giovanni Montemartini's housing report for the years from 1890 to 1901 established that the supply of new housing was not keeping up with demand. Further, more expensive than low-cost units were being built. Using net in-migration as an indication of the demand for new housing, Montemartini divided the migration, like the housing, into workers and bourgeois. He then calculated a ratio of new units of population per new housing units, as shown in Table 4-4. Housing for workers, in particular, had begun to fall into short supply as soon as migration picked up after the depression years of the early 1890s. By 1900, there were four in-migrant workers for each new unit of worker housing. Montemartini ended his report with a call

Table 4-4. Relationship Between Supply and Demand of New Housing Units, 1890–1901

Year	No. of Worker Migrants per Worker Housing Unit	No. of Bourgeois Migrants per Bourgeois Housing Units
1890	1.0	0.4
1891	1.2	0.3
1892	1.7	0.2
1893	1.5	0.8
1894	1.8	0.9
1895	3.5	0.5
1896	4.1	0.5
1897	3.1	0.6
1898	3.5	0.6
1899	2.9	0.8
1900	4.3	0.8

Source: Giovanni Montemartini, *La Questione delle case operaie in Milano* (Milan: Società Umanitaria, 1903), p. 25.

for public intervention, either to build housing or to increase competition in the housing market by other means.[59]

A survey of workers' housing as of July 1, 1903, was part of the general survey on the condition of Milan's working-class population. The goal of the housing survey was to interview all residents of apartments of up to three rooms. They counted 332,841 persons in this category, or 70.2 percent of the population as a whole. These people were distributed according to apartment size, as shown in Table 4-5. Density, in terms of persons per room, was again highest in the outer zones of the city, which were, of course, the working-class neighborhoods.

The yearly rents of workers' apartments constituted 11 percent to 15 percent of their household income. A small proportion (6.5 percent) of families received free housing, a benefit earned by members serving as the concierge or janitor for an apartment house or public building. Housing

Table 4-5.

Apartment Size	No. of Apts.	No. of Residents in Apartments	No. of Rooms per Resident
1 room	37,927	106,222	0.4
2 rooms	86,424	163,273	0.5
3 rooms	48,066	63,343	0.8

Source: Società Umanitaria, *Le Condizioni generali della classe operaia in Milano: Salarie, giornate di lavoro, reddito, ecc.* (Milan: Società Umanitaria, 1905), pp. 50–54.

costs went up with the number of people in the household, but many family members often meant more income because of the greater number of people working. Families did not raise their housing budgets proportionately to their higher incomes, however. Housing was evidently a lower priority in most families than were the urgent needs for food and clothing. Such families thus elected to live in housing that was small, inconvenient, and often unhealthy, even when their incomes rose so that they theoretically could afford more.[60]

A city housing report based on the same data pointed out that rents and housing conditions were finely stratified by occupation. Thus it noted that, "the carpenter was worse off than the cabinet maker; the furrier and the tanner inferior to the saddle or valise maker; in the most lowly occupations—washerwomen, cigar makers, brick makers, and sand quarriers—the proportion of the most miserable housing was the highest."[61] Crowded housing, even crowded beds (schoolchildren were polled on that question), were common. Nearly all respondents complained to the interviewers about the lack of air, light, water, and sanitary facilities. The area to the south of the city center, near the Porta Ticinese, was singled out as being old and dilapidated. But the newer streets to the north of the city were no better. Dilapidated housing, like crowded housing, was more common in the suburban ring than in the city center.[62]

The municipal report also provided a breakdown, by occupation and sex of the household head, of the size and cost of housing for workers in 1903. Households headed by women (these were some 18,0000 of the 96,000 total households) were much more likely to live in crowded quarters. Whereas some 36 percent of families headed by men lived in one room, 50 percent of families headed by women did. (This was not controlled for size of household, however, and families headed by women may have been smaller.) Most female household heads were garment workers, laundresses, or cleaning women, who earned very low wages. Male heads of household in industrial categories like public accommodations, transport, petty commerce, wood and straw, construction work, and metals had particularly high proportions of families in one or two rooms.[63] The report ended on a pessimistic note. There were many, many Milanese who did not enjoy housing "that conformed to the simplest rules of personal safety, whose housing was both uncomfortable and unhealthy, not the hoped-for haven that could regenerate strength, energy, and morals."[64]

In the same period (May 1903) a national law was passed that made it easier for cooperative societies established for building inexpensive housing to find credit and that exempted such buildings from taxation. The law also authorized communes to construct inexpensive rental units, inns, and free public dormitories. An income restriction was placed on the tenants of such facilities.

The old Società edificatrice di case operaie, bagni, e lavatoi pubblici reappeared and built houses on large plots outside the porta Venezia, each equipped with running water and sewage connections. The Società Umanitaria built several apartment houses to the southeast, on the via Solari, outside Porta Genova. The institution's report declared that providing comfortable and hygienic housing was "the most effective" type of support for the poor, "fiscally and morally oppressed by unhealthy, inadequate, and immoral conditions." The Società Umanitaria was both paternalistic and moralizing in its approach to the problem. The first municipal development, with communal showers, reading rooms, and a library, was completed in 1906.[65] Even the one-room units had potable water and their own toilets.

In January 1908, a municipal housing agency opened its doors in response to a continuing shortage of housing in Milan. Now new housing was being constructed on the radial roads rather than in close-in areas. The search for affordable housing was time-consuming and exhausting for workers. Women workers were especially hard put in such a situation because they had both outside work and domestic responsibilities. Landlords with available housing were therefore requested to list their apartments; and persons in search of housing were invited to register.[66]

In the same year, a public wrangle broke out between tenants and proprietors, both organized into pressure groups. The Lega fra gli inquilini (Tenants' league) held a protest meeting in April 1908, in regard to increasing rents. Accusing landlords of unjustly high rents and excessive profits, the league warned that further rent increases could lead to "spontaneous outbreaks of public discontent." It then submitted proposals for improving housing to the city administration.[67] The Chamber of Labor endorsed its petition. The Associazione dei proprietari di casa (Landlords' association) rejected each one of the renters' proposals, and so again the city was forced to take action. In 1908, it established the autonomous Institute for Popular Housing (Istituto per le case popolari e economici), assigned it the titles of previously completed city housing, and voted new funds for housing. Republicans, Socialists, and the Società democratica lombarda joined the protest. After many meetings of the interested parties in 1908 and 1909, some tenants called for a rent strike in September 1909 against the high rents.[68]

New construction did expand the amount of available housing, as did the decision of many poorer migrants to live outside the commune. According to the yearly reports of the housing office in the *Dati statistici*, the number of requests for small housing units fell in 1910 and later, a decline that was attributed to the propensity of new migrants to live elsewhere. A 1910 report on rents shows that workers' housing was considerably more expensive at that time than it had been in 1904. One-room units were 110 to 150 lire, two rooms, 220 to 280 lire, and three rooms, 300 to 380 lire.[69] Also in 1910, building-trades workers found themselves short of work because of a drop in construction activity.[70] Building starts began a long decline. Although the problem was defused for the time being by less pressure for new housing, in 1911, 75 percent of the population of the city lived in apartments of three or fewer rooms.[71]

Public agitation had forced the city to build subsidized housing; the crowding situation was no better; and financial matters had worsened. Housing for workers was still crowded, dilapidated, and unhealthy, but now also expensive.

Conclusion

This brief examination of some aspects of the workers' way of living—the second level of class in Ira Katznelson's formulation—has demonstrated that Schiavi's point about Milanese workers as consumers was well taken. Totally dependent on wages to buy the necessities of life, the nexus of employment–wages–prices was central to workers' well-being. The business cycle greatly affected these relationships. Although the 1880s were relatively prosperous, the years from about 1888 through the early 1890s were a disaster. Then a period of good times accompanied the economic growth starting in 1894. After 1908 there were both rising prices and stasis. The evidence regarding wages and the standard of living suggests that in these matters, most workers were somewhat better off at the end of the period than at its beginning. Overall, the quality of housing probably improved. Although there was more subsidized housing, on the free market, housing was more expensive and just as crowded as it had been earlier. Unemployment and unstable employment continued, but finding a job was easier with the newly systemized worker-sponsored services.

Unlike Germany, with its early paternalistic welfare program (Mary Nolan sees it as an ambitious but ineffectual attempt to de-radicalize the working class), or Great Britain (Douglas Ashford points out that its 1911 welfare scheme was passed under the influence of a group of elite reformers, not organized pressure outside the government), Italy had enacted only piecemeal measures before World War I. A law passed in 1895 required employers to carry government-sponsored accident insurance for their workers. Although an old age and permanent disability fund was established in that same year, it became obligatory shortly before 1914 but only for state employees. A very modest program of maternity benefits for women workers was voted into law in 1907. Protective legislation that set exclusions and work hours for children and women was passed in 1902, the latter being the only welfare law in whose passage the Socialist party played an active role.[72]

In the early 1900s, as earlier, the poorer sections of the working class, including recent migrants, had to devise household strategies—such as sending all able-bodied family members into the labor market—to earn enough for their subsistence. Migrant household heads were often employed in work that was unstable and that required long hours to bring in a living wage. Working conditions in unskilled occupations and the often unhealthy industries in which migrants were employed were usually unregulated. Women and children who worked were paid low wages, as they were treated as part of a family unit, not as self-supporting individuals. Single females who lived alone or who headed households were at an enormous disadvantage in this situation. The result therefore was a family whose primary function was economic survival, with its members tightly interdependent and contributing wages to the household budget to the extent that their home responsibilities permitted.

Recession brought even harder times. Then unemployment or underemployment made family survival strategies less effective, as in the 1889–91 period. After about 1908, another cyclical downturn, fewer textile jobs available for women, and then rising prices demonstrated for all Milanese workers, even the native born, the precariousness of their position. In 1909 and 1913, there were protests over bread prices that would have sounded familiar in the eighteenth and nineteenth centuries. Angry accusations were hurled, declaiming the "rapacity of hoarders" and the "outrageous profits" of bakers. And wage gains were threatened by inflated food prices.[73]

Wage issues dominated workplace politics. In the recession of 1888–91, social questions were introduced to municipal politics by workers, the Partito operaio, the worker city council members, the worker-directed Camera del lavoro, and the Società umanitaria, all of which were supported by sympathetic council members. Problems arising from the state's illiberal policies and repression dominated the years between 1893 and 1899; struggles for fundamental political and civil rights outweighed workers' economistic inclinations. The return of prosperity after 1896 offered an opportunity for new demands for improving workers' economic conditions and for social programs. This chapter discussed the partial amelioration of urban workers' conditions. The ensuing period saw not only a spurt in organization, as we shall point out in the next chapter, but also greater mobilization and collective action on the wage front, as we shall show in Chapters 6 and 7. Overall, despite collective action on wages, local issues like housing, and national protective legislation, political and economic constraints limited the workers' gains.

5

Milanese Worker's Institutions

In 1909, Alessandro Schiavi, then secretary of the Società Umanitaria, provided a capsule history of workers' organizations in Milan in 1909:

> The movement that we call *resistance* and, later, after the first repressions, the organizations euphemistically identified as for *improvement* were preceded and prepared for by associations for mutual benefit that had their origins around the middle of the last century . . . in those earlier societies the spirit of organization and solidarity was cultivated and developed.

With the encouragement of the Partito operaio or the Consolato operaio, "some groups separated themselves from those closed within the mutualist spirit and developed independent sections in order to resist employers' demands and obtain higher wages and shorter hours." What kept these organizations vigorous was their tight link to occupations, "the glue that kept each group tightly associated were the mutual benefits" that endured despite crisis and police repression.

By 1905, a new trend emerged, according to Schiavi, the consolidation of occupations in the same industry—particularly among the metal and machine workers but among others also—into federations with broader bases. Members of some occupations joined colleagues in other cities to form national federations. Schiavi's early interpretation catches the essential outline. He saw these changes as a natural evolution in which differentiation and increasing complexity were the master processes. Although he was certainly aware of the economic changes and political struggles that had swept Milan in the last decades of the nineteenth century, he did not make explicit or implicit connections between these processes and the changing forms of worker organization.[1]

As we have seen, in the last decades of the nineteenth century, capital that formerly had been dispersed in the Lombard countryside, invested in farms, protoindustry, silk reeling and throwing, small metalware manufacture, mechanized spinning mills, and forges, moved toward Milan. Industry boomed in the 1880s, but a depression halted its progress between 1889 and 1894. The Milanese economy grew in scale and complexity. As small-scale agriculture collapsed in the high plains of Lombardy and protoindustry disappeared, the migrants took the road to the metropolis. In the city, workers were divided by occupation and migrant status. Each industry had its own history and typical employment pattern, stratified by age, gender, and place of birth. Economic structural change, migration, and labor markets formed the diverse, cross-cutting context in which worker institutions were born, grew, thrived, or sometimes died. Workers' institution building was a political process, not a natural evolution, as this chapter demonstrates.

In his conceptualization of the third level of class—"formal groups sharing dispositions" —Ira Katznelson deliberately avoids the term *class consciousness* in order to emphasize his "rejection of any notion of degrees of consciousness, with the highest corresponding to the 'real' interests of the working class."[2] This chapter analyzes the process of building workers' voluntary organizations outside political parties as signposts of class dispositions. It asks what the characteristics were of early worker institutions and what circumstances brought them into being. To what degree did these groups represent occupations, cross-class alliances, or workers as such? Under what conditions and how did broad class-based institutions emerge? By whom were they founded and led? The chapter reveals that although the forms of Milanese worker institutions—mutual benefit societies, labor unions, the Chamber of labor, union federations—shared many characteristics with similar institutions elsewhere in Europe, the historical circumstances of their founding and the political actors who shaped the outcomes also made them distinctive.

The Legal Context of Workers' Organizations and Strikes

Italian law shaped workers' constraints and opportunities. Legislation and administrative practice directly affected the conditions of employment, organizations, and the legal status of strikes.

In the early 1880s, Prime Minister Agostino Depretis, representing the historical "Left" that had come to power in 1876, proposed a package of labor reform laws, which came to be known as the Berti project, after Domenico Berti, Depretis's Minister of Agriculture, Industry, and Commerce. These bills included the legal recognition of workers' organizations by means of registration with local authorities, government guarantees for pension funds, compulsory insurance against work-related accidents, and arbitration in labor disputes. Despite some consensus favoring the principle of social legislation, only three weak laws were passed. One law in 1883 gave legal sanction to the National Insurance Fund for Work-related Accidents but did not require employers' participation in the system. Another law in 1883 stipulated the legal recognition of workers' societies, provided that their scope did not go beyond mutual assistance. Workers' organizations, suspicious of the uses to which the law might be put, abstained en masse from registering. Finally, an 1886 law forbade the employment of children under 9; for those in mines, the legal age was set at 10; and for night work, the age was 12. In the last decades of the nineteenth century, then, employers were essentially free to impose whatever conditions they wished for employment.

True, workers were no longer obliged to carry a *libretto di lavoro,* a passbook in which employers would indicate the conditions under which the workers left employment; that obligation had been eliminated in 1865. Both the law abolishing the *libretto* and another in 1889 continued, however, to require that employers report to the police their workers' names, addresses, and places of birth. Furthermore, individual employers were permitted to require that their workers carry a *libretto* in which the period of employment and type of work were recorded as well as an assessment of the worker's conduct on the job. Several strikes in the 1880s and early 1890s protested policies or factory regulations that included similar requirements, such as a certificate of good behavior from the police.[3]

The Sardinian law (Articles 386 and 387 of the criminal code) of the kingdom of Piedmont (adopted by the newly unified kingdom of Italy) recognized a legal arena for "coalition" (to cease or hinder work or to increase its cost) and strikes; they were forbidden only if "without reasonable cause." The criminal law revision of 1889 (the "Zanardelli code," named after Giuseppe Zanardelli, minister of Justice in the Crispi government) criminalized violent strikes and those that threatened violence (in practice this included no more than urging others not to work). Guido Neppi Modona argues that in this negative manner, the code affirmed the noncriminal nature of the peaceful strike and its possible use as a normal tactic in labor relations.[4] Police arbitrary action continued, however. Volker Hunecke concludes from his exhaustive review of strikes in Milan from 1860 to 1892 that in the view of the police, there were no reasonable causes for strikes; state intervention was regular and arbitrary, allegedly aimed at avoiding violence and preventing any infringement of nonstrikers' "right to work."[5]

From 1892 to 1900, matters changed little. Workers' organizations for purposes other than mutual benefit were illegal. Police intervention in workers' meetings and press censorship were commonplace. Public meetings and processions required police authorization, which was frequently denied. The police attended public meetings to monitor speakers for seditious remarks; they could dismiss assemblies if they detected any allegedly illegal statements or discussion. Any action taken by workers or their institutions and press in political or economic arenas subjected them to state scrutiny and perhaps suppression.[6]

Despite the repressive climate, the annual level of strike activity and the number of strikers in Italy climbed rapidly in the years between 1878, the first year reported in the government's strike statistics, and 1885; leveled off from 1885 to 1895, after which they rose even more rapidly to a high point in 1902; and after a five-year setback, the strikes and the strikers resumed their climb in 1908. The distinctive characteristics of these strikes are the large number of agricultural strikes and the frequency with which violence was used in the latter. We shall examine more closely in the next two chapters the pattern of strikes in Italy and Milan as they relate to economic structural change in particular industries and political opportunities. In this chapter the conditions promoting workers' organizations in Milan are the focus. They can be found, I believe, in the opening of new political rights (first suffrage and then the right to organize and strike) to workers, their increased political activity, and the unemployment crisis at the end of the 1880s and early 1890s.

In 1900 Prime Minister Giuseppe Saracco belatedly liberalized the labor law de facto (giving workers the right to unionize without government interference) after the dissolution of Genoa's Chamber of Labor. Contemporaries linked the chamber's shutdown, decreed by the police chief on orders from Rome, to the success of the Genoese longshoremen in negotiating a favorable contract, and so they launched a general strike of protest. Saracco's government fell, under attack by both Socialists and conservatives, to be replaced by one headed by Giuseppe Zanardelli, in which Giovanni Giolitti served as minister of the Interior. The Socialists, following their reformist leader Filippo Turati, supported the government in the hope that Giolitti's reform program would stave off the return of the far Right.

As Neppi Modona points out, however, Giolitti and Turati merely created the illusion of a truce in class conflict, a turn to enlightened reform that was short-lived.[7] Violent police repression of both agricultural and some urban strikes soon resumed. In urban Milan and elsewhere, the legalization of strikes in 1900 liberalized police attitudes and practice and encouraged both new organizations and a wave of strikes, with the strikes occurring at previously unheard-of rates.

The Long History of Workers' Organizations in Milan

A historical review of workers' organizations must consider two types of organization in nineteenth-century Milan. In the first type, the workers participated, but middle-class reformers and ideologues led; examples are most of the mutual benefit societies and the Consolato operaio, a federation of such societies. In the second type, worker leadership predominated; examples are the leagues for "improvement" or resistance, and the Chamber of Labor. The two kinds of organizations were not mutually exclusive chronologically, as mutual benefit societies continued to serve important functions even after leagues were established to press workers' interests more vigorously.

Cross-Class Workers' Organizations

Mutual Benefit Societies

Worker organization had very deep and old roots in Milan.[8] The Pio istituto tipografico, a printers' mutual benefit association that included honorary members and sometimes even employers, was founded in 1804, and the Mutua lavoranti cappellai, a hatter's society, was founded in 1832. Both were among Milan's oldest mutual benefit associations. The unification of Lombardy and Piedmont in the kingdom of Italy in 1859, however, spurred the formation of such groups, often influenced by Mazzinian republicanism. In 1859, for example, L'Associazione generale di mutuo soccorso was founded and it quickly became the largest mutual aid society in Milan, with over 5,000 members in 1864. Like other such associations, it was given funds by dignitaries such as Prince Umberto, who became a life member, and it granted honorary presidential status to others, in this case Giuseppe Garibaldi, in 1861. The Associazione generale offered services like the distribution of free medicine, evening elementary schools, a depository for small savings accounts, and a pension fund for those who were over 65 or disabled. Its fundamental operating principle was always to avoid politics. In 1859 the ribbon weavers' and braid weavers' mutual benefit societies were also founded; in 1860 other associations followed, organized, like the weavers, along craft lines: bronze casters, masons, stonecutters, silk weavers, and printers (new organizations with broader goals than those of the old Pio istituto). Later in the 1860s came organizations of bakers, perfumists, inn and hotel workers, stone pavers, and concierges; in the 1870s followed sand excavators, coachmen, blacksmiths, railroad workers, kiln workers, marble cutters, goldsmiths, tailors, and tripe butchers; and in the 1880s type casters, pastry cooks, hairdressers, leather workers, and glove makers. These societies provided group insurance. Workers paid in a few pennies weekly and drew benefits in the event of illness, unemployment, old age, or death, according to the particular group's criteria. Some nineteen societies established before 1890 were still associated with the Chamber of Labor in 1907.

The inadequacy of mutualism as an instrument of worker politics was summed up by Filippo Turati:

> It was the merit of Mazzini to have planted the first seeds of worker organization. However, the mutual benefit societies moved too cautiously, having declared their avoidance of politics, thereby refusing all contact with ideas and movements of the times, [that is,] with all political life. Although time and ideas moved forward, these societies remained behind. Their abstention from all political activity, their disinterest, the fact that they considered the organized worker an "apolitical animal" and the individual worker a "political animal" served the bosses and . . . their social demagogery and so-called philanthropy.[9]

Thus, although these early organizations provided valuable collective support and benefits for their members, they did not develop assertive positions protecting or promoting the rights of workers.

The Consolato Operaio

The Consolato operaio, founded in 1860 in the aftermath of Italian unification, was the first federation of worker societies. According to Antonio Maffi, one of its leaders, the Consolato came into being because the Associazione generale was "too apolitical." Nearly all the occupation- or craft-based mutual benefit societies belonged to the Consolato, maintaining their own autonomy but participating in the common efforts that it sponsored. By the late 1880s, the Consolato's political program included such goals as autonomous communes, progressive taxes in the place of the regressive *dazio,* abolition of standing armies, right of referendum, juridical equality of the sexes, right to assembly, an end to political police and arbitrary punishment, reform of the awarding of city contracts for public works, free technical education, independence of the courts and election of judges, expropriation of uncultivated land and its distribution to peasant cooperatives, and worker participation in the administration of a reformed system of charity. In the arena of employment, the Consolato called for the right to associate for mutual benefit or strikes, the organization of workers by craft as the best way to economic improvement, worker cooperatives (of production, of credit, and of consumption), equal pay for men and women, protective legislation, and profit sharing.[10] Its activities in Milan included an adult school, a cooperative bank, a workers' library, and clubs.

The Consolato was closely linked with Democratic Radical politicians—for example, Felice Cavalotti, deputy, and Carlo Romussi, editor of *Il Secolo*—and their clients, like Maffi, the worker deputy.[11] During the rise and eclipse of the Partito operaio, (considered in Chapters 8 and 9) this patron–client relationship was challenged by organizations that rejected bourgeois allies in the political arena.

Workers' Initiatives Toward Autonomous Workplace Organization

The first challenger was the Fascio dei lavoratori, an electoral alliance founded in 1889 by the masons' society. (The Fascio's electoral program included many progressive points, such as pay for city council members, open meetings of the council to discuss issues with fellow citizens, elimination of subsidies for schools run by religious orders and extravagant expenses in the city budget, abolition of the *dazio,* and city sponsorship of a chamber of labor.) [12] The vote in administrative (local and provincial) elections had been opened that year to men who could read and write, and consequently the number of eligible voters doubled. The Consolato, hopeful of an alliance of democratic and socialist elements in the November election, began discussions with the Fascio about presenting a joint list of candidates for the upcoming elections. The Fascio insisted on worker candidates and an independent worker electoral program, thus effectively scuttling the Consolato's initiative. The Consolato and the Associazione democratica (the Democratic Radicals' electoral organization) refused to accept most of the candidates that the Fascio proposed, all of whom were workers, men who had recently been their opponents. Further, the Partito operaio militants Giuseppe Croce, Silvio Cattaneo (head of the stonemasons' league), and one Ardigo of the bakers' league were perceived as socialists, hence politically unacceptable. The Consolato did accept printer Vincenzo Corneo, president of the compositors' section of the printers' union, Osvaldo Gnocchi-Viani, and Filippo Turati (running for the provincial council) as candidates. Corneo won election to the city council. *Il Muratore,* the newspaper of the masons' organization, noted bitterly, "Certain democrats speak in their meetings and public lectures to the people about progress and emancipation and liberties, but they are forever seeking ways to deny the workers the position and the rights owed to them." [13]

The Second International's call for a strike on May Day 1890 was the occasion for another contentious meeting of the Consolato operaio and the Fascio dei lavoratori. It started with a challenge to the presence of Carlo Romussi (who in addition to his editorial position with *Il Secolo* served as secretary of the Consolato) because he was not a worker. He replied that he had been active in the Consolato for twenty years and was a worker "even though I wield no hammer." He was permitted to remain, and he proposed a moderate motion, laying out two goals for Milan's May Day celebration: to demonstrate the workers' unity and to demand improvement of the workers' conditions, including the 8-hour day. But, he continued, because of the current industrial recession he opposed any strike, recommending instead public lectures on the evening of May 1 and a demonstration on Sunday, May 4. Bronze caster Alfredo Casati interjected that they were there to plan a demonstration, not a revolution, and moved that they organize a citywide strike and demonstration on May Day. The assembly accepted Casati's motion.

On April 23, a noisy second meeting reconsidered the same issues. Printer Carlo Dell' Avale again introduced the Consolato motion against Casati's version. This time the moderate motion was passed in a close vote. The issue became moot, however, when the police chief, Sangiorgio, citing orders from the minister of the interior, forbade any "public demonstration, parade, or procession and any public meeting or assembly having as its end participation in the May Day workers' demonstration."[14] A joint delegation of workers from the Consolato and the Fascio went to the police chief to protest the ban, but they were rebuffed.

Many workers did strike on May Day 1890. Police and troops broke up several attempts to demonstrate: by workers outside the Pirelli and Elvetica factories who urged their fellows to join them; by a group of unemployed workers that marched from the Fascio's headquarters toward the via Torino; by a crowd of some 4,000 that collected in the piazza del Duomo; and by an assembly of workers at Porta Venezia, near the headquarters of the mechanics' society. There were some arrests, but overall, the day passed without any noteworthy incidents.[15]

Their brief unity against the police behind them, the Consolato and Fascio again clashed in June 1890, as preparations were being made for the municipal election of June 22. The Fascio, now representing eighteen organizations that wished to act independently of the Consolato, met at the masons' society headquarters. Alfredo Casati urged the rejection of any joint list with the Consolato, and the assembly supported his position, agreeing to vote the Lega socialista list, which included Osvaldo Gnocchi-Viani, two other Progressive nonworkers, and two Fascio activists: Constantino Lazzari (bookkeeper and sometimes printer) and Silvio Cattaneo. The Fascio meeting also approved a proposal to establish a permanent committee that would "truly represent worker aspirations" in place of the Consolato.

The Consolato then met and drew up a list of twelve candidates, including Osvaldo Gnocchi-Viani. The representatives from the Consolato's constituent societies left one slot in their slate unfilled, for workers to nominate someone they could support. Needless to say, no such nomination was made.[16]

The night before the election, a candidates' meeting sponsored by the Lega socialista unanimously passed a motion that urged a fusion of the different democratic and socialist groups, permitting "varied opinions and diverse tendencies" to coexist, and settled on a common reform program that would make possible "a complete rennovation of communal life."[17] Eight of the Consolato/Democratic Radical candidates, including Gnocchi-Viani—but none except him from the Lega socialista list—were elected.

On October 12, 1890, the Fascio and the Consolato sponsored a congress before the parliamentary elections that brought together 90 organizational delegates, representing 12,000 Milanese workers. With 1 delegate for every 100 organization members, the 2,700 members of the masons' society gave it the largest delegation, more than one-third of the total. The hope was, once again, a common program and candidates. The two groups reached agreement on a progressive program—based on that of the Partito operaio italiano—but the effort again failed over the issue of electoral coalitions with the Democratic Radicals. The *operaisti* unanimously rejected Dell' Avale's motion for a common electoral position.[18] The Fascio supporters insisted as usual on the workers' independence from bourgeois parties. Casati proposed that if necessary to achieve that goal, the workers should vote with blank ballots. His motion carried, and the Consolato men walked out of the meeting. "The Consolato [delegates]," concluded *Il Muratore*, "who do not move or vote on their own account, but under the influence of certain bourgeois elements, left the meeting once they were defeated in a vote on this issue [worker independence]. . . . [T]he masons' society supports [the Fascio position of] priority of the economic over the political question."[19]

The thirtieth anniversary celebration of the Consolato in November 1890 was an opportunity to reflect, restructure, and retrench. The organization decided to welcome already existing resistance leagues as members and to promote new resistance leagues organized by craft or occupation. The representatives of its associated societies also voted to support the drive for a chamber of labor and equal pay for equal work.[20]

Consolato loyalist Antonio Maffi remarked in his history of the association that in the last decade, its constituent organizations had become more militant: "no more petitions, but protest, agitation, meetings." Most important, he observed, was "the fact that workers had achieved consciousness of themselves; their own collective strength convinced them to fight for their rights, not beg for them." He summed up the difference between the Consolato's approach and those of the Fascio and the Partito operaio. "The Consolato wishes to achieve economic transformation by means of political transformation; hence it uses politics as a means to an end. The tendency that prevailed at our October congress, however, tends toward political ostracism, protest votes, and absolute independence from political parties." He closed with a lengthy and impassioned defense of the vote as the best means for promoting workers' interests.[21]

Planning for May Day 1891 brought Partito operaio members like Giuseppe Croce to a Consolato meeting on March 8. There they argued persuasively that the strike should be on May 1 itself and not May 3, the following Sunday. At a second meeting on March 15, forty-five associations were represented. Several members of nonworker political groups were permitted to speak but not to vote. The assembly agreed to publish a May Day declaration prepared by a unified committee and to hold a national workers' congress. The goal of this congress would be to "coordinate all the forces of Italian workers and to set them united on the path to conquest of their economic and social rights." The declaration, entitled "The Rights of Labor," called for the freedom to organize and resist, legal recognition of craft organizations, the formation of chambers of labor, the 8-hour day, minimum wages, effective protective laws for women and child workers, insurance against workplace accidents, and free and obligatory public education. The program rejected (evaded?) the *operaista* position (represented by Casati and Cattaneo) which refused collaboration with political parties and government reform proposals. The congress planning committee was dominated by accommodationist *operaisti* like Croce and Lazzari (who became more radical in later years). It was this congress, held in August 1891, that opened up the road to worker and middle-class intellectual cooperation in the founding of the Italian Socialist party, as discussed in Chapter 9.[22]

Also in the spring of 1891, negotiations between the Fascio dei lavoratori and the Consolato operaio resumed, with the printers' Società di propaganda as mediator. Because Casati and other intransigent *operaisti* were participants in the process, the agreement more closely approached their position. It emphasized Consolato and Fascio cooperation in the workplace struggle but also stipulated that each organization could maintain its own program and that the Fascio would take part in all future electoral campaigns.[23]

The deepening recession and the unemployment crisis that it engendered, the debates about and eventual formation of a chamber of labor, the eclipse of the Partito operaio, and the founding of a Socialist party signaled the erosion of the Consolato's functions and institutional support. In an effort to build a broad front, the Milan Congress of 1891, which initiated the planning for a Socialist party, appointed Maffi to its constitution-drafting committee. After the Partito dei lavoratori italiani (P.L.I., the original name of the Socialist party) was established in 1892, the *Lotta di Classe* (the party newspaper) declared that the question was one of life or death for the Consolato. "The Chamber of Labor has absorbed the function of representing craft and occupational organizations," it pointed out, "and the Socialist party now provides the political–social function. If it does not join the latter, there will be no reason for the existence of the Consolato." Indeed, a majority (16) of its affiliated societies voted that the Consolato become the party's Milanese section. *Il Secolo* complained: "Gradually, opposition [to the Consolato] grew because many [workers] believed that its objectives were too modest and limited; others believed that it had lost its pragmatic and utilitarian character. . . . And thus we arrived at its 'transformation'." The remnant of its Democratic Radical members and organizations then formed the Tribunato dei lavoratori with the grand goal of seeking "perfect social justice and complete political freedom . . . the abolition of privilege, the diffusion of education and saving, and solidarity."[24]

Autonomous Workers' Organizations

In the 1890s, the mutual benefit societies were supplemented, and in some cases replaced, by more militant leagues, and the Consolato, with its mixed economic and political functions, was displaced by the Camera del lavoro, an independent workers' organization with primarily economic functions, and an autonomous Socialist party.

Leagues for Improvement
and Resistance

The limits of mutual benefit and collaboration with bourgeois allies had become apparent by the 1890s. Craft- or occupation-based societies with "improvement" *(miglioramento)* as a function (in addition to providing mutual assistance) became more common. The program of the Associazione di mutuo soccorso e miglioramento fra i lavoratori muratori, badilanti, manuali e garzoni di Milano (Masons', diggers', laborers', and apprentices' mutual benefit and improvement society), founded at the end of 1890, called not only for subsidies for sick members but also for the "defense of the rights of labor" and the "moral and material improvement of workers' condition."[25]

The constitution (1892) of the Lega di resistenza fra gli operai metallurgici ed affini di Milano marks another modification in the character of workers' craft organizations, an even firmer step in the direction of freedom from bourgeois patronage. It laid out the scope of the organization—the Resistance League of Metal Workers and Affiliates—in five points:

1. To defend its members against bosses' abuses of power.
2. To introduce and implement a wage scale *(tariffa di salari)*.
3. To assist members who were unemployed because of layoffs or disagreements with their bosses.
4. To protect members against violations of the *tariffa*.
5. To promote the schooling and education of its members.[26]

The difference among societies for mutual benefit, organizations for improvement, and leagues of resistance is well illustrated by comparing the activities of the old Associazione generale, the statement of purpose of the masons' improvement league, and the constitution of the metal workers' resistance league. The Associazione generale was interested in the workers' improvement, as, for example, in its provision of evening elementary schools, but its activities addressed individuals, not the group, and the social arena, not the workplace. The masons' improvement league called for defense of the rights of workers as a group and positive action to improve the conditions of masons in the workplace. The metal workers' resistance league aimed to support its members in the workplace and to work assertively for a *tariffa* for the group. The only common concern of the Associazione generale and the metal workers' resistance league was their support for schooling and education. Both the improvement and the resistance leagues expressed a sense of a self-defined group—all workers in an industry or a set of linked occupations—but not a class as opposed to capitalism.

The Chamber of Labor

The first proposal for an independent chamber of labor was made to an assembly of Milanese workers in November 1888. The meeting had been called to discuss the difficulties that workers had in seeking employment, and other current issues: competition to free workers from prison labor; improvement in work conditions (higher wages, fewer hours, abolition of piecework [the *cottimo*], fines, and retention of pay by employers); substitution of children for adult workers; and equal pay for men and women. Giuseppe Croce and Costantino Lazzari, both Partito operaio militants, initiated the call, supported by a large group of labor organizations. Croce opened with a plea for a more equitable and less humiliating organization of labor markets. He reported that the Paris Bourse de travail (labor exchange), funded by the city, offered a place where workers could meet, discuss their interests, and apply for jobs through a system that put their needs first. "Why not in Milan?" he asked.

Unfortunately, Milanese anarchists were on hand to attack the city government and the *operaisti,* accusing the latter of being mystifiers and corrupters of workers, interested only in getting themselves elected to Parliament. The police intervened to stop the attacks, but the combination of audience hostility and indifference prevented any immediate follow-up to the *operaista* initiative. More patient organizing and propaganda were needed before another collective discussion could be attempted.[27]

The labor exchange proposal reappeared in the following year, 1889, when delegation of Milanese workers went to the International Exposition in Paris that summer. There they lost no opportunity to make contact with local workers and visit their institutions, especially the Bourse du travail, and they were impressed by its contribution to labor market stability. At about the same time, Osvaldo Gnocchi-Viani published his brochure, *Le Borse del lavoro,* sponsored by the Partito operaio's executive committee.[28] In it he reviewed the history of the institution in France and Belgium and compared it with interest groups already established in Italy, such as chambers of commerce, agrarian committees, and lawyers' guilds. The existence of such interest groups, Gnocchi-Viani argued, suggested that there would be little opposition to a similar association of workers. Financing for the project, he continued, ought to be provided by the city government. "Italian workers . . . [must by their own efforts] make certain that these new institutions, belonging to *them,* will flourish and prosper."[29]

Circumstances were more propitious than in the previous year. Gnocchi-Viani's lectures on the proposal were welcomed on the Milan circuit. The Fascio operaio added a plank to its electoral program that urged the city to establish and support a labor exchange for the exclusive use of the workers, and it decided to call together workers' organizations to establish a common position. Shortly afterward, the executive committee of the national Associazione fra gli operai tripografi italiani, convening in Milan, sent a circular to its constituent member societies ad leagues urging them to take the lead in establishing labor exchanges in their own localities. These institutions would stand completely apart from electoral politics, promoting instead assistance to unemployed workers, the abolition of private employment agencies and the substitution of worker-controlled ones, the establishment of mediation or arbitration committees, the collection of relevant statistics, and support for social legislation. The printers called for Milanese worker societies to send representatives to a conference on the issue on December 6, 1890.[30] (The Fascio called its own meeting regarding this project, but its leaders accepted the printers' proposal that the groups join forces.)

Seventy-three organizations sent representatives to the conference, which charged a committee of fifteen (all workers, carefully distributed among members of societies affiliated with the Consolato, members of the Fascio, members of nonaffiliated societies, and delegates from the printers' national executive committee) to draft a proposal. Croce and Maffi were members of the committee, and printers Vincenzo Corneo and Angelo Carugati were its chair and secretary, respectively. The group also worked closely with Gnocchi-Viani.[31]

The Unemployment Crises of 1889–1891

The history of the Chamber of Labor's establishment is closely connected with the seasonal and cyclical unemployment crises that rocked Milan's labor markets from 1889 to 1891.

Seasonal unemployment was a long-standing problem in the building trades. It was aggravated in early 1889 by the decision of the *capimastri* (foremen subcontractors) to hire workers from nearby rural areas rather than Milanese workers, in what was perceived (probably correctly) as an effort to find submissive workers at low wages. Leaders of the masons' society protested this policy to the prefect and the police chief. There were rumors of a general strike in the building trades, which the police headed off in February by arresting unemployed workers and sending them back to their villages. Maffi issued a vain protest in the name of all Milanese workers.[32]

A wider and deeper unemployment crisis took place in the early months of 1890. Hundreds of workers in the metal and engineering industry had been laid off in January and February. The Milanese machine-shop owners complained that the lack of government orders for railroad cars and locomotives was hurting their business. The government had erred, they insisted, in not protecting and promoting their industry, which had invested heavily in increasing its productive capacity, through the adoption of new technology and training workers.[33] Calling for solidarity, workers sought to turn out their fellows who still had jobs at the Invitti and Miani–Silvestri factories. Guglielmo Miani—the engineer who headed the latter company and a city council member—urged that they stick to legal means and suggested that they contact directly the government in Rome. In response, a worker delegation, including *operaista* Emidio Brando, went to Rome with Miani to address their complaints to the minister of Public Works, Gaspare Finali. He promised government contracts for Milanese industries. On their return to Milan, Brando and others reported on their discussions in Rome to a meeting of unemployed metal workers. Because it would be some time before the new jobs opened up, the group decided to go to other workers' societies and charities for short-term help for those who had been laid off. Contracts and jobs did come through for Grondona, Miani, Invitti, and Elvetica, and the unemployment situation eased.[34]

By the end of the summer, however, production was again slowing down and layoffs loomed, so a group of workers requested a meeting with the industrialists on ways to avoid another long crisis. A delegation went once again to see Minister Finali, and he again promised contracts to Milan for railroad cars and locomotives. In November 1890, 800 mechanics were reported still to be on the assistance rolls.[35]

In March 1890, when matters were critical in the machine industry, newspapers and police began reporting increased unemployment among workers in the building trades, too. The issue was, once again, the hiring of rural workers for projects in the city. Some masons and their helpers threatened a group of rural workers and frightened them enough that they left Milan. There were said to be 2,000 unemployed construction workers, 600 of them members of the masons' mutual benefit society.[36] The society sent a delegation to the mayor that asked him to use his influence to persuade the building contractors to hire local men. He suggested that perhaps the *capimastri* preferred rural workers because they were more assiduous, obedient, and docile. The *Corriere della Sera* remarked that even the "urban" building-trades workers had originally come from the country to seek work in Milan, though they now lived there permanently.[37]

The *capimastri* agreed to hire 218 Milanese masons, but many fewer were actually accepted. There followed several tumultuous days at the *ponte,* the shape-up at which construction gangs were hired. Two thousand men turned up on one day at 6 A.M., and 1,500 on the next. Police broke up the gatherings, and the masons' chief went, once again, to complain to the mayor. The *capimastri* claimed to have hired 600 men, the police reported, but they refused to limit themselves to members of the Mutua, which they heartily disliked. The *capimastri* joined the call that the city do something concrete about hiring or the crisis would simply keep recurring. The problem became less acute with the approach of spring, and the calls for intervention therefore ceased.[38]

The committee appointed to study and promote the concept of a labor exchange published its report in early March, when mechanics and construction workers were losing their jobs.[39] It declared that "new forms" of organization were needed to cope with the changing social and economic circumstances, that charity and efforts to combat unemployment were inadequate to confront the "social disorder threatened by the antagonisms between labor and capital."[40] A workers' delegation met with Mayor Giulio Belinzaghi. Although the draft of a constitution for the Chamber of Labor had not yet been completed, he assured them of his support, that after the constitution was ratified by the workers, he would recommend it to the city council for approval. His position was surely motivated by his belief that unemployment was a threat to public order and that the proposed institution might alleviate distress. Casero concludes, however, that he also surely underestimated the more assertive aspects of its functions.[41]

Belinzaghi was present on March 30, 1890, when a workers' meeting at the Canobbiana Theater (the hall paid for by the city) began deliberating the chamber's proposed constitution. An acceptable document was hammered out at that meeting, which reconvened on April 20 and 27. The most debated provision was Article 6, which originally stated that only one organization for each craft or occupation would be permitted to affiliate with the Camera del lavoro. Objections were immediately raised, for this requirement took no account of reality; that is, in many occupations and industries, there were several societies with different functions. The article was thereby revised to permit all worker societies organized by occupation or craft to affiliate. It further urged workers to join with one another in such societies but reserved for the Camera del lavoro the right, upon review of any society's constitution, to reject it. Article 2 of the constitution ruled that any societies that joined the chamber must be composed wholly of waged workers. In summary, the chamber's functions were to be, first, serving as an intermediary between the buyers and sellers of labor and, second, promoting the workers' interests in all aspects of life.[42]

In October 1890, the negotiating committee reported that plans for the Chamber of Labor were stalled, that Mayor Belinzaghi had offered a subsidy of only 15,000 lire instead of the originally proposed 45,000 to 50,000 lire. He now argued that both the chamber and the city would be better off with no official connection. Finally, at that moment there was no appropriate and sufficient city-owned office space available. The committee was thus sent back to negotiate further with the city.[43]

Workers and city officials again turned their attention to the unemployment crisis, which again made headlines in November and December. Although new railroad-related orders had been granted to Milan firms, they served simply to keep busy those already employed. A large pool of long-term unemployed still needed assistance. A committee, composed largely of mechanics and chaired by Paolo Zanaboni, was elected to disburse assistance from funds previously contributed by the public; its office was at the Dogana Vecchia, by which name it came to be known.

In January 1891, Gnocchi-Viani, who had been elected to the city council the previous December, posed a series of questions to the council. First, he wished to know what measures, rules, and criteria for addressing the recurrent worker crises the city proposed to adopt. "What do workers want," he demanded, "when they turn to the city government in times of economic crisis? That the city be an instrument of social peace, that the government provide them not charity but their rights. The city cannot close its eyes to these demands; it must act!" He argued, second, that the crises recurred because the economic system was unjust. The commune could help by buying goods and services from worker cooperatives; by encouraging the formation of a labor exchange, which should be a workers' institution with offices provided by the city; by starting a public works program; by promoting shorter work hours; and by providing workplace accident insurance. Continuing in a more diffuse vein, Gnocchi-Viani recommended a city study of rural-to-urban migration and suggested that the mayor meet with local officials in rural areas to discourage migration. His plan would help the poor and increase consumer spending at the same time. The funds to finance it could come from eliminating city subsidies to the La Scala opera company and to religious institutions. Finally, the council should end its silence on the crisis, publicly announce its concern for the condition of the working class, and promptly assist the unemployed with jobs at public works projects. Belinzaghi, while expressing his appreciation for Gnocchi-Viani's sentiments, which he shared, remarked that the issues were too complex for the entire council to address. He proposed that a committee be charged

to study the project. After some debate and general support for a committee, Gnocchi-Viani asked the mayor directly to state the city's concern explicitly and to institute the relief programs that he had recommended. Belinzaghi refused to do so, and the council then voted for the study
committee.[44]

As the fifteen-member city council committee began its study, the workers' planning committee for the Chamber of Labor resumed its activity. The
two committees were linked by Gnocchi-Viani, a member of both. Following the lead of the mayor, the city committee was inclined to recommend
a subsidy for the chamber. It also identified available space in the Castello
Sforzesco; a delegation of workers inspected the location and pronounced
it satisfactory.[45]

Representatives of forty-seven worker organizations met on the evening
of March 2, 1891, under the joint auspices of the Fascio dei lavoratori, the
Consolato operaio, and the Associazione generale degli operai, to constitute
the Camera del lavoro formally and to work out details like the dues to be
paid by individual members.

The city council study committee on the Gnocchi-Viani proposal reported on June 5, 1891. The only recommendation that it fully supported
was that for the Camera del lavoro. The council promptly voted a subsidy
of 15,000 lire, to be reviewed annually, and headquarters in the Castello
rooms already inspected by the workers' committee. The proposal to grant
contracts for city public works to workers' cooperatives was watered down
to a suggestion that the city do so. The other proposals were either declared
impractical and poorly conceived or rejected as too expensive. The committee had reached no consensus on the highly charged political question of
a public statement of concern about the unemployment crisis or immediate
steps to ease it.[46]

The Camera del lavoro, unable to move to its headquarters until late
September 1891, was not involved in that summer's agitation at the Pirelli
plant or the mechanics' general strike (discussed in Chapter 7). It occupied
its new quarters starting on September 17, at which time 23 associations
were formally affiliated, with about 40 more in the process of joining. As
secretary of the chamber, Giuseppe Croce charted a vigorously interventionist policy, frequently acting as a mediator between workers and their
bosses, with uneven results. The institution became a kind of "horizontal
union," writes Guido Cervo; it was more complex organizationally than
earlier institutions had been, appeared to the authorities to be more representative, and avoided electoral involvement.[47]

Contrary to the suggestion of Manacorda and Proccaci in their revised
interpretation of Italian working-class formation, Milan's Camera del lavoro
was not simply a reflection of traditional communalism. The institutional
form was modeled after French and Belgian predecessors and was estab-

lished on workers' initiative with the collaboration (and financial contribution) of the city authorities as a means of combating unemployment and organizing labor markets. It quickly became a mainstay of workers' collective action as well. Although the chamber worked primarily in the economic arena, its leaders viewed workers' problems as closely linked with capitalist development. A political process had brought it into being and continued to shape its vicissitudes.

Growth of the Chamber of Labor and Repression

There was a barely perceptible slowdown in the chamber's organizational activity in 1894, as its member organizations were briefly declared illegal in the Crispi repression of the Socialist party. Nine new leagues appeared in 1894 and 1895. Many leagues, and the Camera del lavoro itself, reconsidered their affiliation with the Socialist party, an affiliation that had included them in the repressive decrees that dissolved the party and its component institutions. The chamber's usefulness was not affected by these changes, however, and a new high—fifty-one—in number of organizations joining the chamber was reached in 1896–97.[48]

In 1897, there were 107 leagues and societies associated with the chamber, with 15,696 members, or 6 percent of the labor force. The distribution of membership in worker organizations by industry in 1897 and 1902 is displayed in Table 5-1. Sand pit excavators, classified under mining, were a relatively small occupational group, all of whose members belonged to their league. After them, the printing and the chemical industries were the most highly organized, followed by the fur and leather, ceramic and glass, metal and machine, construction, and transportation industries. The construction, transportation, and metal and machine industries were large, with many workers, but only construction and transportation had large leagues, representing a cross section of workers in the industry. The metal workers were divided among several leagues and mutual benefit societies.

The chamber's growth was stopped by another government decree dissolving all workers' institutions (and many others, as will be seen in Chapter 10) following the May rebellion in 1898. Progressive Milanese regrouped in 1899, forming a new electoral alliance, L'Unione dei partiti popolari (Democratic Radicals, Republicans, Socialists), for the city council elections in June and December, in which they decisively defeated the long-entrenched Moderates. One of the issues in these elections was the reconstitution of the Camera del lavoro, praised by the press for its "public usefulness" in dealing with unemployment and promoting responsibility in the working class.[49] The prefect refused to give the go-ahead when the chamber's ex-secretary, Croce, petitioned after the June elections for the

Table 5-1. Percentage of Labor Force in Working-Class Organizations, 1897 and 1902, by Industry

Industry	1897	1902
I. Agriculture		
Agriculture	0%	2.6%
Mining	100.0	100.0
II. Manufacturing		
Food	1.1	22.5
Wood, straw, mixed	6.3	17.8
Fur, leather, other	15.4	25.2
Textiles	5.7	14.1
Garment making	1.5	3.5
Ceramic, glass	23.8	70.2
Metal, machines, precision	15.8	26.1
Chemical	29.4	28.5
Rubber	—*	100.0
Printing, lithography	30.0	55.9
Construction	14.0	51.3
Paper, cardboard	2.4	10.6
III. Professions, Commerce, Services		
Professions, arts	0	0
Government	0	8.0
Transportation	15.2	44.0
Domestic service, public accommodations	1.4	5.7
Retail commmerce	2.7	6.0
Total union membership as a percentage of labor force	6.0	16.1

*Rubber workers were organized only in April 1898, several days before the rebellion that resulted in the dissolution of all unions. About two-thirds of the Pirelli workers were said to have joined the league, two-thirds again of this amount being women.

Source: Based on "Quadro delle associazione aderenti alla camera del lavoro sino al 1907," Table in Società Umanitaria, *Origini, vicende e conquiste delle organizzazioni operaie aderenti alla Camera del lavoro in Milano* (Milan, 1909), pp. 72–86, which provides the numbers of members in each working-class organization belonging to the Milan chamber for each year from 1891 to 1907. The years chosen were high points of organization. These numbers were stratified by the number of workers in each industrial category in Milan, based on the population censuses. Numbers were interpolated for intercensal years.

release of its papers and funds so that its former executive committee could start the process of reconstitution. Representatives of member organizations in August took matters in their own hands and declared the chamber open, authorizing the old executive committee to request its former headquarters and annual subsidy from the commune. In hopes of making its actions acceptable, the committee voted to limit the chamber's functions to the employment bureau and arbitration. This effort, too, was rejected.[50] Only in January 1900, after the second decisive electoral victory of the Unione dei partiti popolari, was permission granted to reopen the chamber. On April 15, 1900 (Easter Day), a victorious ceremony marked the return of the Camera del lavoro, with printer Giuseppe Scaramuccia as secretary.[51]

During the festivities, Croce noted the importance of the electoral victory of the Unione for the chamber's rebirth. He proposed no modification of the chamber's functions but praised its role as defender of fundamental civil rights and democracy through coalition politics. Gnocchi-Viani's speech at the same ceremony played down the chamber's political role and emphasized instead its employment function and municipal reform program. The earlier vote of the old executive committee had already attempted to limit the chamber's function to an economic one, in hopes of avoiding any future arbitrary shutdown or elimination of the city subsidy—only to be rebuffed by the authorities. Both speakers hoped for a more openly political role for the chamber in the future, but in different ways. At some level, however, the chamber's earlier economic vision of its function had a political outcome: struggles over workers' rights to organize and strike and over their share of the profits of their labor.[52]

Another spurt of organizational growth followed the chamber's resumption of activity. This growth occurred both in the membership of already established leagues and in the formation of new leagues. Behind the increase lay the improved legal climate for labor organizations. In 1902—the high point for membership in the Chamber of Labor in the period before 1908—there were 168 groups and 44,440 members, 16 percent of the labor force, in the chamber (see Table 5-1 for the distribution of membership by industry). This list combines all types of organization, but by 1902, the proportion of exclusively mutualistic organizations was low. Table 5-1 shows that the total membership in labor organizations increased by 167 percent between 1897 and 1902. Those industries in which the proportion organized increased more rapidly than the aggregate were food, wood, straw, textiles, garment making, ceramics and glass, rubber, construction, paper and cardboard, public accommodations, and transportation. (The proportions organized by industry will be discussed further in Chapters 6 and 7.)

Twentieth-Century Developments

The labor movement developed new forms, especially that of national federations of craft unions. Although the printers and the railroad workers had federated earlier, the first years of the twentieth century saw a great rise in the number of such federations. In 1902, the national federations and the chambers of labor established the Central Secretariat of Resistance, with its headquarters in Milan. In 1906, members of this secretariat founded the

Confederazione generale del lavoro (C.G.L., the General Confederation of Labor).

The balance between political and economic functions changed in the Camera del lavoro as its executive committee took on more duties. The committee was, Giuseppe Paletta explains, "the 'political' soul of the workers' movement." It elaborated union strategies in cooperation with Socialist intellectuals, helping workers develop their analytical capacities and act systematically on their findings. In this process of collaboration the ideas and analyses of reformist socialism were translated into a reformist union strategy.[53]

As spelled out by Alessandro Schiavi, this strategy held both that strikes should be limited to the "lowest" categories, whose economic position was untenable, and that no strikes should occur without organization. The "lowest" categories of worker were unlikely to be organized; thus this strategy actually meant limiting strikes. In place of strikes, or at least supplementing them in an important way, Filippo Turati placed "proletarian politics," which concerned itself with advancing the interests of labor with regard to the state, Parliament, and the municipal authorities.[54] He believed that workers' and Socialists' major efforts should go into building strong organizations and gathering information for statistical analysis.

Beginning in 1901, disagreement in the Socialist Party over the means of class struggle led to a schism between reformists, led by Turati, and revolutionary syndicalists whose theoretical position was shaped by Arturo Labriola, who moved to Milan in late 1902 to become editor of the newspaper *Avanguardia Socialista*. A brief reunification of the two wings engineered by the party directorate failed, and thus in one form or other deep divisions became a byword of Italian Socialism. In Milan, Costantino Lazzari and other former *operaisti* joined with the revolutionaries to oppose the bourgeoisification of the P.S.I. There was a division of labor between the revolutionary allies, however. According to Bonacini, in *Avanguardia Socialista*, Labriola's followers emphasized politics, whereas some former *operaisti* resurrected their workerist exclusivism and economic demands.[55]

In January 1904, an anonymous Milanese worker wrote that workers considered *Avanguardia Socialista* to be "an intransigent Socialist intellectual newspaper," but this time—as Paletta notes ironically—the intellectuals were revolutionaries. The workers' movement had still not found its own voice.[56]

A revolutionary syndicalist slate was victorious in the 1904 elections for the Chamber of Labor's executive committee. It promptly passed a motion making it the chamber's policy "to intervene in all strikes."[57] The number of strikes in Milan remained high in 1904 and 1905, but the number of strikers (except those in the general strike of 1904, a political strike protesting the army's killing of striking agricultural workers) declined precipitously, as did membership in the Camera del lavoro. Reformists recaptured its executive committee in 1906, but the uneasy alternation of reformism and varying forms of radicalism continued.

Conclusion

The formation and transformation of workers' institutions followed a timetable punctuated by political change. Italian unification was a major impetus to the emergence of mutual benefit societies and the Consolato operaio, organizations in which both bourgeois and workers collaborated. Challenges to the Consolato's pattern of patron–client relationships began after the 1889 expansion of suffrage in local elections and continued with planning for the May Day commemorations, starting in 1890. In the economic arena, challenges to mutualism began in the same period, which was marked also by severe economic dislocation.

The succession of different institutional forms was similar to that in other European national states. The importance of the chambers of labor (in Milan and Italy) and their active role in supporting unions in a repressive situation were, however, distinctive. The Milanese chamber developed from a workers' initiative supported by a socialist proposal to the city council. Its activity was cyclical, following the vicissitudes of progressive politics in this period. There was a slowdown with the repression of 1894, a speedup in the intervening years, and a dead halt with the much more comprehensive and severe repression of 1898. The chamber was very active in the economic arena, albeit broadly defined, in the period after its inception, but its leaders retreated after 1898 into the city wide coalition seeking democratic rights. Following the government's extension of the right to strike in 1900, the chamber's activity increased once again. Its substance also changed, as both former Partito operaio workers and syndicalist intellectuals again emphasized the political potential of workplace struggle.

There was little agreement among workers or Socialist intellectuals about how to build institutions and what they should do, for institutional structure was not a simple outcome of changing economic structure. Here Antonio Gramsci's well-known article of 1924, "Il Problema di Milano," provides some insight:

> Why has there never been in Milan—a great industrial city with the largest proletarian population in any Italian industrial center, whose work force alone accounts for 10 percent of the total number of factory workers in Italy—a large revolutionary organization, whereas the workers' movement has regularly been revolutionary? Why has Milan never had more than 3,000 organized Socialist party members? Why have all the Milanese labor organizations—unions, cooperatives, mutual benefit societies—al-

ways been in the hands of reformists or semireformists when the masses
are pulled into the streets by the most revolutionary impulses?

Writing shortly after the establishment of the Communist party and the later
Fascist seizure of power, Gramsci emphasized Milan's importance for national outcomes: "Italian capitalism can be decapitated only in Milan. . . .
The city's problem is thus not a local one; it is a national problem and, in
a certain sense, an international one." His response to his own pressing
questions was that, first, most of Milan's workers were employed in small
units rather than in large factories and the city was full of petty bourgeois
shopkeepers; that, second, the older workers had strong connections to the
Democratic Radicals; and that, third and most importantly, the reformist
socialists had erred drastically in their postwar strategies.[58]

Gramsci was correct about the structure of the Milanese economy; the
move from bourgeois patronage to autonomous labor was a common European urban pattern. But this chapter's look at worker organizations shows
that contrary to Gramsci's assumption, the Chamber of Labor was not a
product of the old mutualist workers, but of the *operaisti* who rejected
bourgeois patronage and developed their own workerist ideology. (From
Gramsci's communist point of view, this was doubtless an outmoded approach, just as the mutualist one was.) Milanese worker "tradition" was
contested by these groups starting in the mid-1880s and, institutionally at
least, was resolved with the founding of the Chamber of Labor. The developments described in this chapter were neither a product of traditional ideologies (communal or democratic) nor unmediated economic structural
change. They were not imposed from above but were the outcome of a
political process that involved workers, intellectuals, and other local actors
such as elected officials. Workers and Socialist leaders, faced with the possibilities opened (or closed) by law and policy, actively constructed the
particular path of Milanese working-class institution building.

The next chapters examine in greater detail the collective action of different groups of workers and demonstrate how workerist ideology and organizations established independent positions in the 1880s and 1890s, only
to join the reformists in a coalition seeking democratic rights in the face of
repression in the last years of the century.

6

Workplace Collective Action I: An Overview and Case Studies

This chapter begins the exploration of Katznelson's fourth level of class formation—collective action—with aggregate and case study analyses of strikes. Although what the nineteenth century came to call the *strike* (Italian *sciopero*)—a collective work stoppage for the purpose of making demands, or resisting the actions, of employers—was not strictly a creation of that century, it did proliferate at that time. Along with the spread of proletarianization, strikes became the dominant form of workplace collective action in western Europe and North America in the second half of the century.

Contemporaries were aware of this process, and politicians, troubled by it, initiated investigations and collected statistics to document it. Francesco Crispi, then Italy's minister of the interior, was alarmed by the resumption of strike activity in the woolen textile mills of the Biellese region (Piedmont) and so appointed a royal commission in 1878 to enumerate and analyze all Italian strikes since unification.[1] The strike statistics collected by the commission and continued by the statistical office of the minister of Agriculture, Industry, and Commerce offer detailed strike-by-strike information about the date, location, and length of the strikes, the grievances over which they occurred, the numbers of strikers and workers in the unit involved, and their outcomes. This evidence has been analyzed on the aggregate level in the last twenty years by social scientists testing three major clusters of causal hypotheses. The same statistics also provide the base for a quantitative description, by industry, of strikes and strikers in Milan from 1878 to 1903.

Case histories of organization and strikes in the printing and construction industries offer a closer look at Milanese patterns, in particular those related to, on the one hand, the timing and form of employers' strategies for controlling workers and, on the other hand, workers' organizations, mobilization, and collective action. These two industries were early to organize and relatively successful in pursuing their interests. They are also richly documented because their organizations produced many records, including newspapers, and were linked to workers' societies in other cities and at the national level.

Chapter 7 presents case studies of organization and collective action for the textile, metal, and engineering industries and for three industries primarily employing women—rubber, garment making, and tobacco. Textile, and metal and engineering workers were early to organize but split in different ways: the first between men and women and also among specialties, and the second among occupations and, more broadly, among skilled, steadily employed, and unskilled workers. The case studies of the industries employing mainly women provide additional information about the problems of women workers in organizing and acting collectively on their interests. Chapter 7 concludes with some general reflections on workplace collective action.

The questions addressed in Chapters 6 and 7 are the following: Under what conditions did the frequency of strikes increase in the short term and over the longer term? What was the relationship between organization and strikes? To what extent and how did organization contribute to the favorable outcomes of strikes? Why did workers in some industries find it difficult to organize or translate organization into successful strikes? What circumstances shaped the timing and low rate of women's labor organization and mobilization?

Three clusters of hypotheses regarding strikes have been examined in aggregate analyses by other scholars. The first theorizes that economic factors are the major cause of variation in strike frequency. One economic explanation is that strikes are more likely when wages and unemployment are relatively low but the cost of living is high; that is, strikes are more common during the upswing of a business cycle. Another version is that strikes increase when collective bargaining breaks down. These perspectives were developed in the United States in the 1950s and 1960s, a period in which unions were relatively mature.[2] They had little to say about historical change and, in particular, the structural conditions under which strike frequency increased or decreased.

The ahistorical nature of the classic studies of strikes contrasts with the second type of explanation, the "breakdown" or "transition" theory. Breakdown theory—a product of nineteenth-century sociologists such as Emile Durkheim and elaborated by structural–functional sociologists in the 1950s and 1960s—argues that rapid social change undermines and ultimately destroys traditional social and economic relations. The outcome of such a breakdown is a period of high social and economic tension and conflict that ends with the reestablishment of stable, functional relations.[3] It was just such an interpretation by Italian public officials in the late 1870s that led to the appointment of the royal commission on strikes.

The ideas of the Marchese Antonio di San Giuliano, who in 1884 sponsored a bill in the Italian Parliament to decriminalize "coalitions" and strikes, exemplify the transition theory. He believed that strikes were the outcome of industrialization, especially the increasing scale of production. Strikes would not be a problem in Italy, according to San Giuliano, "until large corporations develop, preventing direct relations between entrepreneur and worker, wages are high enough for the worker to sustain prolonged absence from work, and large agglomerations of workers increase in number." [4] He was correct that the frequency of strikes was related to large-scale economic structural change, but the relationship was not a simple breakdown of norms (as Italian observers and nineteenth-century sociologists feared), nor was it a disease of transition linked to economic development.

The third explanation, again a historical one, emphasizes politics and organization, seeing strikes as an outcome of union growth and mobilization and an increase in workers' political participation. This alternative also has several versions. The one adopted here starts from the theory that strikes are a form of collective action. Collective action—coordinated action on behalf of shared interests—is the theoretical framework within which we shall look at both workplace activism and other worker politics. Groups that have identified their interests and see the opportunity to act apply what resources they can muster to other groups or to governments. Political power is the positive consequence of the application of resources to governments. Violence occurs when governments or other groups resist or repress the collective action of a mobilizing group, as well as when such a group deliberately chooses violent means. There are five components of collective action: "interest, organization, mobilization, opportunity, and action itself." Time, place, and the issues involved shape the way in which these components are combined. [5]

The five components of collective action suggest ways to specify and concretize the political variable. Organization, in the collective action framework, follows interest and is a source of collective solidarity (organization sometimes is the result of struggle as well). It affects, positively, different groups' propensities to engage in collective action. [6] (The inverse of prior organization is "uprooting," as through recent migration, which, the breakdown theorists believe, propels migrants, disoriented by urban life, into movements of protest, including strikes.) [7] I argue here that the kind of political or social change that is significant in shaping collective action, affecting its timing and the varying readiness of groups to act collectively, is change that affects positively opportunities to act through example, reducing the costs of action, or increasing its rewards.

Italian Strike Patterns and Cross-National
Comparisons

A recent quantitative analysis of Italian strikes and labor conflict from 1881 to 1923 begins by sketching the frequency of strikes. Lorenzo Bordogna, Gian Primo Cella, and Giancarlo Provasi find clear temporal patterns: There was a rapid and steady rise in the frequency of strikes during this period. At the same time there were sharp and irregular variations, most apparent not in the number of strikes but in the number of strikers involved and the number of days lost to strikes. Through 1902, agricultural strikes were the most common. In 1906, however, agreements between employers and the Metal Worker's Federation (FIOM) in Turin's automobile industry structured, for the first time, worker–employer relations and established internal committees in the plants. Thereafter the automobile industry was central to Italy's strike movement. Bordogna, Cella, and Provasi see the spurt in strike frequency in the two postwar years (the *biennio rosso*) as a consequence of increased militancy and also of greater social support. Levels of union membership rose and fell in parallel with the outburst and its abrupt end, which occurred under nongovernmental Fascist attacks starting in 1921 and governmental repression in 1922 and afterward. The three writers' quantitative analysis tests the economic and political–organizational hypotheses and finds support for the latter, with the proviso that in Italy, organization often followed mobilization and strikes rather than preceding them.[8]

Bordogna, Cella, and Provasi, following David Snyder's earlier modification of the political–organizational model that emphasizes certain institutional conditions as necessary for the economic hypothesis to hold, conclude that in Italy these conditions were not present until after World War II.[9] Instead, Italy was characterized (as shown also in Chapters 2, 3, and 5) by diversity and complex stratification among workers, dualism of the national economy with continuing prominence of the agricultural sector, and a political system hostile to workers' aspirations. Before the Fascist takeover, these factors produced "weak and unstable workers' organizations," grouped locally on a territorial basis, that tended to act in the general interest of labor only because national federations either were feeble or did not exist. These factors prompted unions to seek alliances with political parties in order to protect themselves against government repression. The result, they explain, was highly politicized unions and strike levels that varied with the political climate.[10]

How does the history of Italian strikes compare with that of other national states in this period? Addressing one of the central questions of Italian modern history: Was Fascism the result of the capitalist backlash against the workers' "uncontrollable" demands during industrialization? Adriana

Lay and Maria Luisa Pesante compare Italy's industrialization, the frequency of strikes, and their cost with other countries' historical experiences. Theirs is not a sophisticated quantitative analysis; it uses descriptive statistics to construct a case that Italy's development was similar to that of other European states. Further, they decide, the level of class struggle in Italy was not much higher than that in other countries. They find little support for the argument that the workers' excessive militance lay behind capitalists' support of Fascism.[11]

Charles Tilly's six-country comparison of strikes and strikers offers some simple measures of the numbers of strikes and strikers per 100,000 workers in all industries on dates around 1900. Italy's strike rate was 3.8, close to that of Great Britain (4.0), identical to that of France, and much lower than those of the United States (6.5) and Germany (8.6). The numbers of strikers per 100,000 workers in Italy and France were similar, with the rate in Britain slightly higher, and those of Germany and the United States more than double. By these measures, therefore, Italy's strike rate was at the low end of the countries that Tilly considered.

Tilly also discovered commonalities in temporal patterns of strike frequency from 1890 to 1924 in France, Italy, and, to a lesser degree, Germany. The numbers of strikes and strikers in both France and Italy followed an upward trend to the middle or end of the first decade of this century; strike frequency was low during the early war years, more so in France than in Italy, but rose as the war dragged on. In Germany, the rise in number of strikes before the war was less regular; the drop during the war more complete; but the postwar increase paralleled those in France and Italy. The pattern in Great Britain, the United States, and Russia was quite different. In Britain, there was a trough in the number of strikers from the mid-1890s to 1910, accompanied by a less marked decline in strikes and then a sharp postwar peak. The United States saw many more swings around a trend that rose more slowly and only a brief postwar spurt. The number of strikes and strikers in Russia swung sharply from year to year, with a huge peak in 1905, around and during that year's revolution. Italy differed from the other western European countries and the United States, however, in the high proportion of "strike wave" years in the period for which data were available. (Strike waves are peaks in which the numbers of both strikes and strikers exceeded their average of the previous five years by at least 50 percent.) In Russia, the proportion of strike wave years was 30 percent, followed closely by 27 percent in Germany and Italy. In contrast, the figures for France, Great Britain, and the United States were 20 percent, 9 percent, and 3 percent, respectively. Except for this distinction, however, Italy's strike patterns were similar to those of other western European countries.[12]

Strikes in Milan Before 1878

Despite the mutual benefit goals of most Milanese worker organizations and the Consolato operaio's avoidance of class antagonism, workers acted collectively on issues in the 1860s and 1870s that they perceived as pertaining to their interests. Before 1878, two periods stand out because of their high levels of strike activity: 1859–60 and 1872.

Immediately following the unification of Italy came a rush of new organization. "Coalitions" and strikes proliferated in the second half of 1859, under the provisional government of Lombardy. Annalucia Forti Messina counted about thirty "coalitions" and eleven documented work stoppages in the period through 1860.[13] Among the workers participating in these strikes were mechanics, printers, stonecutters, and ribbon weavers. Both printers and stonecutters established new organizations after their 1860 strikes; the ribbon weavers' organization preceded their strike. A close link between organization and strikes thus is evident in this period.

Messina attributes these strikes to economic and social causes, in particular high prices and low wages (a standard economic explanation, as we noted), and not to republican politics, as some observers believed. She also acknowledges, however, the importance of contemporary political change, especially the annexation of Lombardy to the kingdom of Piedmont to form a unified kingdom of Italy. The new regime's openness to free association, she believes, shaped events by enabling workers to meet and discuss their own interests. Although she wisely rejects any political significance, such as that concerning the authorities (that Mazzinian republicans were at the root of the unprecedented rash of strikes), Messina nevertheless provides a political explanation, that the new liberties facilitated workers' organization and demands.[14]

According to Volker Hunecke's assiduous accounting, there were eight strikes in Milan in 1872. That summer, a citywide strike began on August 5 in the foundry of the Suffert company, with mechanics from the associated machine works and other nearby engineering plants quickly joining their fellows. Construction workers left their work sites on the same morning and marched through the city streets calling for other workers to turn out. The movement reached its apogee between August 6 and 8. On August 9 the masons began a back-to-work movement as they returned to the piazza del Duomo renovation project.

Eva Civolani's analysis emphasizes structural and conjunctural economic factors, noting first the increasing density of manufacturing activity in northern Italy, related to the contemporary industrial decline in the south. Second, she points to the greater demand that northern industry enjoyed during the Franco-Prussian War and the German occupation of French territory (when French products were not available in European markets). The summer was also the busy season for construction, and there was a building boom under way in northern Italian cities. Thus it was a favorable time for workers to make demands. At the same time, there was dissatisfaction with wages, which were inadequate to support workers' standard of living in a time of rising prices.[15]

Economic factors seem dominant here; nevertheless, I would point out that the cyclical economic factors that Civolani stresses are linked to political factors: the recent completion (achieved in 1869 and formalized in 1872) of Italian unification, the inflow of foreign capital that followed it, and the war-induced demand for Italian products. Each early "strike wave," then, was related to political changes that reduced the cost of action or augmented the benefits to be gained from it.[16]

An Overview of Milanese Strike Patterns, 1878–1903

The data base for this overview is the strike-by-strike list published by the Italian Statistical Office. (After 1903, the statistics are presented in summary form only.) The completeness of this statistic was called into question by a Società Umanitaria study published in 1904. It reported that the Milan Chamber of Labor's records of strikes for the years 1900 to 1903 yielded about 5 to 10 percent more strikes than did the national-level compilation. Hunecke's search in the archives and contemporary press also found many strikes missing from the published statistics, among them the large masons' and laborers' strike of 1887.[17]

Table 6-1 displays (1) the annual number of strikes and strikers and the average number of strikers per strike listed in the Italian government statistics; and (2) (in parentheses) the number of strikes discovered through data gathered independently by Hunecke for 1878 to 1892 and by the Chamber of Labor for 1900 to 1903. The problem with the additional strikes is that they were often only briefly mentioned, with little information about number of strikers, duration, grievances, or outcome. (This is not the case for all of them, of course; although there is no police report in the state archives for the 1887 masons' and laborers' strike, it was extensively covered in the press. Full information is thus available from other sources about this major strike, at least). No attempt was made to add to or correct the statistical data base derived from the official strike list, which was accepted as a uniform record more satisfactory for comparative purposes than the incompletely corrected records.

The top panel of Figure 6-1 shows the annual total number of strikes in Milan from 1878 to 1903. There are marked steps upward in the number

Table 6-1. Numbers of Strikes, Strikers, and Strikers per Strike in Milan, 1878–1903

	Total No. of Strikes	Total No. of Strikers	Strikers per Strike
1878	1 (4)*	26	26.0
1879	4 (11)	394	98.5
1880	2 (6)	792	396.0
1881	— (3)	—	—
1882	2 (3)	400	200.0
1883	2 (5)	260	130.0
1884	2 (7)	138	69.0
1885	— (4)	—	—
1886	2 (9)	850	425.0
1887	1 (7)	111	111.0
1888	3 (17)	665	221.7
1889	13 (19)	2,235	171.9
1890	6 (13)	765	127.5
1891	7 (15)	3,554	507.7
1892	12 (16)	1,387	115.2
1893	9	991	110.0
1894	14	1,347	96.2
1895	26	1,541	59.3
1896	32	2,399	75.0
1897	20	1,531	76.6
1898	26	3,529	135.7
1899	26	3,707	142.6
1900	25	1,623	64.9
1901	69 (88)†	29,623	429.3
1902	56 (89)	28,869	515.5
1903	37 (52)	15,274	412.8

*Numbers in parentheses from 1878 to 1892 are from Volker Hunecke, *Classe operaia e rivoluzione industriale a Milano* (Bologna: Il Mulino, 1982), Table 6.3, pp. 380–81.

†Numbers in parentheses from 1901 to 1903 are from Società Umanitaria (Ufficio del lavoro), *Scioperi, serrate e vertenze fra capitale e lavoro in Milano nel 1903* (Milan, 1904), p. 9.

Source: Ministero di agricoltura, industria, e commercio, Direzione generale della statistica, *Statistica degli scioperi avvenuti nell' industria e nell' agricoltura* (Rome: Tipografia nazionale, 1878–94, 1895–97, 1896, 1898–1903).

of strikes occurring in 1889, 1895, and 1901. As Table 6-1 shows, the number of strikers, and particularly the number of strikers per strike, did not increase steadily over time, as did the number of strikes. The reason is that even in the early years, there were occasionally very large strikes. There was also incomplete reporting of small and/or brief strikes, as demonstrated by the discrepancy between the official numbers and those discovered in the newspapers. The few small strikes are therefore dwarfed in the yearly means of strikers by the presence of a few very large strikes. The temporal patterns of the number of strikes in the printing, construction, textile, and metal and machine industries depicted in panels 2 through 5 of Figure 6-1 roughly correspond to the aggregate pattern.

Table 6-2 demonstrates that this correspondence is not exclusively based on the fact that these industries contributed the most strikers to the total number. Printing and construction had fewer strikes over the period than did the fur, leather and agricultural products, and garment-making industries. The textile and metal industries held pride of place in terms of numbers of strikes. Fur and leather, metal, construction, transportation, and utilities contributed the greatest numbers of strikers. The largest number of strikers per strike occurred in transportation and utilities and public accommodations. (Domestic service and public accommodations had to be combined in these compilations because of inconsistency in the categories of the published statistics. However, the particular strikes that contributed to

Total Strikes

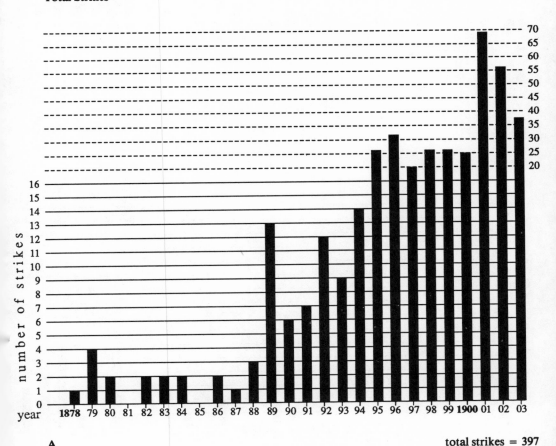

A

total strikes = 397

Figure 6-1 Strikes in Milan, 1878–1903 (*continues*)

Printing

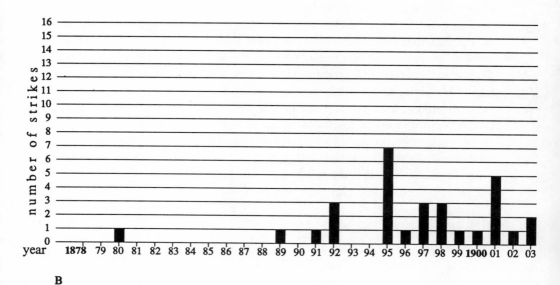

B

Construction

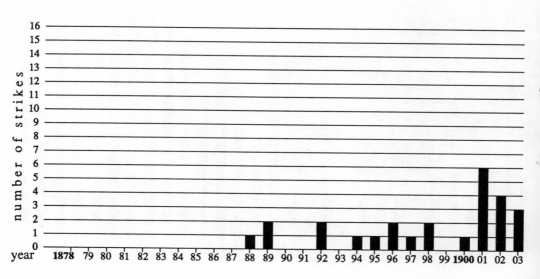

C

Textile

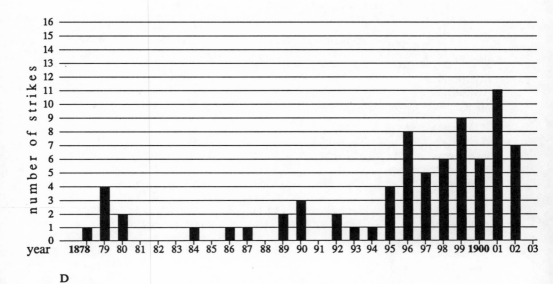

D

Metal - Engineering

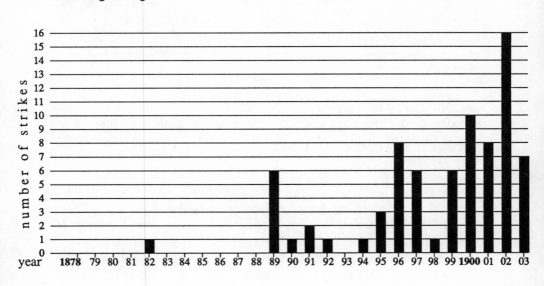

E

Politics and Class in Milan

Table 6-2. Numbers of Total Strikes, Total Strikers, Strikers per Strike, Mean Duration, and Total Person Days by Industry in Milan, 1878–1903

Industry	Total No. of Strikes	Total No. of Strikers	No. of Strikers per Strike	Mean Duration (days)	Total No. of Days Lost
I. Agriculture					
Agriculture	1	8	8.00	1.00	8
Mining	3	410	136.67	26.67	12,888
II. Manufacturing					
Food	10	4,231	423.10	8.20	31,599
Wood, straw, mixed	20	5,052	252.60	10.95	52,516
Fur, leather, other agri. products	53	13,032	245.89	15.38	287,564
Textiles	74	9,868	133.35	7.59	87,325
Garment making	36	3,093	85.92	5.64	19,262
Ceramic, glass	8	472	59.00	9.75	4,243
Metal and machine, precision	77	10,279	133.49	9.45	112,285
Chemicals	15	2,398	154.87	10.73	18,003
Rubber	2	1,220	610.00	3.00	1,300
Printing, lithography	30	3,074	102.47	11.60	96,365
Construction	26	16,001	615.42	10.73	423,659
Paper, cardboard	12	1,620	135.00	9.67	37,133
III. Services, Commerce					
Government	1	20	20.00	4.00	80
Utilities, transport.	18	15,240	846.67	6.00	104,609
Domestic serv., public accomm.	7	8,512	1,216.00	4.43	34,574
Retail comm.	1	29	29.00	1.00	29

Source: Ministero di agricoltura, industria, e commercio, Direzione generale della statistica, *Statistica degli scioperi avvenuti nell' industria e nell' agricoltura* (Rome: Tipografia nazionale, 1878–94, 1895–97, 1896, 1898–1903).

the high number of strikers in this category were restaurant workers's strikes in 1899–1903.) The longest strikes were in fur and leather and printing and lithography. The highest rates of person days lost by strikes were found in the construction, fur and leather, metal and machine, and transportation and utilities industries.

The bottom line of Table 6-3 shows that the majority (more than three-quarters) of the total number of strikers were active in the last five-year period. The involvement of women and children (boys and girls under 16) increased over the period. The total absence of women strikers from the official statistics before 1889 is linked to the underenumeration of small and/or brief strikes, especially in textiles. Women were a majority of strik-

Table 6-3. Strikers by Five-Year Periods, Divided by Sex and Age, in Milan, 1878–1903

	1878–1883	1884–1888	1889–1893	1894–1898	1899–1903
Male	100%	95%	92%	67%	69%
Female	0	5	3	28	17
Children	0	0	5	5	14
% of Total No. of Strikers in This Period	2.0%	1.8%	9.5%	10.5%	76.2%

Source: Ministero di agricoltura, industria, e commercio, Direzione generale della statistica, *Statistica degli scioperi avvenuti nell' industria e nell' agricoltura* (Rome: Tipografia nazionale, 1878–94, 1895–97, 1896, 1898–1903).

ers in textiles in the last two five-year periods (see Table 6-3). The great growth in the proportion of child strikers in 1899–1903 was concentrated in two industries: garment making (primarily girls) and construction (primarily boys). Comparing total numbers of strikers by industry in the periods 1894–98 and 1899–1903 (see Table 6-4) reveals considerable increases in all industries, already observed in Table 6-1, but also an increase from one period to the next in some formerly inactive industries, especially public accommodations. Workers who had seldom or never struck before were mobilized and joined strikes in the first years of the twentieth century. There were also large rises in the number of strikers in transportation and utilities, construction, and the metal industry between the two five-year periods. In these cases, the number of strikers had been higher in the period before 1894–98; hence the increase in the ensuing period is a return to previous high levels rather than a new departure. The three years after 1900 represent a distinct break from earlier levels of strike activity, promoted by the Saracco government's recognition of workers' right to organize and strike. Table 6-5 demonstrates what difference this made in Milan. Almost 50 percent of all the strikes from 1879 through 1903 occurred in the last three years, and they accounted for 63.5 percent of all the strikers in the whole period.

Another measure of strike activity, which relates person days (strikers times length of strike) to the work force in each industry, is shown in Table 6-6 for five-year periods. Again, the great rise in strike activity over time is obvious. There is a wavelike process as workers in different industries began to strike from one period to the next. In the first five-year period, only printers showed much inclination to strike. In the next two five-year periods, first the construction workers and then the metal and machine workers joined the strike movement. These workers are examined more closely in

Table 6-4. Strikers by Industry, Sex, and Age, 1894–1898 and 1899–1903

Industry	1894–1898				1899–1903			
	M	F	C	Total	M	F	C	Total
I. Agriculutre								
Agriculture	0	0	0	0	8	0	0	8
Mining	410	0	0	410	0	0	0	0
II. Manufacturing								
Food	0	0	0	0	2,718	200	405	3,323
Wood, straw, mixed	142	54		196	1,959	26	153	2,138
Fur, leather, agri. prod.	1,365	146	61	1,572	5,941	6,270	369	12,580
Textiles	653	1,892	115	2,660	912	2,979	1,677	5,568
Garment making	269	391	37	697	184	936	722	1,842
Ceramic, glass	0	0	0	0	324	0	97	421
Metal and machines, precision	644	25	61	730	3,919	498	636	5,053
Chemicals	666	29	8	703	1,528	169	37	1,734
Rubber	0	0	0	0	8	12	0	20
Printing, lithography	0	50	132	732	856	7	549	1,412
Construction	566	60	23	649	9,791	50	4,735	14,576
Paper, cardboard	35	154	52	241	378	644	257	1,279
III. Services, Commerce								
Government	0	0	0	0	20	0	0	20
Utilities, transport.	1,308	0	0	1,308	13,460	374	0	13,834
Domestic serv., public accomm.	0	0	0	0	7,356	0	706	8,062
Retail comm.	0	0	0	0	29	0	0	29

Source: Ministero di agricoltura, industria, e commercio, Direzione generale della statistica, *Statistica degli scioperi avvenuti nell' industria e nell' agricoltura* (Rome: Tipografia nazionale, 1878–94, 1895–97, 1896, 1898–1903).

the case studies in this and the next chapter, which show that changes in the organization of production, cyclical factors, and political change facilitated workers' organization in these industries. The growth of association in turn promoted the strike activity.

In the final period, 1899 to 1903, the propensity to strike rose in all industries. Construction again held first place (another big strike in 1901), but the transportation, chemical, and paper workers also were very active. There were big differences, however, in the long-term outcomes of these hectic years, depending on the industrial category. Thus the surge of organization in transportation and utilities turned out to be long lasting, but in the case of paper and chemicals, there were brief but temporary peaks of organization and collective action.

Is there any relationship between migration patterns and strike patterns, as suggested by the "uprooting thesis"? Industries like leather and fur,

Table 6-5. Numbers of Strikes, Strikers, and Strikers per Strike, 1878–1900 and 1900–1903

Year	Number of Strikes	%	Number of Strikers	Number of Strikers per Strike	%
1879–1900	232	50.3	28,062	121	26.9
1901	88	19.1	33,192	377	31.9
1902	89	19.3	30,775	346	29.5
1903	52	11.3	12,213	235	11.7
Total	461	100.0	104,242	226	100.0

Source: Società Umanitaria, *Scioperi, serrate e vertenze fra capitale e lavoro in Milano nel 1903* (Milan, 1904), p. 9. Note that the published strike statistics in Table 6.1 yield much higher ratios of strikers per strike, evidently because of the smaller number of strikes listed in that table and because the strikes that the Società Umanitaria added for 1901–3 involved few strikers.

Table 6-6(a). Propensity to Strike by Industry by Five-Year Periods in Milan, 1878–1903

Industry	Strike Person Days per Labor Force		
	1878–1883	1884–1888	1889–1893
I. Agriculture			
Agriculture	0	0	0
Mining	0	0	0
II. Manufacturing			
Food	0	.58	0
Wood, straw, mixed	0	.36	.01
Fur, leather, other agri. products	.39	0	.48
Textiles	.14	.79	.34
Garment making	.06	.004	.05
Ceramic, glass	0	0	.05
Metal and machines, precision	.03	0	1.89
Chemicals	0	0	.03
Rubber	0	0	1.20
Printing, lithography	16.8	0	.10
Construction	0	.69	.31
Paper, cardboard	0	0	.09
III. Services, Commerce			
Government	0	0	0
Utilities, transport.	0	0	.02
Domestic serv., public accomm.	0	0	.001
Retail comm.	0	0	0

Source: Ministero di agricoltura, industria e commercio, Direzione generale della statistica, *Statistica degli scioperi avvenuti nell' industria e nell' agricoltura* (Rome: Tipografia nazionale, 1878–94, 1895–97, 1896, 1898–1903).

Table 6-6(b). Propensity to Strike by Industry by Five-Year Periods in Milan, 1878–1903

Industry	Strike Person Days per Labor Force	
	1894–1898	1899–1903
I. Agriculture		
Agriculture	0	.001
Mining	54.38	0
II. Manufacturing		
Food	0	2.71
Wood, straw, mixed	.26	3.67
Fur, leather, other agri. products	1.37	2.08
Textiles	1.54	3.87
Garment making	.04	.38
Ceramic, glass	0	1.85
Metal and machines, precision	.91	1.95
Chemicals	.46	10.82
Rubber	0	.05
Printing, lithography	.80	3.81
Construction	.82	41.05
Paper, cardboard	.74	10.12
III. Services, Commerce		
Government	0	.007
Utilities, transport.	.11	7.45
Domestic serv., public accomm.	.002	.66
Retail comm.	0	.002

Source: Ministero di agricoltura, industria, e commercio, Direzione generale della statistica, *Statistica degli scioperi avvenuti nell' industria e nell' agricoltura* (Rome: Tipografia nazionale, 1878–94, 1895–97, 1896, 1898–1903).

textiles, metals and machines, and printing had higher proportions of Milan-born workers than did the Milanese labor force as a whole (see Table 3-2). Construction, utilities and transportation, and public accommodations and domestic service all had a lower proportion. In the numbers of person days lost over the full period (Table 6-2), construction, fur and leather, metal and machines, and transportation and utilities led. Other factors than migrant status seem to be more important, for industries that employed migrants and those that hired predominantly native-born workers both are among the industries with the greatest number of person days lost.

In order to test the effect of organization on the number of strikers, a correlation was calculated between two rates: the number of strikers in each industry out of the total number of workers in the same industry (for 1894 to 1898) and the proportion of the industry's workers in organizations in

1897. The relationship between the two rates was very strong (a product–moment correlation coefficient of .93). The calculation was repeated with the 1899–1903 strikers as a proportion of the labor force in each industry and of organized workers for 1902. The correlation for the second period was weaker (.23) but positive.

This positive relationship of strikers to organized workers by industry was associated with political and institutional change. Such change marked nearly all the upward movements in strike frequency, as illustrated in the following list of these closely connected developments:

1859: Political unification; end of Austrian authoritarian government in Milan; strike "wave" of 1859–60.

1889–91: Expansion of suffrage in local elections; election of worker representatives to municipal council; establishment of Chamber of Labor; increased number of strikes and strikers in 1889.

1891–94: Growth of resistance leagues encouraged by Chamber of Labor and (after 1892) Socialist Party.

1894: Slowdown of organizational activity; consequence of Crispi's repressive decrees against Socialists and leagues associated with them.

1895–96: Political coalition against repression; high points of strike activity.

1896–97: Resumption of growth in labor organization.

1898–99: Dissolution of Chamber of Labor and leagues in May 1898; absence of organization; leveling off of strike activity (number of strikers high).

1900: Resumption of organizational growth; legalization of strikes.

1901–2: New peaks in strike frequency and number of strikers.

This simple chronology reminds us that political factors both promote and inhibit workers' organization and collective action. There were many setbacks, ranging from the government's dissolution of workers' institutions to its surveillance of workers, their organizations, and their strikes and harassment. The orders for this kind of police action, based on the ambiguous legislation concerning strikes (reviewed in Chapter 5), came from Rome. Positive political encouragement came with the various phases of unification and laws that expanded political rights and participation. Another supportive environment for strike activity resulted in 1895–96 from the unified democratic parties' political resistance to Prime Minister Francesco Crispi's dissolution of the Socialist party and affiliated economic organizations (discussed in Chapter 9). These political factors affected the conditions for collective action for all Milanese workers, though only some of them went on strike in each period. To understand the timing by industry and for specific occupations or groups, case studies are necessary.

Industrial Case Studies of Strikes

A close look at strike patterns can reveal causal factors, but the industrial case studies discussed here permit a better understanding of the mechanisms involved. In this chapter we shall examine two well-documented industries—printing and construction, and in the next chapter, we shall consider the textile, metal and engineering, and the female-employing rubber, garment, and tobacco industries. These case studies demonstrate the way in which the effects of economic structural and organizational change were regularly mediated by political factors.

Printing

After the unification of Italy the printing and publishing industry flourished, a kind of *risorgimento* of the book industry, as Ada Gigli Marchetti puts it. The newspaper industry became increasingly concentrated in Milan.[18] A printers' strike played a part in the 1860 strike wave, demanding higher wages and a fixed pay scale, or *tariffa*. The strike ended with the establishment of a conciliation committee and, shortly thereafter, the Società degli artisti tipografi. The old Pio istituto tipografico refused to merge with the new organization. Employers rejected the society's demands for higher wages, but they reluctantly accepted a *tariffa*. In practice, however, it was difficult to enforce the *tariffa* because of the complex organization of work in the industry, and despite a strike in 1863, by the mid-1870s it had gone by the board. The Società then split into sections for compositors and pressmen: The first soon dissolved itself, and the latter reduced its scope to mutual benefits.

As early as 1872, an organization for the introduction and observance of a fixed pay scale (Associazione dei combinatori tipografi milanesi per l'introduzione e l'osservanza di una tariffa) was launched on the suggestion of Mayor Giulio Bellinzaghi. (A delegation of printers had gone to him with their grievances. They also founded at this time a monthly newspaper, *La Tipografia Milanese*.) The organization, as reflected in its publication, was especially concerned with the proliferation of "apprentices," boy workers who were hired to cut costs, not to train in a craft. The first national meeting of printers' societies (held in Milan in 1874) established the Associazione fra gli operai tipografici italiani per la introduzione ed osservanza della tariffa. A committee charged with defining an enforceable *tariffa* for Milan met in 1876 and 1877 and proposed a pay scale. At the second National Printers' Congress, held in September 1878, the Milanese pressman joined what until then had been an organization of compositors only;

henceforward there were compositors' and pressmen's sections of the Associazione. The compositors presented their proposed *tariffa* to all of Milan's printing establishments, but there was no response. Hence, on February 10, 1880, the organization issued an ultimatum: If the publishers did not accept the *tariffa* in one week, it would consider all contractual obligations to be null and void. The bosses collectively rejected the proposed single *tariffa*, and so the stage was set for the strike, which began on February 17.

This strike marked the end of a long process of organization and the beginning of a new period in which the printers sought to match their employers in aggressive mobilization and collective action. The seven hundred compositors developed procedures to prevent scabbing and promote solidarity in the strike. Their newspaper, *La Tipografia Milanese*, called it a war of independence, "a war that no one could evade without being branded a coward." [19] Aided by the solidarity of the pressmen and by financial and moral support from printers elsewhere in Italy and in other nations, the strike lasted until June. (During its course, the entire strike committee was arrested, and the society's president, Vincenzo Corneo, was tried and convicted of interfering with others' right to work.) An *ad hoc* arbitration committee, including workers, publishers, and citizens (chaired by notary Stefano Allochio), agreed on a settlement favorable to the workers, including a uniform wage scale, wage increases, and limits on apprenticeship. Over the next decade, membership in the union increased, as it fought to ensure that employers honored the compositors' *tariffa*, and it forced a similar agreement for the pressmen. [20]

The task continued. Two actions were taken immediately after the 1880 strike. First was the establishment of a "propaganda society" that proselytized printers in nearby cities as well as the founding of nine new sections of the national federation. Second were attacks in the pages of *La Tipografia Milanese* on the Pio istituto and other mutual benefit societies, especially the so-called internal ones that were sponsored by employers. [21]

In 1885, Augusto Dante (1857–1908, a printers' union militant and member of the Partito operaio) spoke to the printers' section of the Lega del lavoro. His message was that competition was again driving publishers to hire youths, thereby causing unemployment among skilled adult male printers. [22] In 1886, the workers at Vallardi, a company that refused to accept the *tariffa*, signed a petition addressed to the police chief of Milan, asking his intervention in enforcing the agreement. Vallardi had hired women in order to cut his wage bill, and he insisted that he would continue to do so. His compositors struck but were unable to change his position. [23]

Although the proportion of women employed in publishing as a whole (as revealed in the population censuses) almost doubled between 1881 and 1901, it was only 10 percent in the latter year. Women worked as book binders and were sometimes employed as compositors, setting straight type (as in books, in contrast with newspapers with narrow columns), but they were very seldom employed as newspaper compositors. That skilled job was reserved for men, most likely by tacit agreement of the workers and their employers. The mechanism for excluding women was demanding equal pay for equal work but maintaining the division of labor by sex. Because the men and women had different tasks, the demand had little meaning. The men seriously claimed both that the ill effects of the lead in type were more injurious to women and that all women were technically inferior to men as compositors.[24]

In 1888, Vallardi tried another strategy. He offered an annual bonus for those who could set 8,000 letters a day, with proportionately smaller bonuses for less rapid work, thus pitting workers against one another in a race for advantage. Vallardi became notorious among Milanese publishers in his unwavering opposition to union demands.[25]

The pressmen presented an ultimatum to their employers in 1888 that was similar to that of the compositors eight years earlier: Either accept the *tariffa* or prepare for a strike. The pressmen had only partial support from the national level or the Milanese compositors, and so they taxed themselves in advance to build a strike fund that would enable them to hold out against their employers. Faced with this resolve and such resources, many employers did agree to the new wage scale. The others were struck by their workers in a series of firm-level strikes in the first months of 1889. A vigilance committee policed the strikes to enforce solidarity.[26]

The industrywide strike of 1892 was well prepared. As early as January 1891, a meeting of printers had called for revision of the *tariffa* of 1880. Workers discussed the proposed demands at length, and a final version—which called for a 9-hour day (8 hours for newspaper printers) and a small increase in their base pay—was approved by a mass meeting on October 28, 1892. An appeal for solidarity, which declared that the motive behind the demand for reduced hours was to help unemployed colleagues, was published the same day. *La Lotta di Classe* reported that the printers' strike fund was four times larger than that in 1880. The employers' Associazione tipografico-liberia agreed to meet with the workers through a joint arbitration committee (four employers, four pressmen, and four compositors), chaired once again by Stefano Allochio. The demand for a shorter working day was accepted completely, but a compromise on the wage demands brought the printers a small increase than they had hoped for. Not all the publishers signed the agreement promptly, but firm-by-firm strikes persuaded them to sign, except for Ricordi and Vallardi.[27]

Not all of Vallardi's workers, perhaps intimidated by his tactics, struck. Some were accosted on the street by several of their striking fellow workers; there was a scuffle in which one of the nonstrikers was hit on the head.

The strikers forced the others to go with them to the Chamber of Labor. There, according to an allegation taken by the police, the strikers pressured them to join the strike, promising them strike pay if they did. The police arrested five strikers, accusing them of threats, violence, and attempts to limit their fellows' right to work. Again, Vallardi was the special target of the association, but again it failed to move him.[28]

The printers moved to integrate their organizations on the national and urban level. A national congress in 1893 changed the name of the federation to the Federazione italiana dei lavoratori del libro, and its journal became *Il Lavoratore del Libro*. In Milan, the pressmen's section merged with the pressmen's mutual benefit society, leaving the Pio istituto even more isolated as a mixed organization.

Workers struck the publishers Wild and Belloni in 1895, attempting to force them to accept the new *tariffa*. The companies then sent their work to Ghidini e Sormani, one of the publishers who had accepted the *tariffa;* the workers promptly struck there, too. These strikes were settled with the help of two printers' union leaders. An appeal was sent to the printers, asking them to refuse to do work from plants that were on strike. The consequence was new strikes. The mixed arbitration committee intervened to adjudicate the tangled issues. In November, Allocchio reported the committee's ruling that although subcontracting was permissible under ordinary circumstances, it was not acceptable under strike conditions.[29]

Strong organization and disciplined collective action did not mean an absence of strikes, but they did mean continuing vigilance and frequent disagreements with employers. There were at least three strikes in 1897— one against Sormani and Ghidini that resulted in the arrest of twenty strikers, and two in 1898.[30]

The printers' strength was demonstrated in the repression following the May events of 1898. All workers' organizations were then dissolved by decree; each was required to apply separately to the prefect to recover its records and funds. The prefect demanded that those authorized to reconstitute limit their functions to mutual benefit. The compositors flaunted their rejection of this demand in April 1899 when they named their reborn organization the Associazione fra gli operai compositori tipografi di Milano per l'introduzioine e osservanza della tariffa, con sussidio di malattia, cronicità e vecchiaia, the last phrase meaning "with benefits for illness, disability, and old age."[31]

In March 1900 the Associazione tipografico-libreria (a syndicate of employers) promulgated a uniform set of work rules requiring that workers submit a certificate of "good conduct" when applying for work, that a week's wages be withheld as a caution, and that workers not smoke or drink during work hours. The printers promptly called a mass meeting of protest at the Chamber of Labor, chaired by Enrico Bertoni, a printer who was a member of the city council. The assembly rejected the proposed rules and voted to demand an upward revision of the *tariffa*. The following January, *La Tipografia Milanese* reported their demands with these ringing words: "that we may free ourselves of economic servitude, restore our dignity as free men, provide effective social benefits that will improve work conditions, offer protection in old age, and political liberty." The *tariffe* of 1901 and 1907 were won by hard bargaining, although a few, like Vallardi, continued to reject the workers' claims. An employers' newspaper reported wryly that printers were "hard workers, possessed good character, were respectful and reasonable, but tended to value their own rights too highly."[32]

Translating into action the logo of their association—three rings standing for mutuality, resistance, and cooperation—the printers had achieved a virtual "closed shop," Giuseppe Paletta concludes.[33] The Società umanitaria study of the 1903 strikes in Milan praised the printers as "98 percent organized. . . . The owners, who know the federation's strength, are the first to advise workers to join."[34]

The printers' success can be attributed to their organizational power and to the industry's general prosperity. They were essential to their employers, who, willing or not, had to deal with them. A major Milanese publisher, Giuseppe Treves, remarked after the pressmen's strike for their wage scale in 1888: "We accepted the *tariffa* because they had knives at our throats."[35] The combination of a booming industry and skilled workers who were in a strong position to make demands was a consequence not only of capital concentration and changes in the labor process. It was linked also to locational decisions by publishers who sought the market of cosmopolitan newspaper-reading Milan and found themselves dealing in the labor market with the skilled, literate, and organized workers who lived there. Further, printers were seldom unemployed; as a police inspector wrote to his chief in the spring of 1892, even when unemployment was looming large in the city, "in this industry, work is stable."[36] The central location and urban experience of the Milanese printers was a factor in the early and exemplary role they played in Milanese, regional, and Italian worker organizations that they led beyond occupational or industrial groups to class-based collective action.

Construction

Construction was a booming industry in Milan in the decades between 1860 and 1889, with the postunification "urban renewal" and then the new park, renovation of the Castello, and residential building.

Among the many workers who struck in the postunification year were the stonecutters *(scalpellini)* and masons. Although a comprehensive history of workers in the building trades would include the stonecutters, this section will examine only the masons, their associated workers, and their organizations.

The masons held a noisy demonstration in late March 1860, along with the railroad workers. The police reported subversive shouts, such as "Viva la Repubblica!" Their strike in July of that year is more fully documented. The masons stayed off the job for a week, holding daily meetings; one worker was arrested for trying to persuade another to stop work, and several for *oziosità* (idleness). A settlement, including a wage increase of 20 percent and an additional hour for meals in July and August, was negotiated by the police, anxious to get the men back to work. The masons resisted— they hoped to get their summer workday shortened—but the authorities forced them back on the job by supporting the *capimastri*. The workers founded the Società di mutuo soccorso dei muratori after the strike.[37]

Masons were also among the instigators of the 1872 citywide strike, although their mutual benefit society disclaimed any responsibility for its members' actions. The masons won wage increases but no relief in their workday hours.[38] In 1877, the Associazione di mutuo soccorso fra gli addetti all'arte edilizie (founded in 1874, replacing the earlier group) bargained for a maximum 11-hour day in the good season, but its major activity before 1886 was sponsoring a trade school for the sons of masons, and an accident fund.

Dissatisfaction with their only organization, the Associazione di mutuo soccorso, which joined bourgeois and honorary members with workers, led the masons to seek other organizational forms. Members of the Partito operaio italiano worked with them, first, to establish the national Federazione degli operai muratori d'Italia and, shortly afterward, the exclusively worker Società mutua e miglioramento fra i lavoranti muratori, manovali, badilanti e garzoni di Milano e provincia (the constitution of this organization is discussed in Chapter 5). Its goal was to improve hours and wages for masons and their assistants. The society grew quickly, reaching 1,500 members in April 1887, when the agitation for better remuneration resumed; its membership oscillated between 3,000 and 1,200 (1896) during this period. Another initiative in 1887 was a masons' cooperative, which, despite a difficult beginning, soon won contracts to build workers' housing at Porta Vittoria, starting late in 1887 and early 1888. Its later success (especially after 1892) was based on city contracts.[39]

Early in September 1887, the masons and other construction workers of Milan met en masse to discuss the refusal of the *capimastri* to honor the wage scale agreed on that very spring. (The builders had agreed to a standard wage per day throughout the year, even though the hours worked per day varied over the seasons). Two cautious, older leaders were called to chair the meeting, and they promptly urged calm. "We do not want socialism," one declared, "but we do demand that no one steal the fruits of our labor. We should not strike—let the *capimastri* refuse us work." Grumbles of protest grew loud. One young man made himself heard over the others. "Let's put the cards on the table," he shouted. "Tomorrow morning, Monday September 5, if my *capomastro* offers me a wage with a 25 or 27 centesimi cut, what should I do?" "Stay home, don't go to work to become a striker," the chairman replied. "So," the fellow yelled, "if it's not soup, it's bread soaked in broth! You advise against a strike, but you say that those who are dissatisfied with the new proposals of the *capimastri* should not work?" The wrangle over words and action continued, but the meeting ended with a vote to refuse to work if the *capimastri* cut wages. A negotiating committee was chosen, and the masons and their helpers went on strike the following day, for the *capimastri* did indeed cut the daily wage offered, by 28 to 30 centesimi.

The strike continued for a month, as various efforts were made to work out a deal with the *capimastri*. In the first week, workers whose bosses had accepted the April agreement stayed on the job. But after the *capimastri* as a group rejected the workers' demands and also a compromise formulated by a committee of worker and employer representatives, chaired by the mayor (the compromise was also rejected by the workers), those who had previously stayed on the job walked out. Thousands of workers met daily at the Arena to consider proposals and counterproposals. The masons both organized their own strike assistance scheme and appealed to the public for support. Much of the press and public were sympathetic to their cause.

Eventually, some *capimastri* agreed to the terms, and workers were permitted to sign on with them. Because most of the workers were still on strike, however, an attempt was made to "tax" those with jobs to help those without them. Defections among workers nevertheless began to sap solidarity. The *capimastri* perceived that support for the strike was dwindling, so became even more determined to hold out. The step-up of military patrols in the city suggested that the authorities also were becoming anxious for a settlement.

Into this impasse stepped Democratic Radical Deputy Felice Cavallotti. In an open letter in *Il Secolo* he wished the strikers well and also suggested to the *capimastri* "as a friend" that a settlement that was fair to labor would be more economical to them than prolonging the strike would be. The *capimastri* promptly proposed Cavallotti as mediator, and the workers accepted him. The settlement that Cavallotti worked out was welcomed by the *capimastri;* the masons' leaders persuaded the strikers to accept a "convention" that gave qualified masons a higher minimum wage: "Common" masons were to receive 2.90 lire per day from April to September, "good" ones, 3.20 lire per day. In the winter months, the equivalent wages were 2.40 and 2.60 lire. Work hours were to be cut year-round.[40]

Il Secolo and the other liberal newspapers congratulated all for the apparently satisfactory outcome, and a huge demonstration of masons saluted the newspapers and Cavalotti. But the moderate newspaper *Corriere della Sera* introduced a discordant note: Some, at least, of the *capimastri* had no intention of accepting the terms of the convention. Further, *Fascio Operaio*, the newspaper of the workerist Partito operaio, quickly pointed out that "although the bourgeois press celebrates, not all workers find [the agreement] so easy to swallow"; the interests of laborers and apprentices, in particular, had been ignored. Almost two years later, the masons' newspaper, *Il Muratore,* disappointed, commented: "So it has been that our sacrifices, our hopes, the struggle itself—it's now obvious—have been replaced by bitter disillusionment. We stonemasons are no better off than before. We were cheated, robbed."[41]

Later in the fall of 1887 the masons' society took up the problems of workers from the nearby countryside who came to Milan to seek jobs. "There [in the countryside] they earn 1.50 lire and eat black bread and polenta; here they earn 2 to 2.50 lire, and they are content because they can eat white bread." The Milanese masons decided to organize rural workers. Branches of the Società were quickly established in Sedriano, Settimo Milanese, Affori, Niguardo, Sesto San Giovanni, and Chiaravalle, all close-in *paesi,* and twelve more branches appeared over the next decade. The opposition of priests and *padroni* made it hard to transform these into functioning branches, however. Only after 1901 and the spread of building cooperatives in these areas, did the organizing effort succeed.[42]

Dissatisfaction in Milan returned as early as the spring of 1888, for many *capimastri* were not observing the convention. At first glance, it would seem to be a good moment to strike, because the busy season and the renovations and new buildings that were part of the new city plan were at hand. The city authorities and the press were hostile to the attempted partial strikes, however, and out-of-town workers were available to substitute for Milanese at lower wages. In the following winter and early spring, massive unemployment and strike rumors brought the police out to repatriate those masons who were temporary migrants. Later that spring, the masons' association supported May Day and the first steps toward creation of the Chamber of Labor.[43]

During the 1887 strike a more politicized leader, Silvio Cattaneo, had emerged. An effective speaker and organizer (he spoke in dialect, so that even the most recently arrived worker could understand him), Cattaneo hoped to extend the society's discipline and solidarity to all workers. Sometimes his efforts were fruitless. In April 1891, for example, Cattaneo wrote to the newspaper *L'Italia del Popolo* to explain an earlier report that the masons' society had voted against joining the Fascio dei lavoratori. He noted that the majority of the society's members who were from Milan had voted for affiliation, but that the votes that rejected it included those of nonresidents who were not fully conscious of class interest.[44]

The Società resumed agitation in September 1891, as the building trades entered their slow season. The police, acting on information supplied by the masons, drew up lists of the *capimastri* who were not observing the 1887 convention and of *cottimisti*, workers who paced their work fast in order to speed up their fellows. But the local authorities forbade the Società to meet. Once the Chamber of Labor was in place, the workers had a greater chance of collectively controlling the labor market. Indeed, Giuseppe Croce, the chamber's secretary, invited the masons and their assistants to use its employment office. Workers themselves went to the *ponte* to tell others about the services available at the chamber. A motion passed by an assembly of workers called for the abolition of the *ponte*, "that indecent spectacle, where workers skilled in the builder's craft are exchanged like slaves. This system is a continuing offense to civility, to the dignity and the interests of laboring men."[45] The masons deplored the behavior of some *capimastri* and welcomed the intervention of the Camera del lavoro. In the winter of 1892, with the help of municipal council members Vincenzo Corneo and Osvaldo Gnocchi-Viani, the masons petitioned the council to devise work projects for the unemployed, favoring residents of Milan over outsiders. The police chief wrote to the prefect on February 9, 1892, reporting a meeting of the masons and their assistants at which they objected to the continuation of the *ponte*, contrary both to their own interests and those of the Chamber of Labor.[46]

La Lotta di Classe, the newspaper established in anticipation of the founding of the Socialist party in the summer of 1892, pointed out in its first issue that promises to the building-trades workers had been ignored. City public works contracts were still awarded to the lowest bidder, not preferentially to workers' cooperatives, and in the current labor market the *capimastri* could count on plenty of willing workers at low wages. Further, the subcontractors were still not hiring at the Chamber of Labor. *La Lotta di Classe* accused one of them of "speculating on the hunger of poor workers."[47]

Yet another effort to enforce the *tariffa* was made in 1893; it too failed. This was an impetus to form a resistance league (La lega di resistenza fra i lavoranti muratori di Milano) among the masons and their helpers (with the assistance of the Camera del lavoro) in the same year. The league was open to those who could not afford the dues of the mutual benefit society, a decision that took straightforward account of the inequality and divisions among workers in the building-trades industry. Native Milanese (or those who had lived and worked in Milan for a long time) were at the top of occupational and wage hierarchies; the newly arrived were at the bottom.

The separation of the resistance and mutual benefit functions thus reduced the cost to poorly paid workers of organizing for resistance.[48]

In 1894, the problems of enforcing the *tariffa* were still foremost. The masons' resistance league protested at the Camera del lavoro that the contractors, who were by then obligated to pay the *tariffa* on city construction contracts, were not doing so. Croce told them that it was their job to make sure that contractors were in compliance. The group voted to continue recruiting new members and to denounce noncomplying contractors and *capimastri* to the police.[49]

The resistance league was dissolved by prefectoral decree in October 1894, along with the other leagues affiliated with the Socialist party. It was not reconstituted until 1901, after yet another repression in which Silvio Cattaneo was convicted *in absentia* as one of the Socialist and other progressive activists accused of involvement in the Fatti di maggio. In 1903, the resistance league joined the mutual benefit society, thus eliminating the differentiation among organized workers, between those who belonged only to the resistance league and those who belonged to both the mutual aid and improvement and the resistance leagues.[50]

The building-trades workers won after staging a large strike (12,000 participated) in 1901, after 27 days off the job. As in 1887, the new wage settlement favored the qualified masons over the much more numerous diggers, assistants, and apprentices. The latter were especially disappointed and continued to be reluctant to join the masons' organizations.[51]

Unlike the printers, then, the masons had to deal with a high proportion of unorganized workers in their industry. Their response was to establish branches in nearby communities. Nevertheless, the organized workers frequently worked side by side with unorganized workers, and relations were not always cordial between the two groups: "Organized masons make difficulties for the unorganized on the job," Alessandro Schiavi commented, "by refusing to help or teach them."[52] The organization operated through social pressure against those who did not join or who, although members, fell behind in their dues, worked too closely with the employer, or scabbed.[53]

Giuseppe Paletta notes that refusal to join the masons' league was sometimes not motivated so much by antiorganization sentiment as by upward mobility (or its potential) in the industry. Masons hoped to become *capimastri* and so adopted a strategy different from that of the printers. They believed, Schiavi suggests, "that the better means of building peace between capital and labor was not through strike battles, but rather by . . . organizing cooperatives for the stabilization of hours and wages that shape the laws of the market."[54] This strategy, Paletta argues, forced the masons' league (the largest in the Chamber of Labor, totaling 10 percent of its membership) to use its strength in relatively conservative ways, for it could not afford to alienate the city government, an important source of contracts for the large masons' cooperative.[55]

Although highly organized (in 1902, over 50 percent; see Table 5-1), the building-trades workers continued to suffer internal divisions between urban workers and others who were recent migrants, willing to take low wages. Regional movements of labor in response to capitalist decisions (in manufacturing, which was urbanizing, and agriculture, which was consolidating and increasing in scale) shaped the Milanese construction workers' capacity to mobilize and act collectively. The associational experience of a relatively small number of masons that promoted organizational initiatives was offset somewhat by the constant problem of unorganized workers and employers' readiness, even eagerness, to employ them. Masons were adroit in their organizing efforts outside Milan and their strategy of establishing a cooperative that bid on city and other contracts. When they struck (always in massive numbers), however, they tended to be satisfied with a skewed distribution of wage benefits, which contributed in turn to a weaker solidarity.

Conclusion

The overview of strikes in Italy that opened this chapter demonstrates that in terms of strike frequency and numbers of workers involved, they were well in line with strike patterns in other western European countries during this period. The timing of these strikes (which included those of agricultural wage laborers) paralleled that of the organization of both urban and rural workers in the 1880s and 1890s and was punctuated by a sharp peak related to the legalization of strikes in 1901. Both Italian and Milanese strike patterns are explained most effectively by a political–organizational hypothesis.

These aggregate findings are useful for suggesting causal connections at a general level. The most important factors here are the extension of political participation and economic rights (to organize, to strike). Internal conditions within industries, however, accounted more fully for the timing of workers' organization, mobilization, and collective action in support of their perceived interests, and for their success or failure in strikes. Following this hypothesis, one need only consider that in both the printing and the construction industries, the timing of the well-organized large strikes demanding fixed pay scales and hours did not coincide with the political turning points that so neatly marked the aggregate strike activity. The landmark strikes in both these industries were the outcomes of long organization and planning, intensive mobilization, and, finally, employers who either rejected workers' demands or reneged on earlier practices or agreements.

If the experience of the Milanese printing and construction industries at the end of the nineteenth century can be generalized in any way, the conditions internal to an industry that first promoted organization and preparation for collective action (hence, the chronological order in which workers in different industries were activated) were changes in the organization of production and the process of proletarianization. The timing of the strikes was linked to a favorable economic climate in the industry (even if not in the economy as a whole). And success in the strikes was linked to the employers' dependence on the workers' regular supply of labor, forms of entrepreneurial capital, the resources that workers could deploy to build solidarity and hold out, and the presence of a relatively privileged worker leaders, with active involvement on a broader scale as well. (The opposite condition—divisions among workers—had a negative effect, of course.)

The chief difference in the cases of the printing and the construction industries contributing to the relative lack of success of the latter was the different forms of capital and the extent of divisions among the workers. The centralization and high fixed capital of the printing business worked against the entrepreneurs, whereas the diffused capital investment exemplified by the subcontracting system in the construction industry hurt its workers. In both printing and construction there was a mix of relatively privileged skilled workers and less-skilled laborers or apprentices. In printing, there was a larger proportion of highly skilled workers; they were able to prevent substitution by the unskilled because it took decades to devise a technology and organization of production to simplify and subdivide their tasks. In construction, however, the unskilled were the majority; they had to be part of any organization and strike. Besides being divided by skill, building-trades workers were also split along lines of place of origin. Milan-born or long-resident skilled workers led the organization, promoted the strikes, received the lion's share of settlements, and often widened divisions in the process. The development of cooperatives as a means to involve all workers more fully in the union movement had the effect of braking their readiness to engage in collective action.

Both these industries shared high public visibility, hence a good probability of intervention by local authorities to keep the peace. Indeed, community arbitration settled the first mass strikes in these industries. In printing it was a victory for workers, but in construction, it was an obfuscation of defeat. The contrasting outcomes favored the winners in later struggles. Although the printers' organization remained powerful (despite losing influence when the industry began to move out of the city in 1910) until it was disbanded by the Fascists in 1925, the construction workers had difficulty organizing their stratified work force and dealing with the equally diffuse hierarchy of entrepreneurs and subcontractors who were their bosses.

7

Workplace Collective
Action II: Case Studies

The Textile Industry

The last quarter of the nineteenth century saw the transformation of the Milanese textile industry from one centered on plain and fancy silk weaving to one with certain silk specialties—ribbons and braid foremost—but even more, mixed-fabric weaving. It was a hard transition for workers.

In 1860, there were 1,500 silk-weaving looms in Milan; in 1864, a Chamber of Commerce report stated that silk weaving could not succeed in cities in which high-waged labor and the high cost of urban life persisted. Giuseppe Colombo detected a "resurgence" of the industry in 1881, claiming 1,000 looms for Milan, but the 1891 Sabbatini report found only 357 hand and 39 mechanical looms in the city.[1]

A strike in 1888 illustrates the consequences for the workers of the decline of plain silk weaving in Milan. Although Ambrogio Osnago, one of Milan's leading silk manufacturers, had signed the *tariffa* negotiated in late 1887 for silk weaving, he reneged on this agreement and cut his workers' wages in July 1888. Three of them refused to accept the lower wage, and a commission of "honored citizens" appealed to Osnago. He rejected their petition, however, declaring, "In my house I am master," and insisting that he had to cut wages because of the low prices he was getting for his cloth. Although the workers struck, they were unable to move him, and the dispute dragged on for months until those whom he had dismissed applied to be rehired.[2]

The ribbon weavers *(nastrai)* were under similar pressure, with increasing unemployment, wage cuts, and threats from their bosses. They had founded a mutual aid society in 1859 and struck successfully for three weeks against the most important manufacturer of ribbons in Milan in 1861; two other strikers occurred in 1872 and another in 1877. Their mutual benefit society, led by Ambrogio Bernacchi, expanded the conditions under which it granted weekly subsidies, giving them to strikers as well as to those who were ill or without work. In 1878 the *nastrai* tried a new tactic: sponsoring successive short strikes at single plants around the city. Although Bernacchi went to jail as a result, they were able to negotiate a *tariffa*.[3] The ribbon weavers fought the feminization that employers pursued in response to this agreement, but it was a losing battle. Wage cuts soon resumed, leading to the collapse of their organization. Only in 1889 did they establish a new improvement league.

Although employers continued to evade the *tariffa* by feminizing their work force over the objections of male *nastrai,* the latter refused to admit women into their league. After several years of serious preparation, a general strike of the category was held in 1893, demanding that employers observe the old *tariffa,* adjusted for changes in the production process. It ended with no more than a "moral victory" for the strikers, with agreement on a *tariffa* that was 5 percent lower for men (25 percent for those who wove on mechanical looms) and 12 percent lower for women. The agreement was welcomed, for the old *tariffa* had lapsed because of feminization, organizational weaknesses, and technological change, and the new *tariffa* was at least accepted by employers.[4] The number of members in the organization dwindled after the strike, however, and in 1894 it was closed, as were others affiliated with the Socialist party.

In 1896, a group of younger members proposed opening the ranks of the league to women, and they prevailed. A new propaganda committee recruited nearly all the remaining ribbon weavers in Milan. The league's membership more than quintupled. Strikes were launched at individual companies in 1898, many of which had favorable results, but the organization was dissolved by government decree in May of that year. Reconstituted in 1900, it suffered from fiscal problems, and hence its strikes were largely defensive and small.[5] By then the *nastrai* were mostly female: In 1881, they had been two-thirds male; in 1901 only 20 percent were men.

Organizations of workers in two small specialties, silk braid weaving and lace making, also existed in Milan during this period. Workers in both industries were in precarious situations similar to that of the ribbon weavers: They found it difficult both to retain members (many of whom were women) in their organizations and to achieve their goals. There had been a braid and elastic weavers' society since 1860, but it disappeared in 1885 or 1886, with the flight of its treasurer, who left with its funds. A new organization composed exclusively of braid weavers was founded in 1894 but disbanded in 1898. It reconstituted itself as a mutual benefit society in October of that year and added resistance to its scope in 1899. The league "was never in a position to achieve its goals," the Società Umanitaria reported in 1909: "Like most of the women's leagues, or those that are predominantly female, the braid weavers, after a period of rising membership, declined once again."

This facile judgment overlooks two factors: (1) Many of the leagues in Milan that were not predominantly female (e.g., the Masons' Resistance League and the Mechanics' Resistance League) disappeared for long periods also, and (2) the parlous condition of the braid-weaving industry, subject to changing fashions, affected its workers' organizational capacity. Simply put, braid trim went out of style, and the industry itself suffered. Further, the women's capacity to hold out in a strike was severely limited by their low wages. Although it was hard to determine an average wage, the report estimated that women's home work wages were 0.8 to 1.2 lire per day, less than a third of what men earned on the looms in the shops. Lace makers founded an improvement league in 1901, but it was plagued by a vacillating membership. Here again, men's and women's wages differed greatly; men who worked on machines earned 4 lire daily; women in the same category earned 1.60 lire, and other women workers in the industry earned even less. The Società Umanitaria report remarked sarcastically that the number of members decreased "whether they had lost or won a strike." Conditions differed among workers; the women's organizational attempts frequently failed because of unfavorable structural conditions.[6]

Police documents provide a capsule history of struggle in other branches of the textile industry in these years. In 1878, Bernacchi, head of the ribbon weavers' organization, founded the Confederazione delle arte tessili, which aspired to bring together workers from all textile branches. Although credited by the police with "good conduct generally," Bernacchi was also reported to be "fomenting discord between bosses and their workers" and trying to promote a general textile strike. Despite the Confederazione's inability actually to do this, the police implicated it nonetheless in a series of plantwide strikes.[7]

One sector that was especially affected was elastic weaving. Hunecke notes a 1877 strike at the factory of one Corrado "Schoetz" (whose name later appears as Schoch). Many of the elastic manufacturers in Milan either mechanized and feminized or moved their factories to rural areas where wages were lower.[8] In 1879, there were a flurry of strikes in the sector that brought an agreement on a *tariffa,* but the strikes nevertheless continued. The first clearly identified all-woman strike occurred in another elastic weaving plant in May 1880. Of the twenty-four strikers whose names and age were recorded, only three were over 25 years old. The police report of each worker's wages in another strike demonstrates quite clearly that they (almost all women) would have to work more rapidly in order to avoid substantial wage loss as their employer reorganized production and the pay system.[9]

Another women's strike against Schoch attracted attention in March 1886. Seventy women walked off the job to protest a wage reduction and took their case to the owner of their factory. He claimed the need to cut costs and pointed out disingenuously that he was cutting wages rather than dismissing workers. Women worked at his power looms from 7 A.M. to 6:30 P.M., with one-half hour for lunch. Their wages had been cut to 18 to 20 lire for two weeks (12 days of work). The *Fascio Operaio's* headline declared "Our Conscience Cries Out:" "Comrades, sisters! Your cause is our cause. We clasp your hands and throw down our gauntlet in response to this insult. Your honor, sisters, is our honor." It announced, further, that it would collect funds in support of the strikers who, supported by the Partito operaio, established a society of women elastic weavers. (It had only an ephemeral existence.) The strike lasted more than two weeks but was lost. Schoch thereupon moved his elastic weaving out of Milan.[10]

A strike of wool weavers entered the police records with the appearance of two workers on January 2, 1880, to file a complaint against their employer, one Mosterts, a manufacturer of flannels. In the previous February, they testified, he had forced all his workers to agree to continue working at the same wage for a year. All had gone smoothly until a few days before Christmas, when his secretary called in twenty workers "of all ages and qualifications" and told them that once they had completed the cloth they were then weaving, they would be dismissed. The weavers discussed the ultimatum among themselves, and a delegation of five went to Mosterts himself. His message was brutally brief: He was moving the factory to Somma Lombarda (near Varese) and would permanently lay off his other Milanese workers in two phases. The workers thus decided to quit at once and seek work elsewhere. However, they reported, they had been unable to find jobs; they asked the police to intervene in their favor but were refused. Some of them petitioned Mosterts himself to regain their jobs, but this request, too, was rejected.[11]

In August 1892, Schoch and his women workers were once again involved in a dispute, this time in a cotton-weaving factory. The Chamber of Labor intervened on behalf of the miserably paid women, claiming that their wages were "inadequate to cover vital expenses and much too low in relation to the number of hours worked." (Their workday had recently been increased from 10 to 12 hours, with no increase in the wage, 1.50 lire. Their wages then were lowered, and the women were assigned two looms each to mind.) The chamber proposed reducing the number of hours, increasing the rate per meter woven, reforming the system of fines, and moving the payday to Saturday (from Sunday, when workers had to report merely to be paid). *La Lotta di Classe* published a lugubrious appeal to their fellow citizens: "They are mothers, almost all of them, . . . widows, wives whose husbands are unemployed. Milan has always behaved in a sisterly fashion in response to the miserable . . . let your heart be your guide!"

One of the authors of this appeal, Osvaldo Gnocchi-Viani, also wrote an open letter to Schoch, pointing to the contradiction when he objected to any intervention in his business (by those who supported his workers) but was happy to call on the police to intervene in his interest. Despite a long strike—in which negotiations failed, the strike leaders were arrested for infringing on others' right to work, and women who applied to other manufacturers for work were rejected—the workers returned in the end to unchanged conditions.[12]

Many more strikes can be documented in cotton and wool weaving. In 1895, a newspaper account (*La Sera,* December 20–21 reported that in Milan eighteen plants were doing mechanical weaving, with some 3,000 workers. Women were paid by the piece, men by the day. Hence the latter "had little reason for solidarity" with the former. Most strikes concerned wage reductions; most of them involved women workers; and most of them were lost.[13] The employers' own economic fragility and the distrust and contempt that they felt for their workers were at the root of the troubled labor relations in Milan's textile factories. These attitudes are evident in the factory regulations imposed in August 1898 at the Strauss mechanical weaving company. The rules called for three months' probation during which workers could be fired without notice; the first week's pay would be held back as a guarantee of satisfactory work; there would be fines for lateness; one day's absence without permission would be deducted from wages as two days; the absence of more than three days would bring dismissal and a loss of the withheld week's pay, and there also would be fines for poor workmanship, talking, and leaving the looms; workers caught stealing yarn from the premises or wasting it through poor work would be dismissed and their withheld pay forfeited; workers would be searched without advance notice as they left work; and finally, workers would not be paid for the extra time they took to mend imperfections in their finished cloth.[14]

A weavers' league was established in 1900, and strikes in the textile industry peaked in 1901, only to drop off steeply in 1903. The union's membership also fell in that year to 100, from 300 to 400 in its first years.

Milan's textile industrialists pursued alternative strategies starting in the late 1870s: the mechanization and substitution of lower-paid female workers for male workers or the abandonment of the city to seek a cheaper work force in the countryside. The character of the industry's work force also changed: It no longer consisted of specialized skilled workers but, rather, unskilled men and women who used machines. Skilled men, in particular, resented and fought the infiltration of women. Once that became inevitable, however, the men were regularly paid more. Even if they did the same work, men were paid a flat daily wage, and women were paid a low rate per piece. According to a newspaper account: "Thus [men] could not—nor did they have any reason to—act in solidarity with [women] strikers."[15] Textile strikes were extremely common in between 1878 and 1903. According to the official statistics the industry was second only to metal and engineering in number of strikes, and the newspapers reported many additional small skirmishes. The textile workers were divided internally, and their organization was discontinuous and tenuous. Their employers were in a weak economic position because of competition from larger-scale manufacturing elsewhere, and—because of the nature of the industry—they were to some extent dependent on the whims of fashion. Because of this combination of factors, strikes were likely to be defensive, and they also were likely to be lost.

The Metal and Machine Industry

The metal and machine workers were active participants in the two early strike waves but were relatively quiescent from 1872 to 1886.

The first problems appeared soon after unification, the evidence being an affidavit (dated June 25, 1859, by Messina) signed by the "Artisti dell'Elvetica." At that time the engineering plant was owned by Giovanni Schlegel, a German. When his workers returned to their jobs following the celebration of the Austrians' departure, they found a note posted at the factory door announcing that their entry was permitted "simply as an act of charity." The police intervened and the manager promised to replace a hated foreman and to continue the workers' 10-hour day. However, when Schlegel returned from Switzerland, where he had fled during the turbulent days of the war, he both cut wages and increased hours. His workers protested, and he was knocked down in a scuffle. There was a brief lockout, and the plant did not open for several days. When it reopened, all the workers returned to unchanged conditions.[16]

In October and November of that year there were "disorders" (the police category) at two engineering companies: Grondona di Miani e Zambelli and Edoardo Suffert, both regarding reductions of hours. Although Miani threatened to dismiss any workers who complained, his accountant negotiated a lesser wage reduction than that originally planned. Suffert agreed to cut hours but reduced the workers' wages as well, leading some to quit their jobs.[17]

In 1872, it was the workers of the Suffert foundry, followed by their fellows in its machine shops, who initiated the strike that eventually spread to a majority of Milan's workers. The metal and machine workers won a 20 percent raise (they had demanded 25 percent) but no agreement on hours.[18] The workers were also forced to accept a new set of factory regulations, drawn up by an *ad hoc* group of employers in the industry and intended to apply to all shops in Milan. The rules required that all workers agree to abide by them before they could be hired and that all applicants for a job provide a document from their former employers certifying their good conduct.

Among the other stipulations were the following: The ordinary workday would be 10 hours, but employers reserved the right to add or subtract up to 2 hours; workers were to arrive in the 10 minutes prior to the opening whistle indicating that work should start; the gate would be reopened after 50 minutes for latecomers, whose pay would be docked correspondingly; taking Mondays off was forbidden; hourly wages would be paid each Saturday; piecework wages would be calculated and paid monthly; workers were responsible for the machines on which they worked, which they were required to clean; smoking, drinking, and spitting were forbidden; insubordination, refusal to work, drunkenness, "disturbing the peace," or breaking these rules would be severely punished; and finally, foundry workers were required to have a locked box furnished with all the necessary tools for their trade, and all other workers were required to furnish their own exact measure.[19]

These severe work rules seem to have been effective in discouraging strikes, for from 1872 to 1887 there were only four brief strikes, all in single factories. Three out of the four involved disputes about the conditions of work.[20] This peace was broken, however, starting in 1888, by a wave of strikes that lasted three and a half years and led to the organization of resistance leagues in several occupations.[21]

Why were the metallurgical and machine shop workers so quiescent in the 1870s and early 1880s? The factory regulations suggest severe discipline, but they also imply a degree of worker autonomy in the labor process, by assigning responsibility to the workers for the machines on which they worked and stipulating the tools that they were required to supply. Also, the rules were not uniformly enforced. This speculation is supported by the fact that beginning in 1887, the peace was broken when employers infringed further on workers' autonomy and enforced some of the old rules.

Metal workers and mechanics were highly differentiated by skill and were often organized into work teams, even though some of them worked in very large establishments. Duccio Bigazzi classifies their many occupations in the following way: (1) The *foundry workers,* of whom there were relatively few in Milan, were mostly employed in the foundry departments of the large machine shops. They were little affected by technological change until after 1900. (2) In contrast, the *firemen and boiler makers (fucinatori e calderai)* were hit by technological innovation and reorganization of the work process in the late 1880s and 1890s, with the men in both occupations organized into teams with their helpers and assistants. Some tasks were simplified by new technology, making it possible for boys or manual workers to do them. Skilled firemen and their work groups can be identified in photographs of the Breda locomotive plant as late as 1908. (3) The number of *machine tool operators,* including metal turners, increased with the greatly expanded use of machine tools in the 1890s and after. These workers, unlike the craftsmen of an earlier era, who (using their own tools) could make many different pieces, were tied to a particular machine and its power-driven tempo. A hierarchy ensued, based on the complexity of the machine used. Again, however, the most skilled workers retained craft qualifications that enabled them to "custom make" needed parts. (4) There were fewer *aggiustatori* and *montatori* (machine fitters and mounters), the *mechanics* of the period, before the widespread introduction of specialized machine tools, who diminished in number with technological change. The *aggiustatori* became specialists rather than skilled generalists. The job of the *montatore* was divided: One part was a simple set of manual tasks, and the other came close to that of a technician or engineer. (5) The *porters (facchini)* and *manual workers* accounted for 10 to 15 percent of the work force. They had the least amount of job security and contractual leverage and the poorest pay of all the metal and machine shop workers.[22]

What was distinctive about the system was the large number of specialized crafts *(mestieri)* and their subdivision along skill lines, with corresponding wage levels. Luciano Davite sees this as a result of the entrepreneurs' differentiation of workers in an effort to reduce the size of the groups with which they had to deal and thus to make resistance more difficult.[23] His evidence is the division of worker organization in the industry; therefore his argument is circular.

I would argue instead that the fragmentation was a consequence of the employers' strategy adopted in response to the business cycle of the Milanese machine industry and its high fixed capital costs at the end of the nineteenth century. Through the 1880s, the machine shops depended on government orders for locomotives and rolling stock; in the 1890s, they began to receive contracts for these products from foreign governments, but the cycles continued. A contract would be signed, and full employment would follow until the work was completed; then the layoffs would begin as the company scrambled for new contracts, as it was critical to keep the expensive plant and equipment in use. Industrialists dealt with the problems posed by the cyclical demand by building a two-tiered work force. A trained, stable core of skilled and loyal workers (or so they hoped) would be assisted by a shifting group of specialists and assistants, or second-rank workers. This was not a new system: The work history of an *aggiustatore* at Grondona shows that he was first hired in August 1877, fired in December 1878; rehired in March 1879, once again fired in July 1883; rehired in August 1883, fired in July 1885; and rehired in February 1885, only to be fired once again in February 1890 (when 140 workers were laid off).[24]

Starting in the late 1880s, however, the engineering companies began to intensify their efforts to reorganize supervision, remuneration, and the labor market.

These workers, their occupations in flux, especially in the depressed 1890s, were employed in the shops already discussed in Chapter 2. Ernesto Breda (who bought the Elvetica in 1886 and immediately began to reorganize it) and Franco Tosi in Legnano (outside Milan) were the most technologically advanced. It was Breda, writes Hunecke, who introduced to Italy industrial rationalization built around new machines and new organization of work. The plant he took over produced "all types of mechanical constructions: bridges and metal-framed glass roofs; fixed and locomotive steam engines; hydraulic and turbine-run engines; railroad cars and harvesters; stationary equipment for railroads, and boilers." [25]

Breda's goal was to specialize in locomotives, and to achieve this he systematized the old, rather modest plant in several phases, adding more steam power and machine tools from abroad. Then, in order to cut labor costs, he added hydraulic nail makers, portable electric drills and augers, and pneumatic drills and riveters. A central control office coordinated planning and supervision. In 1903, the company bought land in Sesto San Giovanni, outside Milan, and built more spacious buildings to rationalize production further. As early as ten years after he took over the business, however, Breda could claim with satisfaction, "There is no longer any trace of what existed before us." [26] Tosi was even more advanced in fixed capital and automatic production processes than Breda, because it built a new plant and started its reorganization earlier.

A second group of very large companies such as Miani–Silvestri and Grondona specialized in railroad rolling stock and other equipment, as well as general machine construction. In these factories, much handwork lingered, not only in mechanical areas but also in woodworking, upholstering, and other finishing processes. Small and medium-sized businesses (each of which employed 50 to 150 men who worked on simple machines with an advanced division of labor) comprised a third group. Bigazzi, whose taxonomy of the industry I have adopted here, concludes that the picture was contradictory: An organization of production and technology comparable to that in more advanced countries coexisted with outmoded forms. He estimates that no more than 20 to 30 percent of all Milanese metal and engineering workers were touched directly by organizational and technical innovations. According to Sabattini, the number of workers in the larger firms of the metal and mechanical industry in the commune of Milan was 11,400, and so Bigazzi speaks here of 2,000 to 3,500 workers. However, his estimate is based on a cross-sectional perspective: The kind of turnover inherent in the two-tiered work force described earlier suggests that over time a much larger number of workers would come in contact with the new ways in the machine industry.[27]

The numbers matter little to Bigazzi, who is making the case for a new type of worker consciousness, not one of pride in craft (*fierezza di mestiere*), but one based on a realization that corporatism or professionalism was an illusion. To support his argument, he quotes *Fascio Operaio* to the effect that the man who does piecework is unable "to bring to his work the pleasure and the artistic passion that are requisite to producing well." Giuseppe Berta offers a similar interpretation, showing that during the period from about 1900 to 1912, this type of industrial reorganization conferred potential leadership on skilled workers, on whose high level of competence the continuing capitalist growth depended. The distance between skilled workers and less skilled ones and laborers was much less than that between artisans and laborers; thus there was more opportunity for solidarity. This appreciation of skilled work was fragile, however; employers renewed their attempt to undermine skill in the second decade of the century.[28]

The image of the man-machine proliferated with the celebration of the skilled worker. An especially apt example comes from the company history (published in 1908) of Breda: "Machine and worker form almost a single organic and intelligent being—who, with the ease of habit turns out hundreds of smooth and polished pieces from the raw, rough, black ones he receives."[29]

It seems unlikely that the owners of the machine industry were reorganizing their shops, job hierarchies, and production processes merely in order to control their workers, as Luciano Davite suggests. Rather, they were making changes aimed at increasing productivity in order to cut costs, which, as they saw it, required greater control of the labor process and the establishment of new social hierarchies on the shop floor. Their chief instrument in this process was piecework, *lavoro al cottimo,* or *cottimo* for short, which formed workers to work more rapidly. This system was initiated in the second half of the 1880s in the Breda reorganization of the Elvetica works. The process of setting the *cottimo* started with a determination of the "normal" (average) pace at which workers in a designated job could produce a given number of units. The company would then offer to pay the worker at a rate 25 to 50 percent higher for any pieces produced beyond the norm. In this way workers were pressed to increase individual productivity, making possible a smaller work force. Similar systems, in which teams, rather than individuals were the production unit, were instituted at Miani–Silvestri and Grondona. In crisis periods, the *cottimo* was speeded up even more, by reducing the base rate.[30]

Higher levels of worker organization followed these changes in the labor process. At the beginning of the 1880s, the only general organization in the sector was the Società Archimede of foundry workers and mechanics, a mutual benefit society. The metal and machine workers' lack of organization was a subject of concern among militants in the better-organized industries. Consider, for example, an article written in 1883 by Augusto Dante, a printer activist, in one of the first issues of *Fascio Operaio*. In the capitalist system, he wrote, mechanics no longer owned machines; they were merely machine builders. "What can be done about this?" he asked rhetorically. His reply: "Organize resistance federations following the example of the printers. . . . [I]n order to defend your labor, take orders for machines themselves, without the intermediary of bosses . . . only then will machines turn to workers' profit, for if they are not with us they are against us."[31]

The theme of a lack of class, as opposed to occupational solidarity, recurred in the same newspaper in 1886. *Operaista* activists inaugurated the Lega lavoranti in metallo in February 1886 with a lecture by Alfredo Casati entitled "Workers and Organization." (A later article, addressing the "metal workers," criticized their reluctance to join the Figli di lavoro, some "because of fear, others because they are deceived, many because of 'egoism'.") The league never got off the ground. But this point was made again in the case of the bronze casters: There was a need for organization "without egoism," but the bronze casters lacked resolve; they were incapable of joining together to protect their interests.[32]

An attempt to establish a league in 1889 bore more fruit. The Federazione di miglioramento fra le arti mecchaniche, affiliated with the Fascio dei lavoratori, was founded by a group of mechanics (including Emidio Brando, who was to play an important role in efforts to ameliorate the mechanics' employment crisis in 1890 and 1891). The Federazione was reported to have 750 members in early 1890. Concerned about "improvement" of the group, the Federazione took a corporatist position alongside the machine shop owners, holding the government responsible for the industry's hard times. Among workers in the industry were dissidents, like Alfredo Casati, who rejected the effort to solicit government concessions and directly blamed workers' plight on the industrialists.

Leaders of the Federazione played an important role in the strike of 1891, but the movement went well beyond them. A new organization, the Lega di resistenza fra metallurgici was founded, which was succeeded by other resistance leagues organized by occupational specialty; the Lega itself disappeared, however, after several years. The first successor resistance league was established by the *tornitori in metallo* (metal turners) in 1892. Others were those of workers in metal furniture (1892), foundries and small machines (both 1894), *aggiustitori e montatori meccanici* (1895), and factory-based blacksmiths (1896). All these leagues were disbanded in 1898 but were reconstituted and joined by others starting in 1900.[33]

More like the textile workers' organizations than those of the printers or building-trades workers, the metal–mechanical organizations were frequently born, combined, and divided. Why? The Società Umanitaria's 1909 study offered this answer: "Due to the low qualifications of those who belong to them and the rapid turnover of workers in the metal industry based on the uninterrupted migration from the countryside of rough elements with their impulsive behavior, there is no experienced nucleus of workers to teach and guide them."[34] A glance at Table 3-2 reveals that workers in the metal industry were more likely in 1881 and 1901 to be native born than was the labor force as a whole, and they were much more likely to be native born than were the construction workers. Although the latter were frequently divided, their league joined the masons, diggers, assistants, and apprentices in 1887. Further, mechanics were less likely to be native born in 1901 than in 1881, yet it was in that very year that the national Federation of Metal Workers (Federazione italiana degli operai metallurgichi, FIOM)—destined to become one of the most powerful labor organizations in Italy—held its first congress. (Discussion of a national federation had begun early in 1898, initiated by the Comitato metallurgico di Milano, but the repression that year stopped the process. It was reopened by the Roman metallurgists in 1899; the organization was built slowly from the bottom up, with the Milanese among the founding groups.)[35]

A high proportion of migrant and unskilled workers is neither a necessary nor a sufficient condition for a late or weak organizational history. The Società Umanitaria was correct in suggesting a shortage of experienced leaders as a factor, however. The lateness of organization in the metal–mechanical sector and its tendency to splinter in the 1890s should be attributed instead to (1) the competitive position of the industry and (2) the continued importance, even as employers initiated the speedup, of specialized workers whose distinctive characteristics were reinforced by the form of reorganization of production common in the late 1880s and the 1890s. That reorganization gave skilled workers more leverage than others had and encouraged them to promote selective (rather than industrywide) strike activity.

The entrepreneurs of the metal–mechanical sector were under pressure because of the cyclical nature of the government contracts on which they depended. Competing internationally for contracts could smooth out the cycles and provide a more steady use of fixed capital. This in turn dictated increasing productivity, and employers sought to achieve this through differentiation and specialization, processes already described. Once they made this decision, employers attacked craft workers' autonomy and resisted their counterclaims. The strikes between 1887 and 1891, Hunecke writes, were likely to concern three issues: "hours, the intensification of work and the *cottimo*, and stricter application of factory regulations."[36] And the Miani-Silvestri and Suffert strikes of 1889 bear this out.

In January 1889, Miani changed his workers' pay period from weekly to biweekly. *La Lombardia* reported (January 9, 1889) that on the previous Monday, instead of working, workers had protested this change. Despite the foremen's promise to refer their grievances to the management, most of the workers struck. They went to the Consolato operaio's headquarters where they drew up a statement of their demands: reduction of the deposit held by the company, from one week's pay to three days' pay; postponement of the introduction of the biweekly pay day until April, so that families could save for the change; company-sponsored group life insurance for workers; establishment of a sick pay fund to which fines collected from workers would contribute; and pay of time-and-a-third for extra night hours or holidays. A "dignified and cordial" compromise was reached, *La Lombardia* reported.

Protest resumed on March 5, 1889, when a one-day strike again ended with the firm ready to compromise. Miani–Silvestri was making new regulations regarding fines, absences, and dismissals. A police strike report remarked that the workers' inclination to take off a day, usually Monday, was no longer acceptable but that the new regulations were favorable to "good workers." The report also recommended that Miani consult with the Consolato operaio for advice on how to avoid alienating his workers.[37]

Miani's strategy of balancing tightened discipline with new (if modest) benefits contrasts starkly with the Suffert policy. There a strike started over a seemingly trivial issue: 158 workers walked out because the buckets in which they heated the water with which they washed before lunch and at the end of the working day had been removed. The company demanded that workers wash up after work hours. The police report speculated that they would be back the next day, for the strike had begun on a sunny Monday, workers' favored "holiday." Later in the day the workers sent a memorandum to the management listing their demands; Suffert himself was out of town. The memorandum protested the changes in the factory regulations: Suffert had posted a somewhat altered version of the 1872 work regulations earlier that month. When he returned later, the same day as the strike, he refused to change anything. The police finally hammered out an agreement on the next day that would restore the old regulations unchanged.[38]

Suffert retreated from his plan to tighten up the work rules. But on April 2, 1889, the police reported another walkout, this time objecting to the newly strict application of the rule regarding lateness. This time Suffert refused to concede and threatened to lay off his workers. Under these conditions, they returned to their posts and, evidently as Suffert demanded, wrote him a letter of apology. He rejected it because they petitioned him in the same letter to rehire a worker who had been dismissed. The workers then apologized with no conditions.[39]

Ernesto Breda's pursuit of competitiveness in the international economic arena led him to use the *cottimo* to increase the individual worker's productivity and then to cut back on his work force. Table 7-1 illustrates his policy. Quite simply, productivity was being driven up and labor costs were reduced. This strategy seemed to pay off over the next five years. In 1894, however, the number of locomotives produced declined by over 50 percent, to 21, and none at all were built in 1895, 1896, and 1897. Breda received other government contracts (for artillery shells) and thus survived (and even prospered) until the eventual recovery of orders for locomotives, which did not get up steam until 1907.

In 1891, a dispute over cuts in the base rate for the *cottimo* exploded at Breda. Forty metal turners refused to accept the cut and were fired on August 15. Their fellow workers gathered at the headquarters of the Federazione delle arti meccaniche that evening to deliberate whether to strike. *L'Italia del popolo* published their statement of grievances the next afternoon. It made two points: (1) The cuts in the base rate of the *cottimo* were unacceptable, and (2) expecting workers to tend three machines led to imperfections in their product; it was unfair to fine them for such errors. The workers elected a committee of citizens (representatives of labor organizations, a socialist lawyer, and a reporter for *L'Italia del Popolo*) who were delegated to present their grievances to Breda.

At noon the next day the committee called on the Breda management. Breda himself was out of town, and the plant engineer refused to meet with the committee or speak with Gnocchi-Viani (then a member of the city council). Later in the afternoon they returned to the factory, where Breda received them. Their demands fell into two categories, he replied. The first—abolition of the *cottimo*, wage increases, and removal of the requirement that workers present a police document detailing their criminal record, if any, before being hired—he refused to consider. The second—rescinding wage cuts for boys and certain fines, among others—he promised to review and respond to in 12 hours. However, he insisted, he would prefer to deal with his own workers, with whom he had bargained successfully before. The strikers thereupon redrafted their demands but made no substantive changes.[40]

Gnocchi-Viani was summoned alone the following day to hear Breda's response: negative on all counts. According to Gnocchi-Viani's report to a strikers' meeting, Breda declared that given the "disastrous conditions of the Italian machine industry, it would be a serious burden to abolish the *cottimo*, for a worker paid a daily wage would no longer be stimulated to work with the intensity required by the company's situation." All the workers' demands, in fact, worked against the company's interests, that is, higher worker productivity and a rational basis for reducing its work force. Breda closed with a threat: His order for forty-three locomotives from the Roman-

Table 7-1. Locomotive Production, Workers Employed, and Wages Paid at Ernesto Breda Co., 1889–1893

Year	Number of Locomotives Produced	Number of Workers Employed	Payroll (lire)
1889	49	1,000	1,000,000
1890	27	900	800,000
1891	13	700	700,000
1892	43	700+	700,000
1893	48	700+	600,000

Source: Carlo Carotti, "L'Introduzione dell'organizzazione 'scientifica' de lavoro in Italia e la prima lotta contro il cottimo," *Classe 7* (1973): 280, based on data from Società italiano Ernesto Breda per costruzione meccaniche, Milano, *Per la millesima locomotiva*, November 1908.

ian government was not yet finalized, and it would be impossible to fill if the workers did not bow to his changes.

A tumultuous meeting received this news. The strikers decided to convene all Milanese machine shop workers to develop a common plan of action against the *cottimo*. They also voted to add several new members to the negotiating committee, including lawyer Filippo Turati and Republican journalist Dario Papa.[41]

On August 28, the strikers, still meeting at the headquarters of the Mechanics' Improvement Federation, heard a report from Gnocchi-Viani on his discussion with Mayor Belinzaghi, who had declared that the city authorities would not intervene in work disputes but had nevertheless offered himself as a mediator. The strikers adopted a proposal made by Carlo Dell' Avale that they convene all the city's metal and mechanical workers in the Arena the next day to vote for a solidarity strike with the Breda workers. Mechanics from the other Milanese factories, as well as Filippo Turati and his companion, Anna Kuliscioff, just returned from the Brussels Congress of the Second International, attended the Arena assembly. The group voted (against the advice of Turati and Kuliscioff, who believed that a stronger organization was necessary) to launch a general strike by all the city's mechanics against the *cottimo*. Groups of Breda workers would visit the various plants to explain their position and call on workers to turn out.[42]

A large analytical chart drawn up by the police (based on daily reports by agents who visited the factories) lists the number of workers in fifty companies and, for each day starting August 31 and ending September 12, the number of strikers in each shop. In the next days, police patrols arrested Breda workers who confronted those at other plants. Most of the manufacturers closed their factories, reopening them only on September 9.[43]

A strike meeting at the Arena on September 8 rejected both a back-to-work motion and another stating that the struggle could not be sustained or expanded because of a lack of resources (which was supported by Turati, Gnocchi-Viani, and Croce). On September 9, a much-diminished group gathered once again to support the struggle. Most of the workers in shops other than Breda returned by September 10, and on the next day the number of Breda strikers had declined to 280. On the following day there were only 12. (Suffert's workers, fought by their employer to a standoff two years earlier, struck for only half a day.) The machine manufacturers' intransigence had triumphed. As *Il Sole* editorialized, "One can understand a strike to increase the rate of the *cottimo,* but not to eliminate it; that would be equivalent to the manufacturers' giving up an effective means of control over the conditions of labor in relation to its productivity." [44]

In the succeeding years, troubled relations between capital and labor continued. The metallurgists' resistance league renewed its call for abolishing the *cottimo.* The Metal Turners' League began a long agitation, with many partial strikes. A full-fledged strike against Suffert in 1895, in concert with its foundry workers, failed to win any concessions but was damaging to the company. It closed its foundry, and to avoid giving in to its striking workers, it hired others, at the higher wage that it refused to give to its own. The same company was struck in 1896 and 1897 over wage reductions and refusal to recognize the league. Suffert again prevailed by hiring scabs. [45]

Other employers reevaluated their unrelenting opposition to their workers' claims in 1896 and proposed the establishment of a *collegio arbitrale* (arbitration board) composed of five industrialists, five workers, and a chair who was acceptable to both groups. (The industry representatives on the board included the stern Breda as well as the more accommodating Miani.) It tackled the knotty problem of the *cottimo* and, after a long debate, finally agreed that "under present conditions the *cottimo* is a technical administrative tool for managing labor that industry is unable to surrender." The board took the position that the system should be seen as a way for workers to maximize their earnings, rather than as a tool of employer exploitation.

The arbitration board's decision caused consternation in the Metal Turners' League. They resolved not to accept this judgment, thus effectively denying the board's legitimacy. In early 1898, a renewed effort by the league resulted in a three-month strike against Stigler, who refused to meet with league representatives to discuss the dismissal of two workers. Stigler laid off his other workers and blacklisted the strikers at all Milanese machine shops. The strikers eventually received some concessions, but their leaders were fired. Industrialists replaced the arbitration board with an organization of their own to coordinate their response to the metal workers' resistance leagues. The following year, the Consorzio fra industriali meccanici e metallurgici di Milano was founded. [46]

The metal and mechanical leagues were closed by government decree

after the May events, later in 1898. But the metal turners' executive committees continued to meet secretly and pay out sickness benefits to its members. It reestablished itself in 1899, with some 50 members, and by 1902
it had expanded to 1,500 members and could boast a well-endowed treasury. There was a severe downturn in the industry from the end of 1901 to
the beginning of 1904. Renewed technological (the introduction of electric
motors) and organizational change lay behind the peak of strikes in 1902.
In 1904, recovery from the recession and the revolutionary syndicalist leadership in the Chamber of Labor encouraged the workers to resume their
struggle. Five hundred of them walked out at Langen and Wolff in a strike
that lasted for 27 days and cost 7,000 lire in strike pay, only to be lost in
the end. Recognizing that the specialized leagues were a hindrance to effective workplace mobilization, the Metal Turners' League moved to merge
with the league of *aggiustatori*. The Società Umanitaria report detected a
moral benefit in the struggle: "the recognition by the industrialists' consor-
tium of the seriousness . . . that drive[s] worker organization." The Società concluded that the mechanical workers' leagues were simply not strong
enough to eliminate the *cottimo;* instead, they must content themselves with
efforts to limit and control its application.[47]

Unlike the printers who organized early and successfully to demand
concessions and a *tariffa,* or the ribbon weavers, who organized early in
defense of their autonomy and economic status, only to be undermined by
their employers' strategy of mechanization and feminization, the metal and
engineering workers organized most vigorously at a later date, in response
to an employer offensive against their autonomy. That offensive, which to
a great extent succeeded, benefited from the division of metal and mechanical workers into specialized leagues that could be dealt with separately. As
Bigazzi correctly argues, these leagues were not so much the result of a
corporative tradition as of the particular situation in which these workers
found themselves. It was not, however, I would contend, a matter of the
"particular necessity of the Milanese mechanics' struggle in this period,
explainable in large part by the process of the transformation of the organization of work," the functionalist explanation to which Bigazzi resorts.
It was instead the result of a rather prosaic aspect of that early stage of
transformation: the grouping of workers by specialty in separate rooms, of
the plant.[48] In these separate rooms, specialists who emerged from the reorganization of production begun in the late 1880s worked in teams with
their assistants. It was not management's control of workers (which was not
fully achieved at this point in the process of reorganization) but the isolation
of the new specialists from one another and the development of collective
relations between each group of specialists and management that hindered
industrial solidarity in the engineering industry.

Women Workers, Organization, and Workplace Collective Action

The conditions of women textile workers and their weak capacity for organization and collective action has already been analyzed. In this section, we shall examine the rubber, garment, and tobacco industries, in each of which women were the majority of workers. The way in which the different industrial situations and work conditions affected women's capacity to organize and act collectively is our focus.

The Milanese rubber industry consisted for many years of the Pirelli company; in the 1881 population census, only 134 rubber workers were listed, half of them women, and only 26 percent born in Milan. In 1901, there were 1,841 rubber workers in the city, 54 percent of them female, and only 22.3 percent born in Milan. Pirelli ran a paternalistic business, in which favors (such as a company-sponsored mutual benefit society) were granted by the company. The workers' gratitude, in turn, was expected to lead them to accept the company's demands.

This concept of mutual obligation was tested by events in 1891. On July 30 of that year Pirelli distributed a flyer to its employees that announced that because of problems in the industry it was instituting a series of severe economies. Wages would be reduced by 10 percent in August, but as soon as economically feasible, it would begin to pay bonuses to "those who accept the current measure calmly and work as hard as they would if their wages had not been cut." A police inspector reported to his superior that the lawyer and anarchist Pietro Gori had been seen in the neighborhood, and accordingly he requested extra patrolmen at the factory gates. On the night of July 31, the workers voted to strike, but nevertheless made a counteroffer to Pirelli, offering to work one less hour daily, with the pay cut that this would require. They complained, however, that the internal mutual benefit society would not supply relief payments for unemployment. Why not, then, dissolve it and divide its funds among its members—themselves? Pirelli agreed to the reduction in hours but reported that there would not be enough in the mutual fund to divide once a subsidy was paid to those he must lay off—twenty mechanics. August 31 was thus set as a strike date after this unsatisfactory response was received.

A workers' committee led by Osvaldo Gnocchi-Viani and Pietro Gori met with Pirelli, and his workers agreed to continue on the job on Pirelli's terms. Dissatisfaction continued throughout August, however. As the police inspector summarized: "It was a foolish idea of Pirelli to reduce the workers' wages when the company was busy filling urgent orders. . . . It created the problem itself, and it has emerged from the confrontation with both its authority and its prestige diminished." He reported that some two hundred workers had established a resistance league. The league was short-lived, however; it is not even mentioned in the Società umanitaria history of worker organization. And there is no mention of women workers in any of the relatively full police correspondence.[49]

As the outcome of an organizing drive led by Carlo Dell' Avale and supported by Anna Kuliscioff, a new rubber workers' resistance league was established on April 3, 1898. This time, women made up 70 percent of the 1,000 workers who enrolled. The Pirelli factory and its environs were the location of the first events of the rebellion that May. The league was, of course, disbanded, but it was reconvened in October 1900, and its membership grew quickly to 1,200 men and 500 women in 1902. Its membership then declined precipitously at the beginning of the following year, only to rebound to 2,670 at its end. In 1905, there were 600 union members (200 of them women) out of the 3,000 workers in the company.

In 1902, the league, which had as its scope "defending and improving the economic and moral conditions of the workers; assisting its members in case of a strike; obtaining reductions in work hours and increased wages," presented a complex set of demands to the company. After long negotiations, Pirelli accepted most of them. Minimum wages were fixed; raises would be automatic and tied to seniority (up to a maximum); the *cottimo* would be modified to distribute more of the benefits of increased productivity to the workers; time-and-a-third wages would be paid for over 11 hours of work and on holidays; regulations concerning holidays, fines, and layoffs were spelled out; workers were required to contribute to the sickness fund; medical expenses would be paid in full by the company (those for women giving birth would be granted only if the woman had worked for the company at least six months); and an internal committee composed exclusively of workers, both organized and nonorganized, would be established.[50] In 1908, Pirelli moved its plant outside the city, near Sesto San Giovanni.

The Milanese garment industry was more robust than the textile sector. It included several specialties, the largest of which was the production of white goods (bed and personal linens, men's shirts and collars, and other small items such as handkerchiefs and underwear). All white goods sewers were women, who often did extremely fine stitching but were nevertheless not considered skilled workers. Most of them worked in their homes. Tailoring involved cutting and fitting to the wearer's body clothing like coats, suits, and expensive dresses and then sewing the garment with curved seams, tucks, and darts that produced a close fit. This specialty employed more men, especially as cutters of men's suits and outerwear. Most of the women tailors worked at home. According to the 1909 Società Umanitaria report, "Homework was generally the rule, not the exception."[51]

A resistance league of male and female tailors was organized in 1890–

91, and one of its goals was equal pay for men and women. As in the case of the ribbon weavers, there is the suggestion here that men were losing their ability to control the labor market through the transmission of skill. Indeed, the flyer for one of the league's meetings at which workers aired their grievances declared that "in our occupation, the introduction of machines and of sales of ready-made garments is generating a sharp shift in the interests of masters as well as workers. . . . [I]ntense competition is damaging workers and small shops, and causing unemployment." [52] The tailors closely observed the general strike of machine shop workers in the same period, and the women tailors sent a colleague to one of the strikers' meetings with a letter of support. [53]

In 1892, there was a strike of one hundred men and women in one garment shop that lasted ten days and ended with a compromise. In 1894 there were some all-women strikes of seamstresses (objecting to abusive foremen and bosses), and in 1896 there was a large mixed strike at the Bocconi shops, which supplied the largest department store in Milan. There most of the leaders were men, but some women activists are named in the police records. [54] A letter from a tailor to *La Battaglia* (September 12, 1896) regretted the weakness of their organization and called on his fellows to banish any thought of becoming a small-shop owner. The league thereafter retreated from the scene.

It was revived in 1899 (as a mutual benefit society), but this time there were separate leagues by sex that were combined only in 1903, on the initiative of the men, who wished to recruit more women. Female membership declined once the active organizing was over, however. A high point of the strikes in 1902 was one against Bocconi in which more than twice as many women as men were involved.

The strike that attracted the most publicity (including an investigation of conditions in the trade by the *Unione Femminile,* a Socialist–feminist journal) was that of the *piscinini,* child helpers in dressmaking and hat-making shops. About 190 of these girls, aged 9 to 12, struck in 1902. Their modest demands were (1) minimum pay of 0.5 lira daily, (2) a 10-hour workday with 1 hour off for meals, (3) additional pay for extra hours, (4) an end to the performance of domestic service for employers in addition to their other work, (5) regular weekly paydays, and (6) Sunday work at double pay. *Avanti!* reported the strike with much sentiment: "The "babies" came to the Chamber of Labor, "the scene of many a struggle by their Daddys and Mamas . . . they fought their first battle at an age when their peers still played with dolls." The girls declared that they wanted a true apprenticeship, and they rejected having to be servants and messengers for the shop. Even after the "successful" strike, however, their wages continued to be exploitative. The *Unione Femminile* helped the *piscinini* organize their own society, *La Fraterna,* with "educational and moral" goals.

Despite some success in organizing women in garment shops for workplace collective action, the Società Umanitaria concluded the economic agitation in the industry was difficult, even unlikely, because the women, in
particular, were "attached to innumerable small shops based on homework."[55]

The women tobacco makers' strikes and agitation attracted attention at
least partly because the industry was a public one. Offering a distinct contrast with domestic service (the largest female occupation in Milan), textiles, or the garment industry, tobacco manufacturing in Milan consisted of
one large factory, employing about 1,300, the vast majority of them women.
It produced cigars and packaged loose tobacco on behalf of the government
monopoly, established to raise revenue. As employees of the state, tobacco
workers had secure jobs and received significant nonwage benefits.

In 1882, tobacco workers unsuccessfully attempted to substitute an independent mutual benefit society for the state-sponsored one. They established their first league, with 600 members, in 1898, after a strike in December of the previous year. The Chamber of Labor negotiated with the
state on their behalf, and the women continued to be active in the chamber
and close to the Socialists. This league was dissolved in May 1898. Through
the efforts of an organizer from the Federation of State Employees and
several male workers, the league was reactivated in 1899. Women workers
flocked to join, and the league had 1,000 women and 169 men members
when it was formally reopened in April 1900. As a section of the state
employees' union, the new league was able to establish its own mutual
benefit plan to supplement the state's sick pay and death benefits.[56]

Even before this league was in place, women in 1891 protested the
dismissal of one of their sister cigar makers, insisting that the action was
unfair (the woman was fired for poor work) because the quality of the tobacco leaves at that time was inferior and it was difficult to work well. The
women threatened to strike, banging on their work tables with their fists.
The director promised to provide better-quality tobacco but refused to rescind the dismissal. Women workers also objected early in 1892 to a male
supervisor's severe discipline. The Chamber of Labor negotiated with the
plant administration on their behalf, and the offender was transferred out of
their workroom. Later that year the cigar makers demanded and received a
raise.[57] Both a half-day strike in 1897 of women who assembled packets of
tobacco and an appeal to the Chamber of Labor for aid in fighting a wage
cut were followed by the formation of their league. A short time after their
"success" in persuading the state to rescind partially the wage cut, however, new machines and low-waged girl operators were substituted for the
women.[58]

In 1901, there was a longer strike in the tobacco monopoly factory. It
began with the women's refusal to work until they had a reply to their

demands from Rome; they demanded that the director of the tobacco monopoly come to Milan to investigate. A woman cochaired the strikers' meeting, which protested police intervention. She continued to serve as one of the two chief negotiators as Socialists and others hastened to turn the strike into a protest against the state as exploiter of women workers. The women returned to work only after they had promises from the Chamber of Labor, the two Socialist deputies from Milan, the prefect, and the plant director that the government would keep its agreement.[59]

In his introductory remarks to the Società Umanitaria's retrospective study of labor organization and workplace collective action in Milan, Schiavi summed up his view of women workers: "Women in organizations . . . do not have the perseverence to pay their dues, the discipline to attend meetings, or the willingness to stick together, and after a brief flare-up of enthusiasm they abandon their leaders—often the single fervent and energetic leader [i.e., they often had only one]—who had begun the movement."[60] To what extent was Schiavi's statement correct?

Women workers usually faced severe constraints that hindered their "discipline" and their "perseverence." Many women workers were isolated from others with similar interests. (The majority of servants and seamstresses, together constituting the majority of women workers, labored in homes, their own or their masters'.) All women workers were poorly paid: Their wages averaged from one-half to two-thirds of those of men. If they lived alone, they lived in poverty or close to it. Most women workers worked in industries with little fixed capital and/or under great competitive pressure. Their jobs clustered at the lowest reaches of the occupational hierarchy, required little training, and offered little possibility of promotion. Under the pervasive and long-lasting system of occupational segregation (the division of occupations—and, even more so, jobs—into those assigned to women and those assigned to men, leading to any given occupation's being primarily male or female), women were eligible for very few jobs; hence they were in constant competition with one another.

Their family situation also put women at a disadvantage. Single women who lived with their parents were usually expected to contribute most of their wages to the family budget. Their ability to pay dues to a league and forgo wages during a strike depended in many cases on their family's willingness to absorb these losses from its budget. Some families were willing to see income lost in the interest of occupational or class solidarity, but many were not. Married women also contributed to a household budget and were subject to familial demands similar to those that single women faced. Their were serious time constraints on married women's participation in organizations owing to their exclusive responsibility for domestic tasks and, if there were children, for child care as well. Unlike men (whose responsibilities to the household were limited to the contribution of their relatively higher wages to the common good), women's position in the household impinged constantly on their ability to earn wages or participate regularly in workers' organizations. Conditions varied, however, in the industries discussed.

The textile industry had the worst combination of factors contributing to women's organizational weakness and defeat in strikes. The industry itself was undergoing difficult times and was seeking ways to maintain competitiveness. Whatever strategy their employers adopted the women lost out. If employers moved their business to a rural area to find compliant and cheap workers, urban women lost their jobs. If they mechanized and substituted lower-paid women for men workers, the women were divided from their fellow workers and were criticized by the men.

La Lotta di Classe (August 20–21, 1892) concluded from the experience of the Schoch strike that year: "We will continue our effort to persuade workers of the need for united action, not impulsive uncoordinated rebellions. Under such conditions strikes are too easily defeated." Yet the opportunities for united action were small. The poor conditions in the industry affected both men and women. The position of women and men was different, however, in regard to both their family obligations and their relations with their bosses. Women workers were being cynically manipulated against men in order to cut wage bills. (Note also that the publicity concerning women workers' plight did not lead to favorable intervention by the government or other outsiders. Textile employers had little reason to stay in Milan if it meant high wages and erosion of their market position, and the authorities did not try to persuade them to do so.) Because of the limited number of jobs for which women would be hired, there was always a ready supply of women workers. And despite their willingness to strike, the women were easy to replace because their bosses did not invest in training them, and because the men did more skilled jobs like fixing looms or elaborate weaving, they were in a stronger position. Once women were substituted for men in low-skilled jobs, the dual labor market enabled employers to buy men's loyalty by granting them better jobs, training, and higher wages. The split between men and women workers was thus a consequence of employers' strategies to maintain competitiveness, reinforced by and reinforcing the household division of labor.

In the rubber industry, gender was less salient to shaping organization and strikes than were the company's paternalism and its workers' recent migrant status. It was Pirelli's policy to hire rural migrants in the hope of guaranteeing a passive work force for his company. As the case of the construction workers showed, rural origins did not necessarily mean passivity once the workers were organized. On the one hand, the predominance of migrants combined with the high proportion of women and child workers could hinder such organization. On the other hand, the fact that Pirelli did bring together many workers in large shops (one for rubber products, the other for insulated wire) suggested, to Socialists and anarchists outside the workplace, that the workers could be organized. The first hypothetical outcome was borne out by experience, for although there was a union, its membership fluctuated widely, and it relied on Pirelli's paternalism rather than workplace struggle to achieve its goals. Some Pirelli workers took a more oppositional stance and acted on it, as we shall see in the discussion of the Fatti di maggio, 1898, but the company's paternalism seems to have been accepted by most of its workers, women and men.

Organization was more effective and strikes succeeded more often among (at least some) women garment makers (although most of them, it should be recalled, were homeworkers who were not involved in these strikes at all) than among textile or rubber workers. The men and women tailors' unions brought these workers together with other workers in class-based action, such as support for other strikes and affiliation with the Chamber of Labor. (Annetta Ferla—a seamstress and member of the Figlie del lavoro [Daughters of Labor], a mutual benefit society—was appointed to the executive committee of the newly formed Socialist party in 1892.) The growth and prosperity of the garment-making industry, supplying Milan's large consumer market, made these workers' action easier, as compared with, particularly, action taken by the textile workers.

Although the women tobacco workers earned less than did the men in the industry, their wages were much better than those of other women workers. Cigar makers in particular were highly skilled and valued workers. Because the tobacco monopoly was expected to turn a profit for the government, it was especially sensitive to its workers' threats to withhold their labor and slow down production. Therefore, although the monopoly, like other employers, attempted to reduce its wage bill by mechanizing and replacing more costly workers by cheap ones, it was much more responsive to its workers' demands than were most private employers. This responsiveness was enhanced by public scrutiny, including that of members of Parliament, who often questioned the minister of finance about strikes and working conditions. Women also were more active in tobacco workers' organizations than were women in other industries, and sometimes they emerged as leaders in strikes. Under such conditions, then, women acted in their collective interests as workers, and in the tobacco industry there was a real possibility of winning their demands.

At times, the women strikers were supported by the *operaisti*, Socialists, some unionists, and the Chamber of Labor; on other occasions they were opposed as workers by their male colleagues. Were there other sources of support for women as workers, members of organizations, and strikers?

Some contemporary feminists, such as Anna Maria Mozzoni, explicitly addressed the woman worker. In her early writings Mozzoni conceptualized wage work as emancipatory, because of its removal of women from the family, the source of sex inequality. She later developed a feminist version of the *operaista* argument that wage earning was a means of freeing oneself of other means of servitude and that such liberation had to be achieved by women themselves. In closing her 1885 essay to the "daughters of the people," she observed: "Remember that you yourself have a mind, a will, and are capable of action."[61] Mozzoni was eager to see women in the work force. She built no organization, however, although she was close to the Socialists and allied herself with Socialist teachers who supported women's suffrage. She respected women workers as autonomous subjects, but she seldom spoke to them and did not seek to organize them.

Among the Socialists, Anna Kuliscioff was the leader most concerned with women workers. In 1890 she read a feminist paper, "The Monopoly of the Male," to the Milan Philological Society, in which she decried men's oppression of women at home and at work and commented that a woman worker is "doubly a slave: on the one hand, of her husband and, on the other hand, of the capitalist." In her work with Turati in the Italian Socialist party, however, Kuliscioff downplayed women's issues, arguing instead that women's liberation was linked to that of the proletariat and campaigning for protective legislation. Yet in a speech to a newly organized men's society in 1893, she stated that it behooved men to collaborate with women, for "as long as the women remain alone and isolated from the general movement, all your hopes to improve your position will fail." She briefly served as editor of a new Socialist women's newspaper in 1913.[62]

Less visible but more practical Socialist women also played their part. The role of the Unione femminile (founded in 1899) in the strike of the *piscinini* has been mentioned, and it also ran organizations similar to settlement houses and sponsored a home for wayward girls. The Milanese Socialist Women's Group also sought to organize women. In 1894 schoolteacher Linda Malnati became the "godmother" of the women's organization Le Figlie del lavoro (an offshoot of the male Figli del lavoro). She accused men of pretending to support women workers. Elsewhere she wrote: "Women are degraded by being forced to compete with men workers. . . . Although women are victims of an odious system that exploits them to the disadvantage of all, they are made to appear apathetic and egoistic. Only organization can save women from this humiliating condition."[63] Despite the sympathetic understanding and willingness of Socialist women, and many Socialist men, to help, barriers to women's organization, and hence to successful strikes in large women-employing industries, were not overcome.

Conclusion

In the textile industry, workers were under great pressure from their bosses to work longer hours for lower wages; otherwise, harsher steps would be taken to cut costs, such as substituting poorly paid women for men workers or moving the business to rural locations where labor was cheap. Both the more-skilled men workers in silk and ribbon weaving and the less-skilled women in elastic and mixed-textile weaving struck in defense of their wages or jobs but lost their struggles. Their strikes attracted little public attention, except from the worker and Socialist press, and the government's efforts to mediate were minimal.

Until 1887, metal and machine workers were divided by craft and disciplined in their workplace but relatively favored in terms of autonomy and wages. An employer drive to increase productivity challenged workers' relative autonomy, and in their struggle against their bosses' impositions, these workers organized broad-based resistance leagues. The leagues were transitory, however, as machine shop workers tended to remain divided by specialty, in the period in which industrialists reorganized the labor process and promoted productivity through specialization. Like printing and construction, the metal and machine industry was the focus of public concern and municipal attention when large strikes were staged. In the latter, though, manufacturers formed their own organization, which left little room for mediation by municipal officials.

The timing of organizational drives and strikes was shaped by long-term organizational and technological change in most industries. Not all workers were able to take advantage of a favorable economic climate to strike. Milan's textile workers were disadvantaged (as were construction workers) by the small scale and vulnerability of the companies for which they worked. Textile manufacturers could shut down and leave the city if pushed too far, and thus they were relatively free to manipulate and exploit their workers. There were very few skilled and privileged workers in the Milanese textile industry, and they were generally unwilling or unable to join or lead their less-skilled fellow workers to organize and strike.

In engineering, employers' fixed capital and dependence on their workers' regular supply of labor were high, as in printing. But they were less in need of highly skilled workers than were publishers. The engineering firms developed strategies for promoting productivity in their competitive industry that kept workers divided by specialty, undermined industrial solidarity, and hindered organization. In the first decades of the twentieth century, engineering workers in Turin were able to build more effective organizations and strike with greater success than they could in Milan. This was due partly to the movement from Milan of the larger plants to smaller cities on its outskirts and partly to the dynamics of political struggles within Socialism and the workers' movement that roiled Milan during this period.

The constraints on women as workers and strikers were not relieved by the support of fellow workers, union members, Socialists, or feminists. Most of the time, Milan's women workers, with the exception of the tobacco workers, were severely handicapped in any effort to act collectively in their own interests. Women workers were not passive but active participants in workplace struggle. But they were structurally disadvantaged, with few resources and only occasional support from sympathizers whose attention was usually directed elsewhere, and so women workers seldom succeeded in building lasting organizations or winning their struggles.

Some of the patterns observed here were similar to those in other European countries. For example, printers and construction workers were early to organize in most countries. Both industries were urban industries par excellence and boomed with industrialization and urbanization. Technological change and growth of scale in printing in Milan were retarded compared with those in other European cities, but their effects were similar. Construction technology and organization of production (including the importance of subcontracting) in other European cities were also similar to the Milanese. But Milan's textile industry was not typical of Italy's, for it was much smaller and more specialized than it was elsewhere. It was even less like that of Europe's huge textile cities, like Manchester, Stockport, or Roubaix. Rather, Milan's textile industry was more like the garment industry in many European cities, in its scale and willingness to follow cheap labor. Overall, Milan's engineering industry was smaller in scale, mixing large plants and tiny ones, than were those of European cities renowned for their machine works (including Turin and St. Petersburg) in World War I and the postwar period. Milan resembled Paris rather than these cities, for both were characterized by suburbanization of the industry after 1900. Women workers' situation in Milan also echoed that in other large cities: To the extent that garment workers sewed at home, they were hard to organize and slow to strike, whereas workers in the tobacco monopoly of France, Spain, and elsewhere in Italy were leaders of both their unions and their strikes. These similarities confirm the importance of conditions internal to industries in the timing, organization, and probability of success in strikes.

State and local government intervention, whether in the direction of repression or mediation, frequently made a definitive contribution to the outcomes of workplace collective action. Politics also entered the workplace struggle in a more fundamental way. Workers' coordinated action on behalf of shared interests, that is, collective action, was part of a political process that began with the identification of common interests, followed by organization, mobilization, and action at moments that seemed opportune. The outcomes largely depended on the weight of resources that employers and workers could deploy apart from intervention by government authorities or other workers organizations. Organizational success predicted effective collective action, and such success favored the winner in the next period of struggle. Similarly, failure fed on itself and conferred a disadvantage on the loser in later efforts. There was a crossover, too, from economic to political arenas: It was those workers with experience in workplace struggle who joined in the last decades of the nineteenth century with intellectuals from their own class or from the bourgeoisie to build workers' organizations for electoral politics, as the next chapters show.

8

Creation and
Repression of a Workers'
Party, 1875–1886

In the aftermath of the May rebellion of 1898 and its repression, Gaetano Salvemini wrote a critical history of Milan's political parties in the nineteenth century. He argued that 1870 was a watershed for both Mazzinian republicans and workers, who up to that time had been closely linked. In that year, the events of the Paris Commune frightened the Milanese Mazzinians, who forthwith allied themselves with the Moderates. The proletariat, according to Salvemini, "sensed for the first time that profound gulf between themselves and the bourgeoisie . . . there began the first currents of the class struggle." Lombardy and Milan became "the great laboratory of proletarian experience." He concluded: "The Milanese administrative [local government] struggles are no less than episodes in, or better, precursors of Italian political struggles. What Milan thinks today, Italy will think tomorrow."[1] Milanese politics was indeed a precursor of Italian national politics, in many ways.

The next three chapters examine that level of class formation concerned with collective action in the Milanese political arena and its relationship to national parties and institutions. The formation of the Partito operaio by workers already involved in political debate and action, its vicissitudes, and its eventual repression are the focus of this chapter. We open with a brief review of Milan's local political history and the progressive currents active in the city in the early 1880s. The chapter's central question is what the conditions were that promoted the creation of a workers' party in the city and those that led to its demise.

Milan's economic importance, the national circulation of its newspapers, and the Milanese base of national political leaders of varying persuasions all made the city central to Italian politics. Workers' politics must be understood first in the arena of local politics, for it was there that their economic institutions like mutual benefit societies, resistance leagues, and urban federations were born and acted. Institution building on the local level and workplace collective action were, however, constrained or facilitated by aspects of national politics such as law and repressive policies. Institutions that broadened the workers' collective action beyond the workplace also originated in Milan. Here the Lombard Workers' Federation, the Partito operaio, and the Italian Socialist party (and many of the actors who shaped these institutions) had their roots.

Two groups of workers were spurred to action by the extension of suffrage in national elections that was granted in 1881: (1) those associated with the Consolato operaio, who with their Democratic Radical patrons nominated a worker to run for Parliament, and (2) the workers and their petty bourgeois colleagues who were active in socialist study groups, the editorial board of *La Plebe,* and certain unions, particularly the printers' resistance league—in brief, workers who were already involved in politics. The latter group founded the Partito operaio and developed its workerist ideology and practice. The disagreements about theory and strategy between these two groups and their leaders underlie the dynamic of worker politics in the region and city in the period up to 1886. During the relatively brief period in which it promoted both worker organization and workplace collective action, the Partito operaio was a political training ground for young workers. The ideas and practices they learned there continued to inform their politics in varying degrees for years after the Partito operaio had disappeared as an institution. The party failed not only because it was repressed but also because it too ambitiously expanded its activities beyond the capacity of its resources and members.

Milanese Politics: Parties, People, and Issues, 1881–1899

The chief characteristic of the men in power in the 1880s and 1890s in Milan was their unrelenting conservatism, the chief characteristic of their politics, an unremitting concern with containing the working classes. After Milan's workers and bourgeois had abandoned Mazzinian republicanism in the aftermath of the Paris Commune, monarchism dominated local politics. The Milanese monarchist conservatives, members of the Associazione constituzionale, are conventionally labeled Moderates, or, collectively, the *consorteria moderata.* The mayors and the *giunta* of *assessori* (a local "cabinet" of the elected municipal councillors who actually ran the city) all were Moderates until 1899, more conservative than the city's parliamentary delegation. The mayors and the dates of their administrations were Giulio Belinzaghi, 1868–84, 1889–92; Gaetano Negri, 1884–89; Giuseppe Vigoni, 1892–95, 1895–99; and Giuseppe Mussi, 1899–1903. (Royal commissioners appointed in Rome briefly directed the city's administration in 1895 and 1899.)

The chief opposition to the Moderate's rule came from the Democratic Radicals, men such as Felice Cavalotti, leader of the party in Parliament, Carlo Romussi, the editor of *Il Secolo,* and Giuseppe Mussi, a lawyer who was first elected to the municipal council in 1868. In Salvemini's view, Milan's Democratic Radicals were "incoherent" and insincere, and their legalism was the outgrowth of their lack of power.[2] The Democratic Radicals' patron–client relationship with the workers' groups associated with the Consolato operaio originated in the workers' alienation from Mazzinian republicanism after 1870.

The main political agenda of the 1880s and 1890s in Milan can be summed up under the headings of setting priorities and directing growth.

The city plan adopted in the mid-to-late 1880s reflected hopes like those of Giuseppe Colombo that the city would preserve its wealthy, residential core and not be transformed into an industrial city. This view of Milan's future was rejected not only by men interested in speculative land buying and building but also by those urging industrial growth, such as the rubber entrepreneur Giovanni Battista Pirelli.

In his maiden speech to the city council (1877), Pirelli claimed that the most pressing need of his constituency in the external *circondario* was "street improvements, particularly the belt roads which are in terrible condition." Better sidewalks and streetlights were also needed for the workers' commute to work.[3] The two big working-class neighborhoods at the north and south of the city were coalescing from the old suburban borghi at the city gates (see Figure 2-2): Porta Tenaglia, Porta Nuova, and Porta Garibaldi to the north and Porta Ticinese and Porta Genova to the south and southwest. Pirelli's words reflected his recognition that paved streets and exemption from the city *dazio* on workers' food and other essentials in the external *circondario* were effectively subsidies for his employees.

The context of Mayor Belinzaghi's 1884 defeat on the issue of permitting land around the Castello to be developed was the continuing difference among the constituency, the interests, and the politics of the internal and external *circondari.* In the center of the city, the wealthier inhabitants, their servants, and some craftsmen were loyal to the conservative *giunta;* in the external *circondario,* voters were concerned with growth and the development of industry and with protecting their favorable position vis-à-vis the *dazio.*[4] They were more likely to support Democratic Radical candidates. For example, in 1889, the council minority included one worker, Vincenzo Corneo, president of the printers' union, elected on that list, and in 1890, an *operaista* and socialist—Osvaldo Gnocchi-Viani (running on both the Consolato's list and one sponsored by the Lega socialista)—won a partial election.

The issue of equalizing the *dazio* burden of the walled city and its suburban areas was both important to consumers and linked to the type of future growth that the politicians envisioned. The first attempt by the conservatives to remove the fiscal privileges of the outer ring was violently resisted in the *micca* revolt," of 1886. Workers who lived in the suburban area carried into the walled city two *micche* (the *micca* was a 380-gram loaf of country bread) for their lunch. Although this quantity of bread was legally subject to the *dazio* when it passed through the city gates, before March 1886, the regulation was not enforced. Mayor Negri then gave orders to collect the *dazio* on the surplus bread in the workers' lunches. (More will be said later in this chapter about the violent reaction to this measure.) Negri's action was but one of several attempts to increase the poorer population's financial contribution to the city. Appropriately, he was labeled *l'affamatore* ("the starver").

As population growth made the expansion of urban services—cemeteries, sewers, hospitals, paved sidewalks, streetlights, and the covering of the *naviglii* (canals)—into political issues, progressive members of the municipal council questioned the commitment of the mayor and his party to these needs. The new streets for bourgeois housing and the park seemed to have first priority with the Negri *giunta*. Opposition grew. In February 1889, the suffrage in administrative (municipal and provincial) elections was enlarged to parallel that inforce national elections. The sanitary services and more schools were eventually financed after the return of Belinzaghi in 1889, carried into the mayor's office by the split of the Moderates into pro-Negri hard-liners and an anti-Negri block calling themselves liberal-monarchists (these included Pirelli and Giuseppe Colombo). Although divided, the Moderates outpolled the Democratic Radicals, who also opposed Negri. Belinzaghi attempted to reduce the level of disagreement in a *giunta di conciliazione*.

New disagreements arose over the subsidy for the proposed Chamber of Labor at the end of 1890, as well as over the slowness in modernizing the tramways. The *ad hoc* group that supported Belinzaghi's *giunta* began to break down, but then Belinzaghi died in 1891, and partial elections were called. Negri refused to return to the mayoralty without a general consensus, and so the council chose his epigone, the engineer Giuseppe Vigoni, as mayor. His program included municipal economies and a cautious modernization of the infrastructure.

In his inaugural speech, Vigoni echoed Colombo when he bemoaned the "disastrous growth" of Milan, which he attributed to the availability of charity in the city. He suggested sending the migrants back to the country to supply farmers with workers and restore fiscal soundness and "decorum" in the city.[5] The council voted to electrify the trams but postponed the debate on social issues, except for the passage of a subsidy for the newly established Chamber of Labor. In 1894, more votes went to the Democratic Radical and Catholic candidates. In a political power play, Vigoni resigned from the council despite a majority in his favor. After an interval during which a royal commisioner governed the city, the Moderates won a clear-cut victory in partial elections, and Vigoni resumed the mayorship.

Negri's failed project to tap more vigorously the financial resource of the ex-Corpi santi was achieved by Vigoni. The need for additional revenue outweighed the interests of suburban industrialists and workers, and the proposed solution—declaring the whole city open and free from the *dazio* rather than making the external *circondario* closed and liable to the *dazio*—was passed in early 1897. There was disagreement, however, on the most equitable and efficient form of municipal taxes to accompany this action. The national law that the city requested in order to levy other forms of revenue-producing taxation was never passed, however, and so Vigoni and his *giunta* offered to resign once again. The May rebellion interrupted these negotiations. The outcome, approved in its aftermath, was to extend the *dazio* to the whole city but to exempt articles of everyday consumption, like food and fuel. A new tax on high rents also was enacted to provide needed revenue.

The Fatti di maggio (the events of May) were an important watershed in Milanese history. Workers' institutions—like mutual benefit societies, unions, and the Socialist party—and dissident newspapers—including the Democratic Radical *Il Secolo,* the Republican *L'Italia del popolo,* and the Catholic *L'Osservatore cattolico*—were closed by royal decree, and their editors and leaders, including several deputies, were arrested. In 1899, op-position parties and workers' organizations gradually resumed their activi-ties. Voters, appalled by the bloody retribution and punishing criminal trials of progressives and, more generally, those opposing the regime, voted in both city and national elections against the moderates. Filippo Turati, the imprisoned Socialist deputy, was reelected twice while in prison. Giuseppe Mussi, whose son had been killed in an antiregime demonstration in 1898, became mayor in 1899, with a council in which Democratic Radicals, Re-publicans, and Socialists held the majority.

In his administration, social concerns like school lunches and housing for the poor came to the fore. At the same time, fiscal demands could not be ignored, and the continuing population growth meant more claims on city finances. Coping with growth and planning for the future continued to be central concerns.

Progressive Initiatives

In 1875, Enrico Bignami moved *La Plebe,* the progressive newspaper that he edited from provincial Lodi, to cosmopolitan Milan. There Bignami and his colleagues began to lay out an independent, eclectic socialist program in *La Plebe*'s editorials and features.

In 1876, Bignami, together with Osvaldo Gnocchi-Viani, founded the Circolo di studi economico–sociali. Both these men opposed the anarchist position, which was being abandoned also by other progressive Italians. Attempted anarchist insurrections had failed in 1874 and 1877, provoking police repression against all progressive groups. Bignami and Gnocchi-Viani argued that Socialists should consider running candidates for Parliament. Andrea Costa's 1879 open letter renouncing anarchism, "Ai miei amici della Romagna" (To my friends in the Romagna), was published in *La Plebe*.

The Consolato operaio, Milan's Democratic Radical federation of benefit societies, also launched in 1880 a Circolo operaio, another discussion and educational group in which economic and social questions could be addressed. It was welcomed by *La Plebe*, which noted that there were then no other organizations suitable for talking over these problems.[6] The Circolo operaio grew rapidly.

Explaining its support of the new Circolo, *La Plebe* attacked the bourgeois political parties, which believed, it claimed, that a "worker was no longer a worker but a serf."[7] The paper called instead for a workers' congress, but this initiative was aborted by the police, who forbade such a meeting. *La Plebe* protested, declaring that "arbitrary rule reigns supreme."[8]

The newspaper again discussed a workers' congress in the fall of 1880, when it published an article claiming that workers preferred a regional, not a national, congress. Further, it continued, this was simply "a matter of form, for many workers hoped for a workers' party, distinct and separate from bourgeois political parties."[9] Later that year *La Plebe* announced that the Consolato operaio had started a movement for a federation of north Italian workers' societies. "Excellent idea!" the article continued, "and excellent, too, will be such a federation if it encourages the progressive development of a working class. . . . Let workers themselves flee the old political parties and constitute their own party."[10]

Gnocchi-Viani and Giuseppe De Franceschi, Bignami's collaborators at *La Plebe*, were Milan's delegates to the 1880 congress of the Federazione dell' alta Italia dell'internazionale the remnants of the Italian section of the Anarchist International. It sought to draft a common program acceptable to both those still faithful to anarchist principles and those, like Gnocchi-Viani, De Franceschi, and Bignami, who had been converted to evolutionary socialism. Although there were only eighteen delegates, this was an impossible task, as the two groups held totally incompatible positions. Gnocchi-Viani and De Franceschi favored socialist participation in national and local elections, and the anarchists insisted on abstention from all elections. Both groups agreed, however, that universal suffrage was no more than a "means of propaganda,'' and they rejected both paternalistic employers' social benefits and bourgeois government-sponsored social legislation. The evolutionary socialists lost most of the votes, but the Italian section of the Anarchist International lost any coherence when the two groups split.[11]

La Plebe, inspired by Milan's progressives and encouraged by Gnocchi-Viani, spelled out a feminist position in the second half of 1880. It cosponsored a new organization, the Lega promotice degli interessi femminili (League for promoting women's interests). A lecture by Anna Maria Mozzoni discussing prostitution as a problem that could be solved only by reforming economic and social relations was reported in the paper.[12] On September 12, *La Plebe* offered an editorial on women: Bourgeois opinion holds that women are ''inferior by nature,'' but socialists do not agree. Rather, socialists believe that women must be emancipated, as they are no less useful in society than are men. Another article claimed that both the civil emancipation of women and the civil and economic emancipation of the male proletariat must be ''infused with a revolutionary spirit, or it risks going up in smoke like academic rhetoric.'' What was needed was ''the fusion of women's cause with the cause of the proletariat.''[13]

Reporting the next month on the founding meeting of the Lega promotice degli interessi femminili, *La Plebe* repeated its injunction: ''Women's emancipation cannot be separated from workers' emancipation: Their oppression is the same in that the liberation of all oppressed groups must be achieved by themselves.'' This conclusion led *La Plebe* to support the league's constitutional provision that only women could vote in its meetings. It was a necessary, ''courageous innovation without which women's cause will always be obfuscated.''[14]

In February 1881, *La Plebe* published the program of the Lega promotice degli interessi femminili. The organization was composed of ''citizens concerned with the inferior condition of women who wish to work for reforms in Parliament, the government, and public opinion.'' The league welcomed ''persons from all tendencies to promote women's interests and an equal and common concept of humanity.''[15] The feminism of early Italian socialism was optimistic and generously progressive.

The Confederazione Operaia Lombarda

Following the Depretis government's expansion of suffrage in 1881, the Consolato operaio issued a call to the first congress of the Lombard Workers' Federation, on September 25–26, 1881. In April, *La Plebe* reported that the proposal had encountered opposition. The old general mutual benefit society, L'Associazione generale, had refused to participate. Milanese workers, according to *La Plebe,* were divided into two groups: those in the

Consolato, true workers, and those in the Associazione, workers in name only, because they were dependent on the notables who were the officers of the society. According to the newspaper, workers ought to rid themselves of nonworker elements in their societies and their federations, "constituting by and for themselves a new party, distinct from all clerical and bourgeois parties." Similar sentiments were repeated in the June 30, 1881, edition of *La Plebe,* in which were discussed both the plan for a socialist party in the Romagna and the project of the Milanese Circolo operaio to establish a Partito operaio.[16]

In August 1881, Andrea Costa inaugurated the Revolutionary Socialist party of Romagna at a congress held in Rimini. The new party was to be a federation of societies and individuals that supported the "principles of modern socialism": in the economic arena, collective ownership, of the land and the means of production; in the political arena, recognition of the civil and political rights of each human being; and in the intellectual and moral sphere, the right to education.[17] It would be "revolutionary . . . in the socialist sense of a revolution that goes beyond merely transforming the roots of social organization." Spreading socialist ideas, organizing socialists in city and country, and inciting the struggle against capitalism through strikes, reforms, or municipal elections would be the means to this end. Gnocchi-Viani, the probable author of *La Plebe's* commentary on the party program, objected to its implied concept of revolutionary dictatorship and its loose concept of class. Nevertheless, in keeping with *La Plebe's* customary eclecticism, he expressed satisfaction with the results of the congress.[18]

The September 1881 congress of the Confederazione operaia lombarda convened eighty-six associations, of which the largest number came from Milan, followed by Como and Pavia. In this congress, 75 percent of the Milan delegations were from occupation- or industry-based organizations. Only 15 percent of the delegations from Como, and 58 percent of those from Pavia were based on occupation. These proportions held in all the congresses (1881, 1882, 1884) for which lists of delegates and participating organizations could be found.[19] The organizational patterns of the delegations represented suggest a preponderance of occupation-specialized worker societies in Milan, perhaps simply a result of scale, but perhaps also indicative of a level of occupational solidarity that was not so common in the smaller cities, where general workers' societies prevailed.

The influence of the Democratic Radicals in the 1881 congress was immediately evident in the fact that the reporter of its constitutional committee was Carlo Romussi, editor of *Il Secolo*. A workerist tone became evident in the discussion about the presidency of the congress when one of the speakers recommended that "workers be nominated as true presidents of a workers' congress." Another suggested that "workers ought to reject the idea of a single president, with its monarchical roots." A presidential committee of five members who would act as president in turn was elected instead.[20]

The planning committee's report declared that the congress reflected "the new historical period that the worker multitudes have entered . . . , [a] period of association replacing the period of the individual." Which societies, Romussi asked, should be permitted to join the confederation? The Consolato's survey of respondents had indicated strong support for a wholly workers' confederation.[21]

The confederation's worker character was clarified further in the report's discussion of politics: "We are not among those who claim that workers' societies ought to avoid politics. . . . [They] should be schools for patriotism. . . . *We do not ask each organization what its political principles are but only if it is a workers' society.*" Manacorda emphasizes that the phrase in italics was "not mere demagogy" but a concession that Romussi—speaking for the Democratic Radicals—willingly made, a concession that he was to regret later but be unable to undo. It was a concession to the idea of *operaismo* (workerism), Manacorda explains, that emerged in that congress "in a self-conscious and combative form."[22] The theme had already been expressed, of course, by *La Plebe*. The newspaper was not always consistent on this issue, but neither was the 1881 congress of the confederation. For example, it voted to permit veterans' mutual benefit societies to join, even though they included nonworker members.

The dominant tone of the congress was democratic. The Confederazione operaia's constitution called for the support of mutual benefit societies, people's banks, cooperatives, and popular and lay schooling. A more radical position was taken in Article 4, Section D, of the constitution, which, as drafted, read that the federation would "encourage institutions designed to achieve justice in relations between capital and labor, helping workers become stockholders in industrial enterprises." This was amended, over Romussi's objection, by adding "and support agitations" between the words *institutions* and *designed.*[23]

The "woman question" was introduced only after the constitution was ratified, and the group voted that it be placed on the agenda of the next congress.[24] The assembly also approved motions (1) favoring universal suffrage, (2) opposing work done in prisons in competition with free labor, (3) calling for a progressive reduction of the salt tax, and (4) stipulating that ordinary congresses of the confederation be held in different cities in Lombardy but that special congresses convene in Milan.

The Partito Operaio Italiano

The first elections under expanded suffrage were to take place in 1882. Because the number of workers' votes could be large, progressive groups met to consider their tactics. In February, the Romagnol Revolutionary Socialist party agreed to run candidates in the election. If elected to Parliament, Socialists were to refuse the oath of loyalty to the monarchy but claim their seats in the Chamber of Deputies in the expectation that they would be removed by force.[25]

In Milan, there were two initiatives, representing contrasting views on workers' politics. The first, launched by the Democratic Radicals of *Il Secolo,* was the candidacy of Consolato operaio loyalist Antonio Maffi. The second was the candidacy of Osvaldo Gnocchi-Viani, running for the Milanese section of the Partito operaio in alliance with the Circolo operaio, the Figli del lavoro, and several workers' leagues and mutual benefit societies, together forming the Unione operaia radicale.

The two candidacies were the outcome of some months of political action and negotiations between the Consolato operaio, on the one hand, and the Circolo operaio and the Partito operaio, on the other. In May 1882, the Circolo operaio had spun off an "electoral section" with the purpose of examining "workers' participation in political [national] and administrative [urban and provincial] elections and cooperating in the formation of a workers' party, independent of all other parties."[26] *La Plebe* extended a warm welcome to the "newborn Partito operaio," and the electoral debate began.[27]

In July, printer Augusto Dante spoke at a crowded meeting of the new party that, he pointed out, was unique in being composed solely of workers. Its goal was to "bring together all Italians in a single camp, in a common struggle, on a single path, demanding their rights and marching toward gradual—but total—change."[28]

The Partito operaio italiano's (P.O.I.) program was published in August 1882. It announced unequivocally that "there can be no political liberty without equal economic liberty." The P.O.I. would give priority to economic over political concerns. Unlike other parties that existed primarily for electoral activity, it was a "permanent party, because exploitation was an everyday occurrence and fighting it required daily struggle."[29]

The P.O.I.'s political program was not completely different from that of the Democratic Radicals and the Republicans: It called for universal suffrage, free secular education, an end of standing armies and state support for religion, communal autonomy, and an end to the regulation of prostitution. The first point of the economic program, however, went beyond that of most of the bourgeois parties in claiming the right to strike. Under the rubric "Action vis-à-vis Capital" it urged the formation of local resistance leagues and their joining in a workers' federation as the "primary goal of our party." *La Plebe*'s next issue reiterated the familiar slogan: "The "workers' emancipation can be achieved only by the workers themselves."[30]

Exactly what the party, the Circolo operaio, and their associated groups would do in the upcoming parliamentary election was still a matter of debate. Some argued for a protest candidate who would be ineligible to sit if elected. From the Consolato operaio came a plea for unity in the effort to send worker deputies to Parliament.[31]

It was only after the Democratic Radicals had agreed to list the Consolato operaio's candidate, Antonio Maffi, that the Unione operaia radicale agreed on the candidacy of Osvaldo Gnocchi-Viani. A special supplement to *La Plebe* editorialized about what a Socialist candidate should be: "the advocate of the little man against the great, of the weak against the rich, . . . a critic of poor people's suffering, a voice that may be smothered, but never corrupted."[32]

The Lega promotrice degli interessi femminile published its own electoral program, which demanded "recognition of the juridical and political personality of woman and her equality with man in civil life." This meant equal education, the right to vote in national and local elections, equal pay in government jobs, and access to all occupations and professions. Feminist Gnocchi-Viani announced in an electoral speech that he supported the programs of the Circolo operaio, the Partito operaio, and the Lega promotrice. He ended with a call for support, in Milan and Italy, for this new social movement.[33]

Although the electorate had tripled with the expansion of suffrage, only 65 percent of those eligible actually voted in the election of October 20, 1882. All four winners in Milan were on the Democratic Radical list, including Antonio Maffi, who came in third, with 10,408 votes. Gnocchi-Viani won only 702 votes. (In the Romagna, Andrea Costa was elected, the first self-proclaimed Socialist to enter Parliament.) *La Plebe* published an open letter to Maffi, stating that it was his historic responsibility to represent the economic and social aspirations of the proletariat. The editorial noted that although socialists were pleased with Maffi's election, they believed that his program would do little for the working classes, and it reprimanded the Democratic Radicals for not understanding that a worker deputy qua worker was simply not enough.[34]

Operaisti versus Democratic Radicals, I: Como, 1882

The struggle between the *operaisti* and the Democratic Radicals for leadership of the Milanese workers became evident during the second congress of the Confederazione operaia lombarda, held in Como in October 1882 (just before the election). There was vigorous debate about abstention from voting. The executive committee had prepared an electoral program for the agenda that closely paralleled that of the Milan Consolato operaio.[35]

The debate at this congress was diffuse, and its electoral program was approved only after long discussion. The 1881–82 executive committee was reelected. Then the group considered its substantive agenda, discussion of (1) pension funds for disabled workers, (2) the juridical recognition (through registration) of workers' societies for mutual benefit and cooperation, (3) the improvement of instruction in the countryside, and (4) the *camere sindicali* (arbitration committees for disputes between capital and labor)—the Depretis government's proposed social legislation. All the proposals were supported, although the last was amended to include *sindacati misti* (mixed unions, i.e., including both workers and representatives of employers). Overall, the tone of this congress was more conservative than that of the first.[36] The *operaista* position made no headway.

With the election over, the Partito operaio, despite its earlier declaration that it would be a permanent party, greatly cut back its activities. For several months, *La Plebe* announced meetings, including one that denounced the Grondona machine works' policy that workers must present a document indicating that they had no criminal record when they applied for jobs as "a humiliating and offensive" procedure. Then, silence. Partito operaio members, however, continued to be mentioned in the newspaper's pages. Costantino Lazzari, for example, was named as secretary of the Lega promotrice degli interessi femminili, which sent a statement of its program to the Chamber of Deputies.[37]

La Plebe was on its last legs, first becoming a monthly and then publishing only occasionally. A debate among Milanese progressives over the appropriate role of the Socialist deputy, Andrea Costa, offered a glimpse of the future. The meeting was chaired by Enrico Bignami, longtime editor of *La Plebe;* Filippo Turati, lawyer and future Socialist, served as secretary. Participants in the debate included Costa, several anarchists, a cluster of *operaisti*—Emil[io] Kerbs, sometime anarchist Ambrogio Galli, Augusto Dante, Giuseppe Croce, Alfredo Casati—Giuseppe De Franceschi, and Paolo Valera.[38]

In June 1883, *La Plebe* welcomed a new arrival to the Milanese newspaper scene—the *Fascio Operaio,* "Voce della Lega dei figli del lavoro"— and in November, it published its last number. The *Fascio Operaio* was the product of a schism in the Figli del lavoro society. One faction, the Democratic Radicals, rejected the other's program that called for daily payments to workers dismissed because they had demanded higher wages or protested wage cuts (in addition to subsidies paid to the unemployed). The second faction became La Lega dei figli del lavoro (League of the Sons of Labor), which sponsored the newspaper and began raising funds for resistance.

In its first issue (July 1, 1883), the founders of the *Fascio Operaio* identified themselves as

> workers in the strictest sense of the word, that is, manual workers . . .
> sons of that immense multitude to whom life is permitted only at the cost
> of being harnessed to everlasting, interminable production.
>   . . . We believe that workers can and must by themselves, and by

themselves only, sustain and protect their own interests, confirming by this fact that great truth: *The workers' emancipation can be achieved only by the workers themselves.*[39]

Operaisti versus Democratic Radicals, II:
Varese, 1883

Two months later, a group of *operaisti* carried their program into combat with Democratic Radical workers and their bourgeois allies in the third congress of the Confederazione operaia lombarda, held in Varese on September 16–17, 1883. The main issue was again the proposed social legislation: bills protecting women and children in the workplace, establishing employer liability for work-related accidents, setting up a pension fund for disabled workers, and requiring legal registration of mutual benefit societies. The congress was the largest yet, with 84 out of the 126 affiliated societies sending delegates.

Giuseppe Croce, the chief spokesperson for the *operaisti,* called for rejecting the government's limitation of workers' autonomy in providing for their own needs. On the issue of pension funds, he immediately suggested that the assembly oppose the government plan. The Democratic Radical majority in the congress, indifferent to workers' autonomy in these matters, voted to amend the proposal by stating a preference for the Confederazione's own plan for regional funds (passed in the 1882 congress).[40]

The next issue was the bill limiting women's and children's work. Most of the delegates supported this bill, and they proposed asking Maffi, the worker deputy, to introduce it in Parliament. But Delegate Bizzozzero, representing the Lega promotrice degli interessi femminili, objected, stating that women should not be equated with children as needing protection: "Women who are themselves mature citizens, free and responsible, are competent to choose their own activity; all legislative tutelage is an unjust limitation of their productive capacity." The Democratic Radical delegates insisted in response that "women are weaker than men and need special laws to ensure that their wage work will not damage their health."[41] They carried the vote, with the provision that enforcement be at the regional level and that women's wages be equal to those of men. The congress also voted to support women's suffrage in national elections (in which men's suffrage had already been expanded) as well as in local elections (in which it had not).

There was greater disagreement on the other bills. A compromise was reached on employer liability for work-related accidents. Croce proposed splitting into two the motion on arbitration boards: one on arbitration and the other demanding the abrogation of Articles 385 through 387 (which criminalized most strikes) of the criminal code. The *operaisti* voted against the first, which nevertheless carried, and the congress voted unanimously for the second. Croce's motion to reject all government intervention in workers' affairs was defeated. He and others argued that the "bourgeois state—representative of capital—cannot also represent workers' interests; hence workers should reject all social legislation projects." The congress agreed to meet soon in special session in Milan, to continue discussion of the Berti project.

Although the *operaisti* had mustered support for their position, none of them was elected to the executive committee of the Confederazione. In this period, *operaismo* was informed by a rough and simplistic consciousness that the central opposition in bourgeois society was that between capitalists and wage earners.[42] This may have been a consciousness of class, but it was held by a small minority of that class.

The *operaista* militants were indeed individually and collectively bold and confrontational, and they sometimes could count on the support of a larger group, as in a tumultuous Milan meeting on expanded suffrage on November 11, 1883. From the start, according to the account in *Fascio operaio* (November 18, 1883), the Democratic Radical deputies Maffi (presiding), Mussi, Marcori, and Bertani, with Carlo Romussi as secretary, were challenged by their audience to be specific about how to expand suffrage further. To the extent that they responded, they were accused of giving evasive answers. Nevertheless, the deputies offered a motion:

> The assembled people of Milan, believing that universal suffrage is a national right [and] that the current administrative system, based on centralized government control, is detrimental to the free development of our country's civil and economic progress;
> Demand the right to the administrative vote for all citizens, regardless of sex, to be exercised in direct voting, and the restoration of local autonomy, following . . . our national traditions.

Costantino Lazzari (one of the Partito operaio's petty bourgeois activists) promptly rose to amend the motion.

There was considerable confusion, as some guessed what he was up to and tried to shout him down, while others insisted that he be heard. His amendment substituted these words for the Radicals' preamble: "Considering that universal and direct participation in the administration of public concerns is a step toward the salvation of the proletariat, which until now has been excluded from participation in any civic function, and is the sole means to achieve a new and better social policy [etc.]." Lazzari criticized the first motion for being full of pretty phrases but empty of any affirmation of the rights of the proletariat.

Deputy Marcori rejected Lazzari's proposal, claiming that it was not an amendment but a new motion. Lazzari countered his objections and shouted in closing, "Proletarians, the amendment concerns our cause; it is in your hands; save it!" The ever-vigilant police stepped in, demanding that there be no further insults of the deputies. The first motion was passed, and the meeting adjourned in a shower of hisses.

Fascio operaio editorialized that the deputies always spoke in abstractions, whereas workers spoke concretely of their interests. They perceived "a great abyss dividing the official democracy, acting in more or less good faith, and the proletarians, concerned with equality, well-being and justice."

Discussion of the proposed Berti project continued in the pages of *Fascio Operaio* in every December issue. In January 1884, plans were announced for a workers' meeting at the end of the month. On January 27, Lazzari made a motion calling for liberty in the broadest sense, rejecting the idea that "any political institution has either the capacity or the competence to pass laws favorable to workers." It was passed by a large majority.[43] The Milanese *operaisti* thus had built a popular base for their opposition to the Berti bills.

Operaisti versus Democratic Radicals, III: Milan, 1884

A special congress of the Confederazione operaia lombarda met in Milan on February 2 and 3, 1884. Eighty-nine associations, thirty-nine of them Milanese, sent delegates.[44] The agenda, as agreed at the previous congress, consisted wholly of the Berti social legislation project: legal recognition of workers' associations, pension funds, compensation for work-related disability, and arbitration of disputes between workers and employers.

Although the president of the congress discussed in his opening statement both the proposed laws and Maffi's "counterproject," Emilio Kerbs insisted that the two—the Berti project and Maffi's alternative—be kept separate and that discussion of the first precede that of the second. Most speakers agreed that the government package was illiberal and ought to be rejected; nevertheless, there were differences of opinion over the extent of that rejection. The *operaista* position of total rejection was not supported by the majority, but overall this congress was more favorable to their position than was the previous one.

The first issue—legal recognition of mutual benefit societies—pitted two motions, both of which rejected the government proposal, against each other. The first called for the grant of "juridical personality" to all workers' associations for the moral or economic improvement or emancipation of the working classes, by means of the simple registration of each society before a municipal magistrate. The second, introduced by Costantino Lazzari, rejected the proposed law or any other type of government registration. It was defeated by fifty-six votes to nine.

There was general agreement on the second issue considered, the right to strike. Vincenzo Corneo, head of the compositors' league, formulated a motion that affirmed the legality of resistance leagues, "the best hope for workers," and rejected the government proposal as "contrary to principles of equity and justice." The motion invited all workers' societies to join a national movement against the project and called for abrogation of Articles 385 through 387 of the criminal code. It was unanimously approved.

A more contentious debate followed the third report, read by Luigi Grando, a Como silk weaver and supporter of the Democratic Radical position on the arbitration of disputes between capital and labor. He pointed to the success of the standing arbitration committee in the Como area, which had successfully settled many disputes. Corneo complimented the reporter but opposed any permanent board, calling instead for *ad hoc* hearing committees. Lazzari, noting that arbitration was possible under the existing law, also rejected the Grando proposal. The final vote was a tie, in which the *operaisti* were joined by several Mazzinians. As delegates left the meeting hall, a member of the Pavia Figli del lavoro objected to government interference in relations between capital and labor and suggested instead that organization and strikes were necessary means for workers' defense.

The defeated Grando later published a perceptive analysis of the split as involving two factions of one party, "one placing more faith in the present social system" and the other "distrustful." He believed that the balance was clearly leaning toward the latter group.[45] The *Fascio Operaio* announced in an exuberant article entitled "Our Impressions" that the congress was a "splendid and solemn victory for liberty."[46]

In March 1884, the newspaper once again floated the idea of resurrecting the Partito operaio. Shortly thereafter, however, it ceased publication temporarily, owing to financial difficulties. Later that year, in August, the Milan Lega dei figli del lavoro left the Consolato operaio and joined some of the provincial societies (the *leghe* of Busto Arsizio, Sacconago, Legnano, and Gallarate) that supported its *operaista* position in a new federation, the Federazione regionale dell' alta Italia del partito operaio italiano.[47]

A hatters' strike in Monza in September 1884 was led by the loca Lega dei figli del lavoro, which exhausted its treasury in paying subsidies to the striking workers.[48] Indeèd, Briguglio reports, eight hundred copies of the *Fascio Operaio* were regularly sold in Monza, more than were distributed in Milan itself.[49]

Operaisti versus Democratic Radicals, IV:
Brescia, January 1885

Operaista debate with the Democratic Radicals resumed at the two Confederazione operaia lombarda congresses that met in 1885.

When the *Fascio operaio* reappeared in September 1884, it fumed polemically at the Consolato worker deputy, Maffi, and at *La Rivista Operaia,* the official newspaper of the Confederazione. The *operaisti* opposed both the employers' paternalism and workers' reformism.[50]

The Brescia congress was postponed from the fall of 1884 to January 1885, because of cholera in the region.[51] The report on pension funds (calling once again for regional organizations) was vigorously opposed by the *operaisti* and rejected by the majority of the delegates.

Arbitration, the question that had divided the fourth congress in its closing hours, was discussed again in a report submitted *in absentia* by Grando, who objected to Lazzari's vision of disputes as collective, involving groups of workers, whereas the problem in Como was often isolated workers, such as hand-loom weavers. Employers were unwilling to convoke *ad hoc* boards for individual disputes. Even in the case of groups, Grando noted, a standing independent mediation board could be more effective in pressuring employers to accept its decisions. In response, an *operaista* spokesperson (writing in the *Fascio operaio* on January 3–4, 1885) attacked Grando for his lack of class consciousness necessary to arrive at the correct position (theirs, of course) on this question. The congress approved a motion proposed by Lazzari and a colleague, which admitted the usefulness of both *ad hoc* and standing arbitration boards for different situations but insisted that workers be free to accept or reject intervention and that government not be involved in the process.

An *operaista* majority was by then evident. It rejected a proposal on pension funds, holding that workers could set up such funds themselves, without government oversight. Bourgeois speakers and Democratic Radical workers were defeated on issue after issue, but they continued to fight. At one point an exasperated Lazzari attacked the "lawyers" who were wasting the time of the congress, at which only workers should have the right to speak. In fact Manacorda points out, the two factions were close on issues such as expansion of schooling and suffrage. The final question was one on which there existed deep disagreement, however: What form should the workers' organizations take? The *operaisti* insisted on resistance leagues, and the Democratic Radicals supported cooperatives. The majority supported the *operaisti*, hailing resistance as the only means to improve the workers' conditions and also to build solidarity between urban and rural workers. They directed the executive committee to plan and convene a special congress in which all workers' organizations would unite in a workers' party. Four *operaisti* (including Giuseppe Croce) were elected to the executive committee of the Confederazione, along with three Democratic Radicals.

The Founding Congress of the Partito Operaio: Milan, April–May 1885

The societies that cooperated in planning the founding congress of the Partito operaio were the five original Leghe dei figli del lavoro (Milan, Busto Arsizio, Sacconago, Legnano, and Gallarate), the Lega of Monza, the lithographers' section of the Milanese league, and two small Milanese workers' associations, those of the upholsterers and the wine shop workers.[52]

Milanese anarchists, headed by Ambrogio Galli, again attempted to resuscitate the Italian branch (anarchist) of the International. They met secretly in March 1885 with other anarchists, mostly from Emilia and Romagna, in Forli. The faction that stood for propagandizing workers for revolutionary action prevailed; the anarchists elected to stay aloof from both the Po valley agrarian strike movement and the socialist agitation led by Deputy Andrea Costa. Instead, they contacted the Milanese leaders of the forthcoming founding congress of the Partito operaio.[53] Gnocchi-Viani, persuaded that a party unifying many tendencies was essential, invited the anarchist colleagues to the April 12–May 3, 1885, founding congress. The anarchists and *operaisti* both shared a distrust of Costa and the Romagnol Socialists. The anarchists, however, saw the *operaisti* as "too exclusive" and "aristocratic." The two groups agreed on the fundamental importance of organization for resistance, but there was a basic difference. The anarchists sought, in vain, to "reinsert protest into the evolutionary process," whereas Gnocchi-Viani maintained that the " 'regime of force' lacked any historical function."[54]

The primary task of the congress was to approve the constitution of the Partito operaio. Worker delegates from Busto and Monza wished to strengthen the proposed first article to make even more explicit the "*purely economic* character of the party."[55] The assembly approved an amendment to the first article formulated by Costantino Lazzari, as follows: "The Partito operaio italiano (P.O.I.) is established to defend workers by means of organizing, craft by craft, . . . to defend one single right, the *right to existence*. . . . The P.O.I. has absolutely no religious or political affiliation."[56] The constitution thus gave itself an exclusively economic function. Occupational and industrial organizations were to be the party's constituent and voting units, resistance its chief activity. Members were to pay dues of 10 centes-

imi per month to build up funds to be paid out as subsidies of 1 lira per
day to striking workers.

The next Partito operaio congress was to be held in Mantua, in conjunction with the sixth congress of the Confederazione operaia lombarda,
at which time it was expected that all worker groups would unite in one
party. Mantua was a city in which both socialism and anarchism had roots,
and it was also the center of the contemporary agrarian labor movement in
the Po valley. The congress voted its solidarity with, and support for, the
strikes of the *braccianti* (day laborers), and it expressed its hope that "the
triumph of our common cause may speed the emancipation of workers from
the tyranny of capitalists and landowners." The *Fascio Operaio*'s publication of this motion led to its first censorship and the trial of the editor
responsible, Giuseppe Croce, "for inciting strikes and class hatred." Croce
was accused of being a "false worker." He replied eloquently in his defense: "I work honestly for a living for my family; because I don't spend
my free time lazily, or in revelry, I am called a false worker, a wicked
instigator."[57] He was nevertheless convicted, sentenced to twenty days in
prison, and fined 51 lire.

The activities of the Partito operaio in 1885 and 1886 were not limited
to urban industrial workers. Its Milanese militants made special efforts in
the agrarian areas of the Alto Milanese. This was a movement, according
to Felice Anzi, who was then a young member, "inspired by the needs of
a class that wanted to take its place in society."[58]

There was a rash of agrarian strikes in July and August 1885 that demanded (1) the abolition of tenants' labor obligations (*corvées*), (2) payment for tenants' work on their landlords' property, and (3) the right to sell
certain farm products in the free market. The strikes, according to Maria
Grazia Meriggi, were linked to the local Figli del lavoro (sometimes through
veterinarians who were members) and to Milanese workers who helped organize them. Several *operaisti*—Croce, Casati, Lazzari, Kerbs, Beretta,
Dante, Fantuzzi, and Galli—were reported by the police to be present at
the organizing meetings. A poster found during a strike at Basiano called
for concrete changes in the tenants' contracts and threatened those who did
not strike or who otherwise undermined solidarity. Meriggi finds the manifesto both "modern" and interwoven with backward-looking appeals to paternalistic values.[59]

The local Figli del lavoro were involved in a renewal of the hatters'
strike movement in Monza in May and June 1885. One of the strikers demands was the employers' recognition of the Lega dei figli del lavoro.
Other demands were limits on dismissal, weekly instead of biweekly paydays, modification of the division of labor, and elimination of the caution
withheld from workers' wages.[60]

In this period, then, the Milanese Lega dei figli del lavoro and its affiliates participated in some of the most visible workers' economic actions in the region.

Operaisti versus Democratic Radicals, V: Mantua, December 6–7, 1885

Between the April–May Partito operaio congress and the "united" workers' congress in December, Giuseppe Croce prepared the way for unification. Most of the bourgeois-dominated associations that were affiliated with the Confederazione boycotted the congress, once the Consolato operaio's last-minute request that the congress be postponed was rejected. However, as one P.O.I. supporter pointed out, the majority of the societies—80 out of 132 present—were not associated with the party. Augusto Dante moved, and the delegates approved, a declaration that "the congress deplores the fact that some societies of the Confederazione lombarda are not attending the Mantua congress. Attributing this abstention not to the will of workers but to bourgeois influence, the congress reaffirms the principle of solidarity." Indeed, the congress delegates were nearly unanimous in agreeing with the *operaisti* on all issues.

Croce read the constitution of the P.O.I., and Lazzari moved that the societies attending "accept the joining of the two associations, based on the constitution of the P.O.I.," and discuss it article by article. All the delegates committed themselves to instill, in their own associations the principles of resistance and emancipation.

The statement that the party was independent and separate from all others was amended to add the phrase "and will participate in the struggles of public life as a distinct class working for its own emancipation." [61] The assembly then upheld the *operaista* position regarding arbitration, prison labor, unemployment, work-related injuries, strikes, the legal recognition of worker societies, and cooperatives.

The congress's central problem, agrarian labor, was addressed the next day. The theater in which the congress was meeting was crowded with peasants and *braccianti*. Several men, speaking in dialect, declared themselves to be the most oppressed class, "unable to educate or organize ourselves. Our bosses and authorities react furiously to our every attempt to emancipate ourselves, and the police dissolve our associations." Another added, "Meanwhile, unemployment spreads, and a *bracciante* who has work makes at most 300 lire yearly." [62] *"Mi son contadin ca lavora sempar"* "I am a peasant who works all the time," another shouted in dialect, "but I cannot make a living!" [63]

Various *operaisti* then spoke, urging the peasants to organize, as the only way for workers, urban or rural, to emancipate themselves. The assembly overflowed with declarations of solidarity and brotherhood. Lazzari proposed a motion demanding "the reclamation of the soil as the common property of those who work it." The motion was passed unanimously, to shouts of *"Viva i contadini!"* [64]

The congress turned then to questions about workers and parliamentary elections. Opinions varied, from that of Brando, who strictly opposed nominating a worker candidate for deputy, to those of Croce and Dante, who believed that such a deputy could help the workers in economic matters. The discussion ended with the passage of another motion by Lazzari: "That the Partito operaio enter the political struggle in ways that promote the emancipation of its class and that individual sections retain the freedom to participate in national elections, in response to opportunities based on local conditions." [65]

The congress, as Manacorda summed it up, was a "true triumph." The Partito operaio had laboriously gained leadership of the workers' and peasants' movement in Lombardy, and its distinctive workerist ideology prevailed, at least for the time being. [66] In the coming months, the network of associated societies grew more dense in Milan and throughout Lombardy. New Leghe miste (mixed leagues, i.e., not made up of single occupations or industries) dei figli del lavoro were founded. The movement spread to Piedmont and Ligury, put down roots in Emilia and Romagna, Tuscany (Livorno), and even in the Mezzogiorno, in Naples. In many places, socialists abandoned their debates with anarchists and accepted the P.O.I. as the vehicle for class organization and propaganda.

A year-end summary report for 1885 on the *spirito pubblico* in Milan, from the police chief to the prefect, remarked on the continuing growth of the "Partito operaio italiano, its sections the Figli del lavoro, and its resistance societies." The police chief claimed that the ideas and practice of the P.O.I. had already made an impact on the formation of leagues and strikes, and he predicted that its influence would spread with the workers' greater mobility. The economic agenda that the party planned might also lead to events that would be impossible to resolve peacefully, he warned. [67]

The Milan Festa del Lavoro

Similar forebodings informed the police response to the large celebration planned for March 27, 1886, by the Milanese Lega mista dei figli del lavoro. In the afterglow of the Mantua congress and the rapid growth of the P.O.I. sections (including its first women's section), the league invited delegates from throughout Italy, including organizations that had boycotted the congress. [68] Some eight hundred were expected; there would be a parade and a banquet so large that it would be held in the tram station. The *Fascio Operaio* reported the plan to celebrate the new Partito operaio with a short history and discussion of its customary mottoes: Resistance, workers' emancipation, and in union is our strength. The week before the *festa* (March 20–21, 1886), the *Fascio Operaio* published "Il Canto dei lavoratori: L'Inno del partito operaio italiano" ("The Song of the Workers: The P.O.I. Anthem"), with words composed by Filippo Turato. For its chorus, Turati borrowed the motto of the Canuts of Lyons: "Live working or die fighting!"

The same issue of the *Fascio Operaio* reported in its weekly column "Cose di città" (Around Town) that the price of bread had gone up 2 centesimi and "as though that were not enough, the *dazio* will now be imposed even on the small amount of bread that workers carry with them as they enter the city daily to work. It's shameful . . . organized workers are preparing to force the city to cancel this wrongheaded policy."

Three days before the projected *festa* the police chief prohibited the celebration as a public gathering. *La Lombardia* expressed the hope that the socialists would act sensibly and that the police, "who are not obligated to have good sense, will learn from them." [69] The committee quickly changed the *festa* to a private gathering, with admission by ticket only. The police made extensive preparations to see that nobody stepped beyond the limits they had drawn. Plainclothes agents attended, and a patrol stood at the ready outside. Delegates from workers' societies in Lombardy and Piedmont took part in the uneventful celebration. The Garibaldian hymn was sung. The banner was dedicated; hanging from a mace topped by a sickle and hammer, it portrayed a bare-chested worker turning from his task to hail the rising sun: the emancipation of the worker. It was embroidered with the quintessential *operaista* motto: "The workers' emancipation can be achieved only by the workers themselves." Combative and conciliatory speeches were delivered, including one that referred to the injustice of the city plan for imposing the *dazio* on the workers' bread. [70]

The day before the *festa*, the *Fascio Operaio* published an intemperate attack on the authorities under the headline, "Declaration of War," reviling the "organized bourgeoisie." [71] It was censored, and some of the copies were seized by the police. According to one report, however, the printers hid most of the press run and distributed the copies clandestinely. The following week the newspaper reprinted parts of the offending article, and it also announced extra coverage for the upcoming parliamentary elections. Once again, the "Around Town" column commented on the *dazio del pane*, warning that it was gathering opposition at the gates of the city. [72]

The Revolt of the Micca

In the last days of March, Mayor Negri, along with his *assessore* in charge of food supply, Gaetano Vimercati, decided to enforce the *dazio* collection on the excess weight in bread that workers were carrying into the city. Workers could either pay the duty or bring in less and buy the rest—at a higher price—inside the walls.[73]

"Subversive" flyers, signed by "many workers," decried the "general poverty, lack of work, and higher cost of necessities" and attacked the municipal administration for its "perfidious" act, ordering the *dazio* guards to collect duty on the second *micca*. They closed with a call for a demonstration in the piazza del Duomo, Thursday, April 1, at 9 P.M.[74]

The first effort to collect the duty came on the morning of April 1 at the porta Tenaglia; it particularly affected building-trades workers who lived in the external *circondario*. A crowd gathered around the kiosk of the *dazio* guards, determined not to pay.[75] One stone mason with two *micche* was stopped and held by a guard. He threw one of his loaves across the barrier, where someone in the crowd caught it. He shouted, "Help! Free me! They've grabbed me!" The crowd responded with shouts of "Let him go! Let him go!" As the guards locked him in their kiosk, they were pelted with stones and loaves by angry bystanders. Five other workers were seized and locked up, as well as one Domenico Leonardi, identified in police reports as a singer and the chief agitator.

City police and municipal assessor Emmanuele Greppi were called to the scene. The guards asserted that the protest had been prepared in advance. Greppi and the police assigned extra agents, *carabinieri,* and two companies of *alpini* (an army unit) to guard the city hall in the piazza della Scala and be ready for the anticipated demonstration in the adjoining cathedral square. In the evening, curious bourgeois crowded the square. A stream of workers came in the gates to the north and proceeded toward the city's center, where, *La Lombardia* remarked, "they [usually] were rarely seen." The workers shouted, "Bread! Bread! Down with the municipal government!" Bystanders began to move away, and merchants shuttered their shops. Rocks were hurled at streetlights and unprotected windows. When they saw the police and troops moving into the piazza del Duomo, however, the demonstrators dispersed, abandoning the square to the troops.[76]

One cluster of *alpini, carabinieri,* and urban police proceeded south on the via Torino, shouting to all to go home. Those who persisted were arrested, eighty-five in all. Others attacked isolated police with rocks, freeing some prisoners. *La Lombardia,* although it opposed the city effort to enforce collection of the *dazio* and expressed the hope, on April 2, that the previous day's demonstration would enlighten the authorities, claimed that "true workers" would see that violence could only hurt their cause.

On the second day, there was a repetition of the troubles at the *dazio* kiosk, and in the evening, there were more demonstrations and fights. Mayor Negri issued a declaration, dated April 2, 1886, pointing out that the issue was not a new tax but a simple equalization of the burden between the internal and the external *circondari*. He insisted that the city had the good of all classes at heart and that its policy was equal treatment. Unimpressed, demonstrators again filled the center of the city that evening. There were about forty arrests.

On April 3, the municipal council met in a special session to consider what action should be taken. Mayor Negri exploded in outrage:

> The deplorable disorders just past are in large part the fruit of the incredibly violent language to which we have been subjected . . . they forget all we have done for the public good in the short time I have directed the city administration . . . when a man has dedicated his life to his country . . . it hurts, it wounds the heart, to be falsely accused.[77]

The council, perhaps also surprised by the vehemence of the public reaction to Negri's measure, formally requested him and the ruling cabinet to permit free entry for workers' meals on the job of up to 800 grams of bread. Posters rescinding the policy were immediately printed and posted.[78]

The headline of the *Fascio operaio* was "La piazza ha vinto. [The crowd has won.] The bourgeois cry 'vandalism' when there is damage to property but are silent when workers' bread is taxed."[79] The edition was promptly confiscated, and the newspaper offices were closed. Croce was again arrested and eventually sentenced to two months in prison and a 100 lire fine.

The lists of the men arrested and held (many of them were then soon released) show that all were employed, most in working class jobs.[80] Stonemasons entering the center of the city through the Porta Tenaglia to work on the New Park and the Castello sparked the first day's demonstrations; indeed they also were overrepresented among the arrestees in the three days of demonstrations that followed. The mutual benefit society of stonemasons petitioned the city to rescind its policy, but only on April 3, the same day that the city council met.[81] In that evening's demonstrations in the city center, the occupational distribution of arrestees, who included Felice Anzi, then a young Partito operaio member, was a more typical cross section of city residents.

Analysis of a consolidated list of persons arrested in this four-day "revolt" shows that the arrestees were very young. All male, they ranged in age from 12 to 65, but 44.4 percent were between the ages of 16 and 20. (Of the male labor force, only 12.4 percent were in that age range.) Of the 100 arrestees whose home addresses were given, about 36 percent lived in the external *circondario,* quite close to the proportion of the city's population reported in the 1881 census as residing there. Those arrestees whose birthplace was noted in the police lists were born in Milan in exactly the same proportion as was the male labor force as a whole. The arrested construction workers, however, were more likely to be urban born than were the construction workers as a whole. Demographically, then, the arrestees resembled the male working population of Milan, except for their youth.

The building-trades workers who took part in the events of the *micca* were part of a broad-based protest against a city policy that penalized in a petty way the residents of the external *circondario* while spending lavishly on bourgeois urban renewal in the center city. The pattern of arrests does not suggest collusion among the stonemasons or a coalition of stonemasons with other workers or with the Partito operaio. Indeed, the arrestees claimed not to be members of any organization, and both the Consolato operaio and the Milan section of the P.O.I. claimed no knowledge of the demonstrators.

The *Fascio Operaio* wrote (on April 3–4, 1886) that the more formal April 3, 1886, demonstration was an outburst of discontent against the city policy but that it lacked "a line on how to proceed." The democratic newspapers were generally sympathetic to the demonstrators' grievances. Both the *Fascio Operaio*'s use of the word *popolo* rather than *worker* and the broad cross section of persons involved, many of whom lived and worked in the neighborhoods of the demonstrations (downtown) rather than in the external *circondario* primarily affected by the new policy, suggest a community rather than an occupational or class base for the demonstration.

The demonstrations grew out of the political activity of the workers' organizations but went beyond rhetoric to confrontation. A larger population was involved than the leadership that originally questioned the city's policy. The escalation of tactics was due at least in part to the reaction of the police to the protest. In the short run, the crowd made its point—that the policy was unjust and arbitrary—and saw its wishes heeded.

In the aftermath, a group including the Consolato operaio, mutual benefit societies, and the Partito operaio protested the city's policy (which by then had been rescinded) and the police repression.[82]

The Parliamentary Elections and Partito Operaio Conflict with the Democratic Radicals

The Partito operaio decided to run its own candidates in the May parliamentary elections. "In opposition to the privileged coalitions of capitalists and to the division of society into classes, we proclaim the rights of man," the electoral program of its Milan section announced. "In opposition to bureaucracy, we preach free associations, organized and federated in the public interest, and in opposition to the brutal domination of capital, we demand the rights of labor."[83] Its platform put forward a long list of demands, including an 8-hour day, a minimum wage, equal pay for equal work for both sexes, the abolition of prison labor, the right to strike and to "coalition," universal suffrage, the abolition of political police, and the right to divorce. It was, as Guido Cervo sees it, "a purely agitational and propagandistic program." The *operaisti* were not interested in sending representatives to Parliament in order to fight for workers' interests but, rather, to act as "privileged propagandists," enjoying parliamentary immunity.[84]

The radicals again nominated Antonio Maffi. A Consolato operaio leaflet proclaimed that his term as deputy had demonstrated "how a worker can, along with other citizens, participate in governing. The election of a worker ought to have the character not of opposition to other classes but of appreciation of one's own class." It supported the "party of democracy that stands for government of the people."[85]

The *Fascio Operaio* replied in an article entitled "The Two Democracies." One was "official" democracy, and the other was "*true* democracy, which embraces popular aspirations. The P.O.I. is its new symbol. . . . It is not a special interest. . . . [T]he interest of the majority is the interest of civil society and progress. . . . [I]n this case the workers' electoral battle is also a class struggle."[86]

The *Fascio Operaio* found the election returns encouraging, even though the Democratic Radicals won as a whole. The Partito operaio candidates had received 5,451 votes in Milan Province. Winning was not the point, the editorial pointed out; making propaganda was.[87] The Democratic Radicals and especially Felice Cavallotti, however, believed that their candidates had been hurt in some constituencies by the Partito operaio candidates.

On May 30, 1886, Cavallotti asked in the Chamber of Deputies whether "well-financed anarchist agitation was related to the persecutions undergone by the Italian Radical party." He accused the government of financially assisting the Partito operaio in order to run its candidates against the radicals. The *Fascio Operaio* attacked the Democratic Radicals for their undocumented accusations. Under the headline "La democrazia vile," they bitterly defended themselves. There were huge press runs of the ensuing issues, which protested Cavallotti's "calumnious and gratuitous insinuations." Printers Augusto Dante and Flaminio Fantuzzi, both Partito operaio members, were fired by Sonzogno, the publisher of Cavallotti's mouthpiece, *Il Secolo*. The party accused Cavallotti of defamation, but at the same time its militants tried to change his mind by showing him their financial accounts. Cavallotti retreated somewhat from his accusations.[88]

On June 23, 1886, by order of the prefect, the Partito operaio was dissolved; its books and banners were confiscated; and the *Fascio Operaio* was closed. Milanese *operaista* leaders were arrested, accused of conspiracy and inciting to civil conflict. Cavallotti then went further, insinuating in *Il Secolo* that Prime Minister Agostino Depretis was, in this manner, ridding himself of those who had served him well in the election campaign.

Lawyers Filippo Turati and Luigi Majno, interpreting the P.O.I.'s plight as
due to the lack of political rights, volunteered to defend the *operaisti*.

Conclusion

By 1885, the Partito operaio, deliberately eschewing electoral engagement,
had become an effective institution for promoting workplace struggle and
spreading its workerist approach to politics. It was founded and led by
several petty bourgeois intellectuals, such as Bignami, Gnocchi-Viani, and
Lazzari, who were allied with a band of hardy workers like Croce, Dante,
Kerbs, Galli, Casati, Corneo, Fantuzzi, Brando, and Beretta. The printers
among them—Dante, Fantuzzi, Galli, and Corneo—had been active in their
own union and its struggles and had reached out to other workers to talk
about the value of organization. Structural change in printing—particularly
its growth in scale and the employers' efforts to cut labor—determined the
timing of their activism. Their individual characteristics, such as skill, lit-
eracy, and Milanese birth, prepared them to welcome ideas and mobilize
others for action. Casati and Croce, sons of manual workers and manual
workers themselves, were self-taught independent thinkers who were, with
Gnocchi-Viani and Lazzari, the strategists of *operaismo,* with its insistence
on workplace struggle rather than electoral politics. They rejected patron–
client relations, whether in the local arena with the Consolato operaio and
the Democratic Radicals or in the national arena through government-granted
social legislation. The party chose to build exclusively workers' institutions
to extend its influence. P.O.I. activists, many of them artisans or skilled
workers who themselves had fought proletarianization, acted according to a
strong class outlook and simple ideology. They carried their program to all
workers, including agricultural proletarians, and they enjoyed modest success in this effort.

The party's weakness lay in the very ambition of its challenge. Although it focused on the workplace and led workers to fight for their rights
there, its resources were limited. In particular, its rejection of alliances and
bourgeois members limited it financially, politically, and strategically. A
party with a program so challenging needed supporters and allies, at least
at the level of protection of rights. The *operaisti* tried to act alone. Even
when they were fighting about wages (as shown in Chapters 6 and 7), the
chances for their success were closely linked to the resources that a group
of workers could bring to the strike. They were less able to mobilize people
and resources for their political programs, and their failure to develop a
strategy for political engagement was critical. Although their fiery rhetoric
offered little political danger to conservatives, it did provide an excuse for
repression.

The Partito operaio was not a natural outgrowth of economic structural change in Milan and its region; rather, it was constructed by workers who shared some experiences and developed, along with some middle-class progressive theorists, original ideas and practices. The timing of the party's birth was shaped politically by the 1881 expansion of suffrage that enabled some workers to vote. The party that these men created as an instrument for workplace action was influenced by the leadership struggle between the *operaisti* and the Democratic Radicals. The issues of party independence and the relationship of workplace and electoral politics continued to be central to workers' politics long after the Partito operaio was gone.

9

The Formation and Repression of a Socialist Party, 1886–1894

La Tipografia, the printers' league newspaper, called Milan "the city of grand initiatives" when it discussed (in 1889) the efforts then under way to establish the Chamber of Labor.[1] Indeed, Milan was the home of the worker and intellectual founders of the Partito operaio italiano and the initiators of the Camera del lavoro movement in Italy. The formation of the Partito dei lavoratori italiani (as the Socialist party was called in its first year) also grew out of the Milanese-based politics of some of the same workers and intellectuals, with and against other Socialists, Democratic Radicals, Republicans, and anarchists from Milan and other regions. The process began when it became clear that the Partito operaio, repressed in 1886, was unable either to resume its former level of activity or to move forward in new directions. It climaxed with the creation of the new party in 1892. In 1894, the party encountered new government repression and began to embrace the reformist ideology that characterized one of its major tendencies for decades.

This chapter looks at this great initiative, another aspect in Italian experience of Katznelson's fourth level of class, mobilization and collective action. It examines the failure of the Partito operaio and the conditions of the new party's birth, the circumstances that led to new government repression, and the early signs of what became the party's characteristic reformist policy. The differentiation of function between institutions like the Chamber of Labor and the Socialist party was partly the result of inclinations of those who founded these organizations and partly an outcome of interaction in the Milanese and Italian political arena among *operaisti*, anarchists, Democratic Radicals, and Socialists. It was a division of labor that made possible the emergence of a Socialist party in a country whose workers were highly heterogeneous in their propensity and capacity for economic or political struggle.

Filippo Turati, Milanese Intellectuals, and "Democracy"

Turati's willingness to undertake the legal defense of the arrested *operaisti* did not entail any sudden change of heart. In May 1886, before the election, he had published an article in which he reflected on the concept of democracy: "A democracy that turns inward and rejects new struggles born of the people is contradictory. . . . True democracy . . . modern democracy, does not fear the people asserting themselves."[2] In keeping with these principles, Turati defended the *operaisti* in letter to Cavallotti, dated May 30, 1886.

In June, Turati was the first to resign when Cavallotti's Associazione democratica refused a public discussion of the disagreement between the *operaisti* and the deputy. In a letter to *L'Italia* announcing his resignation, Turati proposed a new Circolo di studi sociali, in which a more modern concept of democracy could be constructed. In it nonworker activists, "conscious of the importance and moral value of the new independent popular movements—like the Partito operaio"—could support and collaborate with them.[3]

Two days later, after a divisive debate of the Associazione democratica in which Cavallotti's supporters prevailed, another group of members, including Anna Maria Mozzoni, resigned. A *dichiarazione d'onore* (declaration of honor), which denied Cavallotti's charges against the Partito operaio, was published. The declaration, which also pledged solidarity with the *operaisti,* was signed by Socialist intellectuals and union leaders, in addition to the resigned members of the Associazione democratica.[4]

The *Operaisti* Regroup

Within a month of their party's dissolution, those Milanese *operaista* activists who were not in prison organized the Unione operaia mutua istruttiva (Worker's mutual educational union), which Anzi later labeled the "legal version of the Lega dei figli del lavoro." They sought to bring the party's scattered and demoralized members back together and rebuild links to branches elsewhere.[5] The imprisoned leaders were released in September, and the *Fascio Operaio* resumed publication in October. Its first editorial was indignant and unbowed: "Our great hope is *the redemption of all oppressed people*. . . . Let us join together, our hope is our union. Another article denied that the Partito operaio was "a party of troublemakers and conspirators" (as accused) and once again took up the cudgel against Cavallotti. The prefectoral decree forbade the reconstitution of the party or the use of its name or program. But, the article demanded rhetorically, "has it disappeared? Can the need of the oppressed for freedom disappear; the need for improvement disappear for proletarians? No."[6] This feisty reprise reflected the determination of the *operaisti,* undiminished despite the imprisonment and upcoming trial of their leaders.

The accused *operaisti* were dignified and calm at their trial in January 1887. They stuck to their principles, claiming that they had acted within their rights and according to their sense of duty. The legal defense, led by Turati, described the party as a "coalition of workers' forces promoting freedom of contract between capital and labor."[7] The accused *operaisti* received light sentences of two to three months' imprisonment for incitement to strike, except for Casati, who was convicted of inciting to civil conflict; his prison term was to be nine months. (He was pardoned and released in June 1887.) The *Fascio Operaio* drew a simple lesson from the trial: "One does not reason with the bourgeoisie because one cannot."[8]

An article in the *Fascio Operaio* (attributed by Cervo to Casati) expressed gratitude for the support offered immediately after the dissolution of the party. It noted that Costa's position as a member of the Chamber of Deputies made more effective his defense of the *operaisti.* The party could benefit from elected representatives who would uphold their cause in the privileged arena of Parliament, the anonymous writer wrote: "The last elections demonstrated to us that if we do not protect our rights, we are disarmed . . .; only a tenth of our comrades have the weapon of the vote. Don't let it happen again; let us seize the electoral weapon." It had been a rude awakening for the *operaisti.* They had naively believed that they could safely nurture their movement. In the crude reality of the Italian political system, however, that was impossible.[9] In Milan, the Circolo socialista that brought together "evolutionary socialists, *operaisti,* and anarchists in a stormy household" was founded in early 1887.[10] It shared headquarters with the Unione mutua operaia istruttiva.

The Pavia (Third) Congress of the Partito Operaio Italiano

Determined—although still weakened from the repression—the *operaisti* met in Pavia on September 18 and 19, 1887. The party's treasury was depleted, and some of its members were so poor that they had to travel on foot from Milan (where any such meeting was forbidden) to nearby Pavia. There, Casati's argument that experience had demonstrated the danger of too much centralization inspired the transfer of the party headquarters to Alessandria, in Piedmont. The *Fascio Operaio* would continue to be published in Milan.

The anarchists attempted to force a vote on whether the Partito operaio ought to be socialist. (By this they meant anarchico–revolutionary–antilegalitarian–abstentionist socialist, their own orientation). The Milanese *operaisti,* supported by Deputy Andrea Costa (who was attending in hopes of uniting his party and the Partito operaio), opposed any such revision of their program, which would lead, they believed, to divisions in the membership. For them, the party was "the struggle of a whole class, which does not have a program for governing but one for improvement and emancipation." The anarchist proposal was rejected,[11] and so too was Costa's motion to broaden the criteria for membership in the P.O.I. to include "citizens who believe in and accept its program." On the question of participation in parliamentary politics, the *operaisti* occupied the center ground between the anarchists, who preached absolute abstention, and Costa, who supported full electoral participation. A motion by Lazzari affirmed that the Partito operaio italiano, having no political program, was not a party, and thus its sections were free to accept or reject electoral politics as long as they preserved a class base for political, economic, or moral participation in public life.[12] The congress closed with a discussion of ways to improve the financial base of the *Fascio Operaio,* the designation of Bologna for the next P.O.I. congress, and a declaration of solidarity with the striking construction workers of Milan.

The Milanese anarchists did not abandon their efforts when they were defeated in the Pavia congress but continued to fight in the Circolo socialista. In February 1888, the P.O.I.'s central committee issued a warning to its sections against "efforts to adulterate the true nature and the life itself of the Partito operaio" by certain groups that wish to make the party a "chapel [sect] of doctrinaire socialists." In July, it evicted the Circolo socialista from their common quarters. At the same time, party militants continued to disagree publicly with Cavallotti and the Democratic Radicals, not only in words, but also in action. On May 25, 1888, for example, there was a fight between the two groups at a lecture by Cavallotti in the Milan Public Gardens.[13]

Alfredo Casati reiterated the majority *operaista* position in his unsigned essay "Fra due fuocchi" (Between two fires) published in the *Fascio Operaio* on July 14–15, 1888. The two fires were those of bourgeois democracy and anarchist socialism. Casati reserved his most biting and angry words, however, for the anarchists, whom he called "new priests" who were crying excommunication at the *operaisti,* "workers who have organized a class party in order to wage the struggle against capital." The Partito operaio, Casati continued, had for seven years fought alongside workers and peasants, to "raise their consciousness of their interests, to improve their miserable condition, and to organize and make them into socialists, not *sentimentalists* [emphasis in original]." History will show, he insisted, who had worked more for their cause, who were the more honest and sincere socialists. "Enough sectarianism! . . . Go your way, and let us go ours." Although he invited the anarchists to attend the upcoming congress to debate their position, Casati hoped by his polemic to strengthen the antianarchist forces.

The Bologna (Fourth) Congress of the Partito Operaio Italiano, September 8–9, 1888

Matters were not to be so simple. The Bologna congress faced restructuring problems because of financial as well as ideological and strategic exigencies.

The chief order of business was the proposed communal and administrative electoral program, which the *operaisti* regarded as nonpolitical. Politics for them was the sterile and/or corrupt bourgeois party struggle at the state level. Electoral participation at the communal and regional level could be organized by economic associations. Indeed, the first point in the administrative program was "removing any political character from the communes and reducing them to simple administrative, economic, and moral organisms." There was a hint here of an older corporatism and, even more, of what came to be known elsewhere in Europe as municipal socialism. This political practice assigned value to local-level relationships because of their closeness to popular control. Those services that the *operaisti* were unwilling to demand of the state—such as aid to the aged, subsidized housing, and popular schooling; regulation of sanitary conditions, housing, and workplace health standards; and reform of the *dazio* regime, of subsidies for church-related institutions, and of the awarding of municipal building contracts—they were willing to ask of the commune.[14] Nevertheless, some *operaisti* rejected the notion that participation in communal elections was a means to forward the workers' interests. This position was even more rigidly abstentionist than some anarchists (who called for seizing power in the communes) admitted.[15] A compromise permitted autonomy based on local conditions.

On the issue of women and children working, the delegates agreed that to the extent that they would compete with men workers, such labor was not desirable; for them women workers were victims of the bourgeois system. Rather, their program declared that the solution to women workers' problems lay not in laws limiting their wage labor but in the women's organizing as adults for equality in the workplace. This anti-social-legislation, do-it-yourself position echoed the *operaista* principles drawn up by and for male workers.

Thus the majority—or at least a large minority—of the delegates were even more economistic and antipolitic in their outlook than was the Milanese group that edited the *Fascio Operaio*. Nevertheless, a majority finally voted to urge members to work with all their strength to restore concord among the "revolutionary forces."

The ensuing period was no more peaceful for the Partito operaio. In May 1889, most of the Milanese *operaista* leaders were arrested and charged once again with incitement to civil strife. Lazzari and the publishers of the *Fascio Operaio* were convicted of misdemeanors related to publication of the newspaper.

New Organizational Initiatives

In an effort to regain some momentum, Lazzari, Casati, and Turati joined other workers and intellectuals in founding the Lega socialista milanese in July 1889. The *operaisti* hoped to build a coalition against the proanarchist current among their fellow P.O.I. members. Briguglio and Manacorda both quote Turati's statement that the league welcomed the "socialists" of the Partito operaio, excluding the anarchists. But they differ on the Lega's activity: Briguglio points to continuity with the Partito operaio's efforts to organize peasants, and Manacorda emphasizes Turati's theoretical work.[16]

It is likely that the *operaisti* and Turati had different goals in mind in founding the Lega socialista. Lazzari and Casati saw it as an institutional base, not completely separate from the Partito operaio, but distant enough from the party's internal differences between members and longtime leadership to permit going forward in order to solve some of the P.O.I.'s internal contradictions. This interpretation is supported by the emergence at around the same time of an *operaista*-sponsored electoral federation, the Fascio dei lavoratori (discussed in Chapter 5), to push their municipal program in 1889 and 1890.[17]

Turati had begun his voyage toward socialism when he founded the Circolo di studi sociali after his resignation from the Associazione democratica in 1886. The twin influences of Anna Kuliscioff and Antonio Labriola, a Marxist professor with whom he corresponded extensively, were critical to his socialist education. The Partito operaio's municipal socialism appealed to Turati because of its democratic tone. Turati's name was placed by the Fascio dei lavoratori on the combined electoral list of progressive candidates in 1889, and he won a seat on the provincial council.

Carlo Dell' Avale's ambitious effort in the fall of 1889 to unify the P.O.I., the Consolato, and the Fascio dei lavoratori failed, and—the election over—the Fascio merged with the Partito operaio.[18]

The years 1889–1891 were not only a period of institutional uncertainty and the multiplication of organizations but also a time of economic recession and high unemployment (as described in Chapter 5). Milanese workers' organizations were overwhelmed with practical problems as they fought for programs like assistance and public works projects for the unemployed. They also were deeply involved in planning the Camera del lavoro and the political struggle to implement it.

There was more consensus on economic issues between the *operaisti* and the Democratic Radicals in the Consolato operaio than on political issues, such as electoral alliances, which regularly failed. Casati, especially, routinely opposed electoral coalitions, even that with the Lega socialista in the 1890 city elections, under which his colleagues Lazzari and Gnocchi-Viani ran for the city council. Casati's position was that there should be "no worker coalitions with bourgeois intellectuals or democrats." Lazzari's readiness to run for office was based on the belief that the Partito operaio needed to regain energy through action and, further, that coalition was the only means to protect the workers' organizations from repression.[19] Lazzari and Croce, for the Fascio operaio, supported a June 1890 proposal to establish a permanent workers' electoral committee, including the independent workers' organizations led by Carlo Dell' Avale. Casati abstained, neither supporting nor opposing the proposed committee.[20]

A meeting of workers' organizations considered an agenda for the fall parliamentary election that was *operaista* in tone, even though the old *operaisti* and the Fascio leaders did not attend. It agreed on a common program, but again there was no single slate of candidates. Within the Consolato operaio, dissent among the constituent workers' organizations increased. The *operaisti* led an "abstentionist procession" on election day, in which five hundred workers marched, urging others not to vote. The Democratic Radicals lost ground in the election, with both Giuseppe Mussi and Antonio Maffi being defeated in their bids for reelection as deputies.

Mazzinian Splits and the Paris Workers' Congresses of 1889

The Mazzinian (republican) Società operaie italiane affratellate met in Naples in June 1889. The group was increasingly divided, with an emerging minority that upheld the socialist ("collectivist") positions. Although their national congress avoided a formal split, the debate continued after the meeting had ended, especially in the Romagna, where cracks finally become visible in June 1890.[21]

The Mazzinians sent a delegation of workers to attend the International Exposition in Paris, celebrating the centennial of the French Revolution.

French workers were not unified at this time either; there were two workers' congresses (one "Possibilist," the other "Marxist") in Paris in the summer of 1889. Turati, invited to Paris as a representative of Milanese socialists, delegated Costa, who was heading an invited group from the Revolutionary Socialist party, to stand in for Lega socialista milanese delegates. They were instructed to work for the unification of the two congresses. The Partito operaio sent Croce to the Possibilists' congress. Other Italians who attended included two workers from Livorno, representing local organizations, and three anarchists. Because the attempt to unify the two Paris workers' congresses failed, the Romagnol and Livornese delegates (with the exception of one who chose the Marxist congress) attended them on alternate days; Croce stuck to the Possibilists, and the anarchists opposed both congresses.

Manacorda assigns great importance to the Paris congress experience. There, Italian delegates discovered the European workers' movement as a "great force," he writes, a force capable of making its case against bourgeois governments in the world arena. They were especially impressed by the German Social Democrats, fresh from electoral successes in national politics. In October 1890, the Lega socialista milanese and the Partito operaio sent a joint message congratulating the German Social Democratic congress at Halle. The letter was composed by Antonio Labriola and Filippo Turati, who were by then thinking about and planning a similar united Socialist party for Italy.[22]

Although three congresses of Italian progressives—those of the Partito socialista rivoluzionario, the anarchists, and the Partito operaio—took place in the next three months, none of these could bring into being the new party that Labriola and Turati had in mind.

The Workers' and Socialist Congresses of Fall 1890 and Winter 1891

The congress of the Partito socialista rivoluzionario italiano in Ravenna in October 1890 concentrated on electoral issues (Costa was running for reelection), over the objections of the anarchists. It was not a focus calculated to build bridges to the *operaisti*. Despite reciprocal messages of support between Costa and the P.O.I., and pious hopes for unification, the *operaisti* scheduled their own congress for the next month. Turati did not attend the Ravenna congress but sent a message urging the Revolutionary Socialists not to identify themselves with bourgeois democracy by emphasizing their own electoral ambitions. Turati also confronted what Manacorda calls the thorny central problem of Italian socialism in the period: the position of the socialist intellectuals in the workers' movement. On the one hand, wrote

Turati, were the "workers' organizations devoid of any emancipatory conscience; on the other hand, the [Romagnol] Socialist party, reduced to a
debating society." Labriola, his theorist colleague, saw the necessity for
the intellectuals to participate with the workers in a Socialist party, but he
also distrusted any socialism not firmly linked to the workers' movement.[23]
Turati suggested that the independent workers' organizations had passed
through their "primitive period" and would continue to be the propulsive
force of social struggle but that higher understanding would come from
elsewhere, perhaps from the bourgeois intellectuals.

The congress of Ravenna, Manacorda concludes, was a regional electoral "petty congress" of no national significance.[24]

The fifth and last congress of the Partito operaio italiano met in Milan,
on November 1 and 2, 1890. The local negotiations between the Fascio
operaio and the Consolato operaio over electoral alliances had again broken
down, and the anarchists moved to amend the agenda to consider the problem at the congress. But their motion was rejected by the majority.[25] Turati
tried to interpret to this congress the historical process under way: "In your
movement. . . . I see the modern socialist base that is no less than scientific consciousness . . . of the social drama in which you are the protago-
nists and the heroes."[26]

After a spirited discussion, the congress amended the party's constitution to specify that "resistance to capital is central to the Partito operaio
and that by this means only can workers defend themselves from exploitation" and to require that each section insert the principle of the strike into
its constitution. Fiscal autonomy of the sections was reaffirmed, removing
all except propaganda functions from the central committee. Although the
first chambers of labor were still being established, the congress voted its
support for such organizations and agreed to work for their dissemination.
The third item on the agenda was May Day and the 8-hour workday. Casati
read a report on the issue, waxing eloquent (for two hours) on its moral,
economic, and health advantages. His successful motion declared that the
shorter workday must be achieved through the workers' struggle in a class
party (i.e., not through bourgeois social legislation).

The Partito operaio italiano, its functions having been appropriated by
other institutions, was on its last legs. The *Fascio Operaio,* which had
resumed publication only in March 1890 (after its 1889 police shutdown)
ceased publishing permanently in January 1891. The party was no longer
viable, although its ideology and practice continued to flourish in resistance
leagues and local federations and soon developed further in the chambers
of labor.[27]

Italian anarchists (among whom there were relatively few Milanese activists) held their congress in early January 1891 at Capolago (in Switzerland, just across the border) to found their own party, the Partito socialista

anarchico–rivoluzionario, destined to have a very short life: Government repression and arrests after May Day 1891 crushed the fragile organization.[28]

Intellectuals Seek a Path to a New Party

Following up on his convictions regarding the role of intellectuals in a new Socialist party, Filippo Turati, along with Dario Papa—the Republican editor of *L'Italia del Popolo*—and others, founded a new organization, the Unione democratico–sociale, in December 1890. The Unione was not intended simply as an electoral committee but, rather, as a circle for political discussion that brought together social Republicans, the left wing of the Democratic Radicals, and socialists. Moderate workers, like printer Carlo Dell' Avale of Genio e lavoro (a mutual aid society) and others of the Consolato operaio, joined Turati and Carlo Tanzi, a young socialist lawyer, on its executive committee. There were other worker members, including Giuseppe Croce, but most of the members were intellectuals: professionals and students interested in progressive politics.

Along with the Unione democratico–sociale, Turati started a new journal, *Critica sociale*, a transmutation of *Cuore e Critica*, a journal of culture and politics, edited by Archangelo Ghisleri. In the past few years, *Cuore e Critica* had become an influential voice for positivist sociologists from all over Italy. Although it had not explicitly discussed politics, it had laid the groundwork for democratic politics with a deep concern about social problems in an autocratic state.[29]

Turati's goal, through both the organization and the journal, was to construct a clear and consistent progressive political position. Thus the journal was not to be militantly socialist; its task was to create a "socialist consciousness" not only among its student and intellectual readers but also among the workers, in particular the *operaisti*. In the first issue of *Critica sociale*, Turati discussed the conditions needed in order for workers to achieve socialist consciousness: greater development of the industrial proletariat and a "little more clarity regarding its strength and their destiny." He and Anna Kuliscioff hoped to carry socialism to the workers' movement and to unify the various socialist currents under the banner of scientific socialism, "a historical and scientific point of view . . . derived from the necessary continuing evolution of the great economic and moral forces that comprise history."[30]

On the practical side, Turati worked in the Unione democratico–sociale in a familiar way: by excluding local anarchists, including Pietro Gori. Later, after the 1891 administrative elections, Turati also guided the group's break with the Mazzinian—noncollectivist—Republicans. In the same period, Turati and Kuliscioff took the lead in crafting the program of the Lega socialista milanese. Severely criticized by Manacorda for its Lasallian holdovers and its neglect of issues of landholding and agrarian workers, the Lega socialista program was published in the *Critica sociale*.[31]

The Romagnol Socialists decided in January 1891 to sponsor a national congress of socialist unification. In contrast with Turati's imaginative and energetic multilevel efforts, the Romagnol socialists' organizational incapacity was stunning, and not surprisingly the enterprise sputtered to a halt in June 1891.

Milanese Workers' Drive for Unity

On its thirtieth anniversary in the fall of 1890, the Consolato had already reduced some of the distance separating it and the *operaisti* in the Fascio dei lavoratori, as discussed in Chapter 5. What remained were issues regarding politics as a means and an end, with Maffi accusing the *operaisti* of abstentionism, not unreasonably, considering the behavior of many of them in that fall's election.

The planning for May Day 1891 brought most workers' groups together again, as the Consolato supported *operaista* Giuseppe Croce's proposal that the workers' holiday be celebrated on May 1 itself, not on the following Sunday. In the process of drafting a joint May Day declaration, the groups decided to convene a national workers' congress that summer "to coordinate and unite all Italian workers' forces to conquer their economic and social rights" and to name national delegates to the Brussels congress of the International.[32]

The basic task of the congress was to consider the most effective means for workers' organizations to achieve their program within bourgeois political institutions. Its agenda included (1) effective national protective labor legislation and the means to enforce it; (2) rights to organize, strike, and establish cooperatives; (3) the position of the working class vis-à-vis militarism, and workers' duty in this matter; (4) the 8-hour workday; and (5) the establishment of a workers' newspaper.[33] Although both the agenda of the congress and the May Day 1891 declaration (summed up in Chapter 6) were relatively economistic, and hence close to the *operaista* position, both documents also contained more heterodox elements. Most important was the demand for government-sponsored protection for workers, minimum wages, and maximum hours. These implied the kind of political campaign that the Partito operaio had always rejected. For Casati especially, demands of this sort threatened the subordination of the Italian workers' movement to bourgeois socialist intellectuals. Silvio Cattaneo shared his concerns, but Croce, Brando (with his recent experience leading the Milanese machine shop workers through recurring unemployment crises), and Lazzari were not persuaded. And it was they who were to play the much greater role in the national workers' congress in August 1891. Casati took comfort in their agreement that the congress was to be "wholly worker in nature," but he was largely excluded from its planning. Croce proved exceptionally effective, and in the end, the call to the congress put no restrictions on the occupation or class of the delegates.[34]

Casati was more involved in negotiating the "alliance pact" between the Consolato operaio and the Fascio dei lavoratori, in June 1891. The printers' propaganda committee mediated the process. The agreement focused primarily on common economic action to be undertaken by the two groups: working collaboratively on all types of workers' economic struggles, especially strikes, and promoting and assisting the formation of resistance leagues, producer and consumer cooperatives, workers' schools and study groups, and mutual benefit societies. In addition, the Fascio agreed to take part in both administrative and national electoral struggles, and the Consolato, to support unconditionally the worker candidates put forward by the Fascio. In this reiteration of the primacy of economic issues, Casati was probably speaking for the majority of Milanese Partito operaio members.[35]

That same month, the progressive electoral coalition (a joint list of the Unione democratico–Sociale and the Lega socialista) was defeated in the administrative elections by the moderates; only one Republican on the joint list—and he was also on that of the Democratic Radical Associazione democratica—was elected.

Although some of the Milan operaisti still hoped that the national workers' congress would unify their movement along lines of economic action without bourgeois participation, they did little to prepare for the congress. Copies of the P.O.I. constitution and program were their only weapons in their attempt to shape the new institution in its image. At this congress Casati's opponent would no longer by the wily and opportunistic but clumsy Maffi but, rather, Turati and his operaista allies, Croce and Lazzari.

The National Workers' Congress, August 2–3, 1891

The context of the national workers' congress, as Cervo points out, was the strong presence of worker reformism in Milan, whose leaders at this time were Antonio Maffi and Carlo Dell' Avale. Within the Partito operaio, the proponents of municipal socialism and reform politics on the local level agreed on many issues with the democratic reformist workers. A coalition of these groups was the base on which the workers' movement was unified.[36]

On August 2, some 250 delegates representing 450 Italian organizations (resistance, mutual benefit, and cooperative, with many delegates from larger centers representing organizations from smaller cities), mostly from the north and center, gathered at the headquarters of the Consolato operaio. The inclusion of nonworkers was immediately evident, for Filippo Turati played a central role.

And indeed, Turati reported on the first agenda item (and also most of the others): the position of the Italian workers' movement vis-à-vis labor legislation. His motion explicitly demanded from the state an 8-hour workday, one day of rest each week, inspection of health and safety conditions at work, social insurance, and laws forbidding child labor (for those under 14), and the provision of popular schooling. He declared that workers' collective action was needed to bring about passage of such laws. The anarchists quickly objected. Some *operaisti* supported them, but Casati stayed silent, calculating that the opposition was very weak. Turati had not made a theoretical or abstract case for his motion, but one concretely anchored in workers' experience, and it made sense. The motion passed by a large majority. Thus, right at the start, Manacorda writes, the Milan congress proclaimed the necessity of the workers' political action (against the *operaista* and anarchists' practice) and specifically proletarian politics (against the Mazzinians).[37]

There was even less opposition to Turati's second motion, calling for the end of legal restrictions on strikes, government recognition of unions, and affirmation of workers' solidarity with strikers and blacklisted workers. Third came Lazzari's report supporting antimilitarist and antinationalist propaganda, which passed with no opposition and little discussion. Croce and Turati were appointed as alternates to Lazzari and Camillo Prampolini (both of whom declined the nomination) to attend the Brussels congress of the Second International.[38]

Bigger battles were fought on the second day of the congress. Carlo Dell' Avale (a worker who supported electoral participation) was scheduled to open the debate on the form of Italian workers' organization. Before he could speak, however, Anna Maria Mozzoni took the floor. On the chairperson's podium at the front of the hall, a pile of copies of the constitution and program of the Partito operaio italiano had been deposited. Mozzoni proposed that "in order to avoid useless discussion," these documents be debated as a draft constitution of any new organization. Lazzari and others quickly objected that this was not on the agenda of the congress and that in any case, Dell' Avale's report must be heard first.[39]

Dell' Avale made a confused case for a Partito operaio–socialista whose membership would be exclusively by ways of workers' organizations and whose activities would be "tightly linked to economic questions." Objections were raised from all sides. Maffi disliked the word *socialist* in the proposed party's title and substituted "Partito dei lavoratori italiani." The anarchists and some *operaisti* called for broader membership (not simply members of workers' organizations but individuals, too). Angiolo Cabrini complained that Dell' Avale's motion was not a program and proposed that they accept the program of the Partito operaio for the new organization and modify it through discussion. The reformists strongly opposed this motion, which Turati also viewed as dangerous. He suggested instead that the session be suspended and that a small group be designated to formulate a plan for organizing the new party. Cabrini thereupon withdrew his motion.[40]

Those who drafted the new motion included Turati and Maffi but not Casati or Dell' Avale. In it, Maffi's preferred party name was accepted. The motion declared that the new party would seek "the emancipation of workers from the political and economic monopoly of the capitalist class. It would take part in public life as an independent class institution, and it would conduct its struggle against the capitalist monopoly through solidarity, resistance, and propaganda."[41] Membership would be open to waged workers, urban or rural, who belonged to affiliated organizations, as well as individuals who did not. Those who were themselves supervisors or exploiters of others' labor were excluded. Affiliated organizations ordinarily would be wholly run by workers, but exceptions could be made by the central committee. A committee would be named by this congress to draw up within a month a draft constitution and program. Manacorda sees this formulation as a clever move on Turati's part, eliminating from consideration the Partito operaio's constitution while at the same time preserving some of its more acceptable features. Because the previous day's session had voted in favor of social legislation, Turati did not insist on an explicit reference to political struggle.[42]

Casati asked once more that the P.O.I. constitution be debated, as Cabrini had proposed that morning. An angry exchange between Croce, who was chairing the session, and Silvio Cattaneo, who wished to speak in support of Casati's request, led to Croce's stepping down, and Maffi's taking his place as chairperson. But Maffi was even more authoritarian, and the *operaisti* protested loudly. The congress itself seemed threatened, and indeed, one of the men of the Consolato operaio talked of a walkout if the Partito operaio's constitution were accepted as the basis for debate. Lazzari stepped in and asked Casati to give way in the interest of unity, telling him that it was a "matter of form, not substance." Casati fell silent, but the debate raged on, as the anarchists proposed an amendment permitting organizations to choose whether they wished to enter political struggles. This time it was the members of the Fascio dei lavoratori who threatened to leave if the motion were passed as read. Turati relented, and the motion was amended on several counts. A new section on organizational autonomy was added, reading "Each section and federation will preserve its autonomy in all matters except those that are essential to the interests of the party," and the charge to the constitutional drafting committee was enlarged, giving it the function of a provisional central committee and the organizer of a second congress within a year. Its members were to be Bertini for the printers, Maffi and Cremonese for the Consolato operaio, Croce for the Fascio dei lavoratori, Cattaneo for the masons, Lazzari, and Mozzoni. Dell' Avale was nominated but disqualified himself as the head of Genio e lavoro, an organization with nonworker members. A proposal to establish a weekly newspaper was also adopted.

Although Turati's evaluation of the congress was positive, he regretted that a true Socialist party had not yet been born. He had avoided divisive debate and wisely under the circumstances, had not even mentioned the explicitly Socialist program hammered out earlier in the year by the Lega socialista milanese. The Italian workers' movement had decisively broken with the patronage of bourgeois parties. And some of the Milanese *operaisti* had taken the first steps toward unifying workers and intellectuals on a national level.

Operaista, Socialist, and Worker
Maneuvers in 1892

As the constitution-drafting committee went to work, the mechanics began their strike of August–September 1891 (discussed in detail in Chapter 7), and Giuseppe Croce began his tenure as secretary of the Camera del lavoro. This institution, rather than being simply an employment exchange, as the city officials had conceived it, quickly assumed an assertive role in strikes and labor struggles. Croce's energy and his interventionist philosophy contrasted with the Consolato operaio's limited conception of its role in labor struggles, and they reflected his own experience in the Partito operaio.

In the spring of 1892, May Day was celebrated with a citywide strike, planned by a united workers' committee. A banner on which was written "Eight hours of work and abolition of the *cottimo*" hung at the entrance of the Chamber of Labor offices. This demand meant, Turati explained in his speech, "increased wages, an end to unemployment . . . and also eight hours available for individual self-cultivation and exploration of ideas."[43]

The Consolato followed with its own "congress," a series of meetings over a three-week period, at which its members rejected the secretary's report as being too conservative. At the next session a majority of the organizations present declared their support for the principle of class struggle. Maffi was chosen as the Consolato's representative to the Mazzinian workers' congress in Palermo. In their last two sessions the Consolato members discussed and approved the motions passed by the national workers' congress.[44]

The Consolato operaio, the Fascio dei lavoratori, and the independent workers' organizations all agreed on a list of candidates for the June administrative elections, and these candidates ran together with those from the Lega socialista and the Unione democratico–sociale. A single-issue electoral newspaper entitled *La Lotta di Classe* (class struggle, published by "the democratic socialist workers' organizations of Milan") was published on June 18, 1892. It urged other groups, including the bourgeois, to vote with the socialists. During the campaign, the anarchists attacked the candidates, especially Croce, who in his capacity as secretary of the Camera del lavoro had refused to permit Pietro Gori to lecture on its premises.

The results of the combined effort were disappointing. Not only were none of the candidates elected, but the total number of votes cast for the group was smaller than that of the previous year. Turati nevertheless expressed satisfaction in *Critica sociale* with the clarity of the program (excluding the anarchists) and the refusal to ally with the Democratic Radicals, two cornerstones of the strategy that he was pursuing in order to educate workers in socialist principles.[45]

The call to the founding congress of the P.L.I. in mid-August was limited to organizations that had accepted the positions of the national workers' congress of the previous year: collectivism and action in the political arena. Some Milanese groups perceived this exclusion as evidence of the probable orientation of the draft constitution and program, and accordingly they issued a call to "all workers' societies, irrespective of position," to a meeting on July 23 at the Consolato operaio. Croce, reporting on behalf of the drafting committee, spoke at length about the special train tickets and other less relevant matters, but his listeners became irritable and noisy, and so he ceded the floor to Dell' Avale. The latter promptly introduced the central question: admission to the Genoa congress of representatives of the anarchist workers' organizations and of political associations. The motion that was passed left little doubt that there was a large element (the majority at this meeting) that supported the exclusion of intellectuals and political organizations from the upcoming congress and from a central role in the new party. Its points were (1) the congress would admit representatives of any and all workers' societies; (2) Socialist workers' associations were to be composed exclusively of workers organized by craft; (3) any party newspaper was to be managed and edited by organized workers; and (4) workers' economic interests were to be the criteria for considering all policies of the future party. The old objection to politics as such seems not to have been an issue for the majority at this time. Antonio Labriola (in a letter written to Friedrich Engels after the Genoa congress) called this meeting "an uprising of the 'pure workers' against the political Socialists."[46]

Hastily, Turati and his colleagues put out the long-delayed national socialist newspaper that had been projected a year earlier: *La Lotta di classe*.
Camillo Prampolini (editor also of *La Giustizia* [justice] of Reggio Emilia)
was named its editor, but in reality it was written and published by Turati
and the Milanese group around him. The first issue of *La Lotta di Classe:
Giornale dei Lavoratori Italiani* came out on July 30–31, 1892. Manacorda
notes that only in Milan could such a truly national Socialist newspaper be
published, that from its beginning it had a broad outlook and at the same
time considered local interests and spoke in terms understandable to the
workers.[47] *La Lotta di Classe* declared in its first issue that since the demise
of *Fascio Operaio*, "the silence of the Milanese working class has been
like the silence of an entire nation. . . . [The *Fascio Operaio*'s significance was] its affirmation that the problem of labor is not simply a local
question, limited to one occupation or one federation . . ., not concerned
merely with hours, wages, work rules. [It] . . . is a social question." A
new workers' newspaper was urgently needed, and *Lotta* would fill that
void. The party's draft constitution and program were published in the same
issue (July 30–31, 1892).

In the next two weeks, analysis and critique filled the pages of *Critica
sociale* and the other Milanese newspapers. In an effort to emphasize the
socialist character of the party, *La Lotta di Classe* noted on August 6–7,
1892, that its goal was the "economic and political expropriation of the
capitalist bourgeoisie and the parasite classes, ending the private ownership
of the means of production . . . and the transformation of capitalist production into production by workers on their own account." The means to
achieve these ends were the freedom to organize, political struggle, labor
legislation, and organizations like the Chamber of Labor, all carried forward through a socialist workers' party.

A dismissive critique (attributed to Turati) of the draft constitution and
program appeared in *La Lotta* on August 13–14, on the eve of the Genoa
congress. The program was vague, he wrote, it was expressed in terms of
commonplace verities like equality, popular sovereignty, and emancipation.
It needed specifications, that is, an end to wage dependence, collectivization of the land and means of production, and the conquest of power. The
means to achieve these ends were class organization and struggle independent of and against all bourgeois parties. Both the program and the constitution ought to be equally informed by the Socialist worker principle:

> The Partito operaio becomes a Socialist party and ceases to be simply a
> corporation. The corporations remain for the defense of occupational in-
> terests and as the prime material of the party, but the party raises the
> level of action and concentrates forces. Members are no longer printers,
> bakers, or metal workers . . . they are then conscious workers and So-
> cialists.[48]

The article's author, however, did not have full faith in the workers' ability to do this by themselves. Rather, the role of intellectuals was the keystone of Turati's strategy. "The Socialist intellectuals, in fact, were the only element by means of which the unitary efforts of the Italian workers' movement could be coordinated and brought to fruition, beyond regional diversities, different experiences, and specific local political situations."[49]

Casati was the one *operaista* leader who was determined to uphold the minority position (Angiolo Cabrini, an intellectual, wrote in support but abandoned him at the congress) against worker unification under intellectual leadership. An unpublished manuscript by Casati from this period is entitled "Salviamo il Partito operaio dal devenire un partito borghese o semiborghese" (Let us save the P.O. from becoming a bourgeois or semibourgeois party). In it he wrote that "the political concept, or rather the way it expresses itself in the bourgeois setting, divides us; . . . we must first organize on an economic base and then enter into struggle."[50]

Casati reported on an assembly of the organizations belonging to the Fascio dei lavoratori (of which he was secretary) in a letter to *La Lombardia*, published on August 11, 1892. The group had defined his mandate as delegate to the Genoa Congress as (1) to uphold the Partito dei lavoratori as a pure "class" party, distinct from religious or political parties; (2) to insist that all functions of the party be staffed by workers only; and (3) to ensure that only delegates of workers' organizations, and not political associations, be admitted to the congress. He concluded, based on these criteria, that the party ought not be called *Socialist*. This report was entitled "The Partito Operaio Does Not Wish to Be Called a Socialist Party." It prompted a quick response from Turati, published on August 12, in which he reminded *La Lombardia*'s readers that the Fascio dei lavoratori was no more than a small part of the P.O.I., itself about to be absorbed into the Partito dei lavoratori italiani. The earlier report had been the personal idea of Casati, Turati noted in closing, "my good friend, shopkeeper, and poet." (The last was a cruel allusion to Casati's earlier efforts at writing poetry, an activity from which Gnocchi-Viani had gently dissuaded him.)[51] In yet another letter, published on August 13, 1892, Casati repeated his opposition to having bourgeois intellectuals in the party. The Partito operaio that Turati wanted to bury was not yet dead, he explained.

The Founding Congress of the Partito dei Lavoratori Italiani

The founding congress convened in the Sala Sivori in Genoa the morning of August 14, 1892. There were delegates from all over Italy, including Sicily, where urban and rural workers were organizing and mobilizing in

Fasci, a movement that grew increasingly challenging in the next year and a half. Representatives of the Sicilian Fasci had been important to the move to socialism in the recent congress of Mazzinian workers' organizations. Leaders of *operaismo* and all varieties of socialism and anarchists in red shirts and ties were present, but Kuliscioff and Turati, writes Manacorda, were the "principal artificers." [52]

Discord quickly became evident. Casati objected to Kuliscioff's nomination of four men as rotating chair, instead insisting that all officers of the congress be authentic workers. Despite a stormy debate, her motion was passed by a large majority. It became obvious that Casati and the anarchists Gori and Galleani were conducting an obstructionist campaign against the Socialist majority. When Maffi rose to open the discussion of the draft party constitution and program, one of the anarchists moved that the discussion be postponed to the next day because not everyone had yet had a chance to read and study it. Shouts and protests rose from the majority. Turati yelled, "Down with tyranny; out with the despots!" Prampolini spoke more calmly, repeating the opinion that he had already published in his newspaper, *La Giustizia*. He urged the anarchists to leave and to hold their own congress, for they were another party. Forget the sterile conflict and permit the Socialists to establish their party, he admonished them. As Cervo notes, Turati and Prampolini did not hesitate to push the group toward schism. Pietro Gori replied for his colleagues, that they insisted on making their point within, not without, the Socialist congress. [53]

Turati and others decided that if the anarchists would not leave, the majority would. That evening they arranged for the congress to reconvene the next morning in the assembly room of the Carabinieri Genovesi, via della Pace, welcoming all those ready to support electoral politics as a means of achieving power. The anarchists continued to meet in the Sala Sivori, with Casati and the diehard *operaisti* of the Fascio dei lavoratori. A group of delegates from the Veneto, Ligury, and Romagna (including Andrea Costa), objecting to the high-handed manner in which the matter had been settled, decided not to attend either congress.

Most of the Milanese chose the congress at the via della Pace, where the party was finally constituted. The very first session of the congress had revealed to them, Cervo argues, that the intellectuals were the necessary glue for a workers' political party, given the "profoundly differentiated workers' realities." Those who joined the congress at the via della Pace were ready to sacrifice class autonomy in order to achieve political unification of workers and intellectuals. [54] Among the Milanese groups who broke with the Fascio dei lavoratori on this issue were the Federazione meccanica, the Unione dei figli del lavoro (male and female sections), and the Lega muratori. *Operaismo* lost its identity in both congresses, however, as Casati's coalition with the anarchists meant surrendering autonomy, just as accepting the formation of the Partito dei lavoratori italiani did.

The anarchists and *operaisti* crafted an alternative program and constitution that incorporated the autonomy of party sections vis-à-vis politics and made political action in both national and local arenas problematic. It thus questioned recent *operaista* majority policy positions. The anarchists dominated the central committee of the party as constituted, but it collapsed in several months.

At the via della Pace, opposition came from Maffi, whose contribution to the draft constitution and program (writes Manacorda) consisted of a "Democratic Socialist hat placed on the head of the *operaista* constitution," and from Lazzari and Cabrini in the name of *operaismo*. The draft documents were substantially modified along the lines of the critique published in *La Lotta di Classe*. According to the official minutes, the congress "clearly [affirmed] what the workers' party stands for—socialization of the means of production—and to obtain this—workplace struggle and the conquest of public powers. For the first, resistance leagues, chambers of labor . . .; for the second . . . a class party." Non-Milanese delegates were getting restless about what seemed like an internal disagreement among Milanese. Both they and the great majority of the Milanese (convinced that the battle had been lost at the start) voted for Turati's amendment. In the afternoon session, the section of the constitution limiting membership to "pure and simple workers" was amended to apply only to organizational affiliation. Thus *operaista* worker exclusivism was preserved in labor organizations, even though the party was open to nonworkers. The party sections would be autonomous only in administrative matters.[55]

In its self-congratulatory report (August 20–21, 1892), *La Lotta di Classe* praised the birth of a "disciplined" party. The central committee, whose members were Enrico Bertini (Associazione tipografi), Giuseppe Croce (Camera del lavoro of Milan), Carlo Dell' Avale (Genio e lavoro and Associazione tipografi), Annetta Ferla (Unione mutua figlie del lavoro), Giuseppe Fossati (Federazione meccanica), Costantino Lazzari (who brought his long experience in the Partito operaio and Milanese worker politics), and Antonio Maffi (ex-deputy, representing the Consolato operaio of Milan and the Mazzinian workers' societies)—all Milanese—would continue to sit in Milan. The workers' movement had created a party that would be a political tutor, spokesperson for popular demands in the current Italian conditions that restricted political liberty, and promoter of worker associationism. For most of its members, the party's goal was the improvement of workers' everyday life through class struggle.[56]

The debate about the intellectuals' role in the party was primarily a Milanese one; elsewhere in Italy intellectuals and professionals were ac-

cepted as organizers and leaders. The experience of the Milanese workers
in the Partito operaio and in resistance leagues had acquainted them with
the ways of organization and collective action, as Robert Michels asserted
many years later. It also gave them the opportunity to develop independent
opinions and autonomous practices, often in opposition to the bourgeois
"Democratic Radicals. Socialist intellectuals were, for them, colleagues and
friends and sometimes defenders. In adopting the Turatian version of the
Partito dei lavoratori italiani, they accepted a subordinate relationship with
the intellectuals for the Italian workers' movement as a whole in order to
establish a unified party. The *operaisti* who made this decision continued
to be active in the Socialist party, but later, workers (as Michels also pointed
out) could seldom be found in the party leadership.[57]

In Milan, the Fascio dei lavoratori followed Casati into schism, but the
provincial organizations of the Partito operaio joined the P.L.I. Most of the
independent Milanese workers' organizations joined the party, and the Consolato operaio became its Milan section. At the time of the congress of
Reggio Emilia, a year later, 10,000 Milanese members were reported, but
in June 1894, *La Battaglia* reported only 7,000 Milanese members.[58] The
printers' league also refused to join the party, insisting that there had been
too many compromises. *La Lotta di Classe* reported that its vote against
affiliation had been engineered by a motley coalition: "anarchists, Mazzinians, moderates, and apolitical printers, trying to create out of their occupation a labor aristocracy."[59]

Individuals like Gnocchi-Viani affiliated quickly, despite initial concerns. He wrote an open letter to *La Lotta di Classe* (published on August
27–28, 1892) explaining that his worry that workers' organizations might
be absorbed by the new party disappeared once he read the report on the
Genoa congress, as well as the constitution and the program of the new
party. He urged those who might be hesitating to examine the evidence and
join also, in order to continue the struggle for social renewal. Andrea Costa
was less friendly, mostly because he disapproved of the way in which the
schism had been provoked. His party and his Jacobin democratic ideology
now ceded its claim to represent Italy's Socialists to Lombard Socialism,
which had a stronger worker base than did the Romagna.[60] The process of
reconciliation between Costa and the P.L.I. took about a year.

Several years later Casati himself joined the party. The exact date and
circumstances are unknown, but it was sometime before 1895. In the year
immediately following the Genoa congress, Casati lectured occasionally in
Milan to hostile audiences and was aggressively questioned by Socialists.
For example, in early September 1892, Casati declared in a public lecture
at the Consolato operaio that the true Socialist congress in Genoa that August had been the one in the Sala Sivori. He tried to differentiate himself
from the anarchists: What he wanted, he explained, was that bourgeois be

friends and advisers of the socialists, but not party members. No one supported his position. Turati declared that the important thing was that anarchists and Socialists now recognized that they were two parties with different positions.[61]

Socialist Politics in Milan, 1892–1893

In the fall of 1892, parliamentary elections came too soon after the Genoa congress for much preparation, but the Socialists nevertheless ran a full slate of candidates in Milan. *La Lotta di Classe* wrote, "In order to affirm ourselves, we must distinguish ourselves, not . . . play the game of others." It disapproved when *Il Secolo* recommended that Democratic Radical voters cast their ballot for Osvaldo Gnocchi-Viani instead of Giuseppe Colombo. The results were not encouraging: The six Socialist candidates won a total of 1,326 votes (Gnocchi-Viani alone got 630 of them). It was the only "logical and dignified" way to "prepare for the true victories [to come]."[62]

Although the Milanese Socialists were strictly opposed to alliances, they were also faced with the prospect of Gnocchi-Viani's losing his seat on the Consiglio communale in the June 1893 partial administrative elections. Their solution was to run a serious campaign with fourteen candidates and, again, to look to the future. The issues, as *La Lotta di Classe* saw them, included "popular schooling, protection for labor, an end to private profit on communal projects, a brake on the ever-increasing debt . . . no more spending public money to protect bourgeois interests, [and] the encouragement of workers' organization and struggles." By entering the electoral struggle themselves, the Socialists hoped "to push the bourgeoisie to constitute itself a party . . . whereas it has ordinarily gone about its own affairs in the commune without publicity or opposition."[63]

The Socialist administrative planks called for the referendum, the freedom of religious and political opinion for municipal employees, and the collectivization of all public services. On the economic side their demands included facilitating workers' cooperative bids for city contracts, the 8-hour workday, and minimum wages for workers on city contracts. And in fiscal matters, they urged the abolition of the *dazio* and the end of communal budget items for theater subsidies, receptions, monuments, and the like. The Socialists were defeated, but the number of votes per candidate increased. Turati received 1,500, and other candidates, between 1,200 and 1,400. As *La Lotta di Classe* observed, the results illustrated the "decadence of the democratic bourgeois party faced with the growth of the Socialist party." It was apparent, it continued, that "in losing, one can win and that in winning, one can lose."[64]

The second congress of the Partito dei lavoratori italiani was scheduled for September 8, 9, and 10, 1893. An anticipatory column by Ezio Marabini in *La Lotta di Classe* brought up the question of changing the name of the party: Because not all the members were workers, the current title was misleading; further, it lacked any statement of principles. The party, he wrote, was about the "redemption not only of the working class but of all mankind." The party thus ought to be called "Socialist." [65]

In the next month, several suggestions for additions to or clarifications of the program were made. The Mantua section proposed complete equality between the sexes and civil and political rights for women; others called for modifications in the party's position on politics and struggle. [66] The most comprehensive set of proposed revisions, however, came from the Milanese section (the Consolato operaio). It accepted the addition of the word *socialist* to the party's title, as well as a change in the dues structure proposed by the central committee. It proposed motions on the questions of both electoral alliances and the role of deputies and other officials elected under the party banner. Its position on alliances (supported by Turati) was unyielding: "In elections, the party must act as a party independent of all others, running its own candidates, who are party members and have unconditionally accepted its program." As for the elected deputies, "they are delegates of the party," and their position on various questions should be set by regional and central committees and ultimately by the party congress. In Parliament, the deputies were expected to introduce legislation that conformed to the party's minimum program, with two goals in mind: propaganda and persuading the bourgeois parties to accept the program. Socialist deputies were not to support bills introduced by members of other parties, but they were permitted to amend them. [67] *La Lotta di Classe* emphasized that the congress's decisions regarding the party's tactical program would be "the [regulatory] code for the organization and discipline of our party." [68]

The Congress of Reggio Emilia, September 1893

The Partito dei lavoratori italiani was in good shape and growing when its second congress met. The most striking development of the past year had been the continuing expansion of the Fasci dei lavoratori (rural and urban organizations of workers and peasants) in Sicily. Demonstrations were sweeping the island, and already some had been repressed militarily. Garibaldi Bosco, a delegate from Sicily, carried greetings from 65,000 workers.

The Milanese delegation came to the congress with the most comprehensive position, having submitted motions in advance on electoral and parliamentary tactics as well as on lesser issues. Its position on the first, despite support during the discussion, was voted down, and a more flexible one put forward by Croce and others was passed. The approved motion stated that

> political action for the conquest of public powers must represent the will of the party to act independently of all other parties, running its own candidates who have unconditionally accepted the party's program in both political and administrative elections and repudiating any combinations or compromises that, while taking account of local conditions, might detract from the principles and conduct of the party's own line or be in contradiction to it.

This wording, Briguglio points out, called for the repudiation of compromises based on local conditions but implied that they might be acceptable if they did not undercut the party's line.[69]

On parliamentary tactics, Lazzari offered an *operaista*-style motion that described parliamentary discipline broadly but precisely. Opening with the declaration that "political power is none other than the class organization of the bourgeoisie by means of which it carries on its struggle against the proletariat," it directed the Socialist deputies to form a disciplined faction. The motion also insisted that the deputies "support only Socialist proposals voted by the party for purposes of propaganda and affirmation."[70] Turati and Kuliscioff raised strong objections. Turati claimed that to forbid deputies from voting for other parties' or individuals' bills would prevent their conducting a "logical struggle in Parliament." Lazzari refused to budge on this point, and so Croce adopted the motion and permitted removal of the antivoting section. Prampolini amended it to add that "in no case may the Socialist parliamentary faction vote confidence in the ministry," and then it was passed. Briguglio, who is ever alert to continuing relationships of Socialists and *operaisti,* quotes Lazzari's *Memorie* on his pleasure at "having contributed to the creation of a large and invincible force for action and struggle that would continue the tradition of the Partito operaio without being deceived by parliamentary politics or dragged down by the ambitious will of personal influence."[71] This was but one of the increasingly frequent differences of opinion between Lazzari and Turati.

The congress also delegated responsibility for agrarian cooperatives and resistance leagues to the regional and provincial federations, which would "oversee and direct" the party's economic actions.[72] The proposal for equal pay for equal work and political rights for women did not come up for discussion at the congress. (Kuliscioff, who had supported women's rights at the Geneva congress of the Socialist International, did not press for them in Reggio Emilia, apparently because she felt that it would be too divisive here.) Finally, the congress approved a new title for the party: Il Partito socialista dei lavoratori italiani. The delegates and their local supporters marched into the streets in an "immense procession . . . a broad river of men, marching irresistibly on the road to conquest."[73]

La Lotta di Classe congratulated the congress, remarking that "the party emerged from it more disciplined, ready for victories, [and] unified in its will and method." Bourgeois newspapers, it pointed out, could not help but recognize the importance of the congress, the consequences of which would be the displacement of the "parties that have up to now monopolized Italian public opinion." In 1921, Michels concluded that this was the true "constitutive congress of the party." Its goal was no longer limited reform—he wrote, hyperbolically—but radical transformation, as Socialism had "invaded the factories, winning over proletarian support."[74]

The Repression of the Sicilian Fasci

Matters were degenerating, however, in Sicily, where the continuing demonstrations of local Fasci dei lavoratori were being met by mounting police repression.

There were public meetings in Milan to express solidarity with the Sicilian Fasci (most of which were nominally at least in the Socialist party) at the end of September and in October. Three hundred persons attended a protest at the Consolato operaio on September 30, at which Alfredo Casati accused the "chiefs of the Partito dei lavoratori" of being "slaves of legalism" and abandoning their mission to lead workers in a revolutionary direction. He proposed the establishment of a committee that would work with the other "subversive parties to organize an intense and extensive agitation for the political rights of Sicilian workers." Lazzari, Dell' Avale, Croce, and others denied that the party had ever thought of renouncing its revolutionary mission. After a vigorous debate, Casati's motion and another radical one were rejected. The assembly stated instead that class war was being waged in Sicily, urged the Sicilians to continue their "dignified resistance to arbitrary government," and called on "all Italian Socialists to demonstrate their solidarity with Sicilian workers."[75]

Bank scandals brought down the liberal Giolitti government at the end of November, and in Milan the Socialists demonstrated, singing the "Inno del Lavoratori" and shouting "Down with the thieves! Long live Socialism!" Giuseppe Zanardelli was the first approached to succeed Giolitti but then was vetoed by King Umberto. The king preferred Francesco Crispi, who held out the promise of a strong government that could both control the rising tide of protest in Sicily and pursue Italy's colonial expansion. The Democratic Radicals supported Crispi at this time, but their Milanese representatives soon regretted their compliance.[76]

Internal surveillance intensified under Crispi, as exemplified by a letter from the minister of the interior (dated December 6, 1893) reporting to the prefect at Milan that a secret "action committee" had been set up by the central committee of the Fascio siciliani meeting in Palermo. A delegate had been sent to the "Directive Council of the Partito operaio" in Milan for consultation where he (his name was not known) had allegedly been advised by the Milanese to keep calm and act prudently, in anticipation of "better organization and preparation and a more propitious moment." The minister asked local officials in Milan to investigate and send any additional information, in particular the name of the Sicilian who had come to Milan. A draft reply in the Milanese Archivio di stato reports that they had no knowledge of such a visitor but noted that Turati corresponded with Garibaldi Bosco, one of Fasci leaders.[77]

In the first days of 1894, Crispi formally dissolved the Fasci and imposed martial law in Sicily. On January 4, demonstrators gathered at the gates of Milan and marched toward the piazza del Duomo. "The increasingly serious news coming from Sicily . . . brought crowds of workers and Socialists into the streets, crowds that increased gradually, alternately singing the 'Inno dei Lavoratori' and shouting 'Viva il socialismo! Abbasso i massacratori!' ", La Lotta di Classe reported. Two thousand persons sang the anthem in the Galleria, "making a solemn impression." Police and carabinieri broke up the crowd, which calmed down after the repressive forces passed by. Police reported thirty arrests, among them Enrico Bertini, aged 39, a printer and proofreader and member of the central committee of the P.S.L.I. Bertini was roughed up by the police, and Filippo Turati intervened to protest their brutality. He in turn was promptly seized by the carabinieri. Later when he was released, he went directly to the police chief to protest Bertini's arrest.[78]

Sicilian Socialists urged calm and the Socialist parliamentary delegation decried the protest as "spasms symptomatic of profound and ancient injustices" and contrasted the bloody Sicilian riots with the conscious workers' organization in their class party. (They underestimated, however, the contribution of the forces of order to the violence in Sicily.) There were new demonstrations in the piazza del Duomo on January 5 and 6, and for several days the police continued to report isolated protests and the appearance of subversive flyers.

Turati did not share the Socialist parliamentary delegation's eagerness simply to disclaim all responsibility for the disturbances. In its announcement of a meeting of the Milan section of the Partito socialista dei lavoratori italiani on January 16, 1894, La Lotta di Classe criticized the Statuto (the Italian constitution) for its grudging concession of liberty, which it perceived as the cause of the lamentable events in Sicily. Turati's purpose in convening his comrades was to reach consensus on the section's position vis-à-vis the condition of the Sicilian workers and the government's "shameful action.'' *La Lotta di Classe* reported that although the session was ''private,'' it was crowded with police agents, *provocateurs,* that is, ''everybody except the police chief as honorary chairman.'' Turati's speech ridiculed the weakness of the bourgeois parties and announced that the Socialists would prepare to assume the struggle for collectivism. Socialist propaganda in Milan, he observed, was being greeted by apathy, but other sections were more active. An anarchist denounced Turati for having called aspects of the Sicilian protest *vandalism;* what was needed, according to the speaker, was ''action, not chatter.'' Although Ettore Cicotti argued that the Socialists needed to hasten the dissolution of the bourgeoisie, Casati insisted that it was no time to be criticizing one another. There would never be a time when the whole population was in revolt, and so the party should profit from this insurrection to force a ''turn toward Socialism.''

A motion by Turati was approved that claimed that the Sicilian repression was a ''manifestation of the fierce struggle of the bourgeois class [to retain power].'' ''Full and active solidarity with the oppressed of Siciliy and the world who struggle for proletarian emancipation'' was necessary, it advised, and it pledged ''to redouble efforts . . . to hasten the Socialist revolution.'' *La Lotta di Classe* congratulated the meeting because ''good sense, or better, the practical sense of the situation, had triumphed.''[79]

Kuliscioff, Turati, and Engels: The Question of Alliances with Bourgeois Parties

When the Sicilian rebellion was joined by one in Lunigiana (an insurrectionary strike by marble quarry workers), Turati began to regard it as pre-revolutionary. Accordingly, in an article in *Critica sociale* entitled ''La Sicilia insorta'' (Insurgent Sicily), he wrote not about bourgeois reaction but about mass revolt. At the same time, however, he did not see the democratic parties as potential leaders or shapers of the movement. Whereas most Socialists believed that the members of the Fasci were primitive rebels, far removed from modern scientific Socialism, Turati—and Kuliscioff—were developing an interpretation that saw Socialist support and solidarity as a necessity.[80]

At the end of January 1894, Kuliscioff decided to write to Friedrich Engels, asking his advice about the situation and recruiting his authoritative support. In her letter, she described Italy as ''two-thirds medieval'' country whose discontented masses were a mixture of peasants, the unemployed, ruined smallholders, and petty shopkeepers, all of whom lacked any consciousness. The crisis was economic and moral, but the corrupt government responded by imposing a state of siege. The bourgeoisie was not like that of 1789, ready to seize leadership; rather, the Democratic Radicals and Republicans had neither organization or consistent politics. Socialists could not make a revolution in such a country. The possible emergence of a republic seemed to be the only positive aspect of the crisis (to Kuliscioff, at least; it appears that at this point she was speaking for herself, according to Lorenzo Strik Lievers). Turati added a postscript, indicating that he shared Kuliscioff's concerns.[81]

Engels's reply argued first, that the classes that Kuliscioff had described as lacking consciousness had the capacity "to furnish the combatants and leaders of a revolution." Second, Engels reasoned that although the Socialists should take part in the agitation, the preparatory initiatives and direction of the movement ought to be set by the democratic parties. The Socialist party should maintain its independence, ally tactically with the Radicals and Republicans, but be ready to go into opposition with a new regime. What concerned Engels was that the Socialists not assume the tasks of the democratic parties. Turati pursued his critique of the democratic parties, insisting on the necessity of the various sectors of the bourgeoisie unifying. Engels's letter was published without comment in *Critica sociale*. Although individual issues of the Socialist press were censored in the winter, spring, and summer of 1894, most government repression took place outside Milan, thereby permitting something of "politics as usual" throughout the period.

In the supplementary parliamentary elections of March 1984, the Democratic Radicals did not support Gnocchi-Viani in his run against Colombo. They were more concerned about curbing the growth of Socialist votes than welcoming progressive candidates. The Socialists interpreted this position as an opening for a potential alliance of Moderates and Radicals. "That is indeed the local solution of the situation of the Milanese parties," wrote *La Lotta di Classe,* "one that . . . we anticipated."[82]

In the partial administrative elections in June, the Socialist candidates ran on a slogan similar to that of 1893: "No alliances, a single Socialist list." The proportion of votes won by the moderates declined, whereas those of Catholics, Socialists, and Democratic Radicals increased. Mayor Vigoni resigned and refused renomination, and so a royal commission was appointed to govern the city until February 1895, when full local elections were scheduled. The Socialist interpretation was that there was a polarization under way between left and right, Socialist and bourgeois, with fewer differences among the bourgeois parties. Maurizio Punzo points out that the opposite interpretation is equally possible, that the polarization of left and right was splitting the bourgeois parties, moving the Democratic Radicals, Republicans, and Socialists closer together and leaving the Catholics and Moderates at the opposite pole. Lorenzo Strik Lievers believes that Turati and Kuliscioff were beginning to withdraw from their antialliance position, as evidenced in their remarks about the large divisions of opinion among the various sections of the bourgeoisie, and its structural backwardness. Indeed, in June, the Democratic Radicals broke with Crispi. Turati responded to those who insisted on the intransigent policy of Reggio Emilia: "The coherence of tactics consists of modifying them according to changing circumstances."[83]

Parliament granted the government additional exceptional powers against all types of political activists, but particularly against the anarchists, through a "revision" of the electoral lists in July. In August, the prefect of Bologna prohibited the Socialist party congress, scheduled for September 7–9, from meeting, and on October 16, the prefect of Milan dissolved the Partito socialista dei lavoratori italiani and all the Milanese organizations affiliated with it, including the Consolato operaio and the constituent leagues of the Chamber of Labor. Then on October 22, Crispi extended the decree against Socialist organizations in all of Italy.

La Battaglia reissued its number 30, changing its subtitle from "Giornale del Consolato operaio: Confederazione del partito socialista dei lavoratori" to "Giornale dei socialisti milanesi" (newspaper of the Milanese socialists) and continued to publish. It printed in full the decree of dissolution issued by Antonio Winspeare, prefect of Milan. In it, he pointed out that the minutes of the congress of Reggio Emilia revealed the party's subversive ends and that its influence was evident also in the revolutionary movements in Sicily and Lunigiana.

The Democratic Radicals, Republicans, and Socialists promptly organized the Lega per la difesa della liberta (League for the defense of liberty), which would be "above parties, schools, personal and group tendencies. Its goal is to defend civil society." *La Battaglia, L'Italia del Popolo, La Lombardia, La Lotta di Classe,* and *Il Secolo* joined in accepting contributions and memberships.[84]

This was the moment of Turati's *svolta* (turn) toward cooperation with the Democratic parties. He recalled Engels's letter in an essay on his new position but explained that he was not seeking an alliance for a joint progressive offensive, as Engels had recommended, but in order "to defend the most elementary liberties."[85] Socialists and Democrats had been pushed into cooperation by the government's repression.

Conclusion

The political process of the Italian Socialist party's formation was shaped by the disappearance (to a large degree attributable to repression) and arrival of other institutions and by changes in individual and group strategies and goals. Economic structural change had forced some workers into forming new relations with the capitalists. Whereas before, many workers had been satisfied with patron–client relations with Democratic Radical or Republican bourgeois, the Partito operaio had offered new approaches to deal with new problems. Although the P.O.I. (and independent leagues as well) had made successful innovations in the economic arena, organizing and mobilizing workers for workplace struggles, it was less sure when it ventured into the political arena. Its leaders disagreed about the forms and arena of political action. They were challenged on the extreme left by anarchists who objected to the slightest move in the direction of politics. Equally important, however, was the fact that the government (and other bourgeois parties, including the Democratic Radical) distrusted and feared independent worker ventures into politics. Government repression, combined with its leaders' differences on the value of political action, weakened the Partito operaio enough that it was unable to agree on a strategy and act on it. At the same time, new institutions like the Chamber of Labor and occupation-based resistance leagues entered workplace struggles and eroded the party's former role. Partito operaio loyalists like Gnocchi-Viani, Lazzari, and Croce all were active in this process, which may be more precisely described as a centralization and specialization of function rather than a substitution.

The 1886 government repression had the effect of modifying progressive strategies and ways of thinking about politics. Turati began to study Socialist principles in the Unione democratico–sociale and the Lega socialista and to incorporate them into his democratic political philosophy. His goal became the development of a coherent progressive politics, and his strategy, that of building a Socialist party of conscious, participating workers that could turn Italy toward a more democratic and just political path. To be sure, his colleagues Kuliscioff and Labriola influenced him in this process, but it also made sense in terms of his own previously held ideas about democratic decentralization.

In the first years of the 1890s, Turati (and Antonio Labriola) were meticulous in their concern that any Socialist party be firmly linked to the workers' movement. Certain worker leaders agreed. *Operaisti* Lazzari, Casati, and eventually Croce joined Turati in the Lega socialista. Only Croce followed the path into political socialism with little hesitation. Both Lazzari and Casati balked and protested at times; Casati, most strikingly, joined the anarchists in 1892 in refusing to accept the Partito dei lavoratori italiani. Workers connected with the Consolato operaio, like Maffi and Dell' Avale— to varying degrees still influenced by the Democratic Radicals—moved closer to the Fascio dei lavoratori (another Partito operaio successor organization) on economic issues, municipal socialism, and decentralized reform. These reform socialists, pragmatic and not always consistent in their beliefs, were ready to go along with the Socialist party projected in the 1891 national workers' congress and, with compromise, prodding from Turati and his colleagues, and a sense of urgency based on illiberal contemporary politics, with the party's formation in 1892. In the end some *operaisti* and reformist workers chose political unification with nonworkers who supported their rights to associate according to their interests, over full class autonomy.

According to Turati, Socialist intellectuals were a necessary prerequisite to party formation because of the diversity of the Italian workers' experience and the unevenness of their consciousness. Indeed, despite their autonomous collective experience, the ideology that the *operaisti* developed in practice, and their considerable success in economic struggles, it is unlikely that once the party was founded, the workers could have captured even the limited national political ground that the intellectuals did. It was not only a case of more coherent theory, as the next chapter shows, but their control over party publications and the kind of experience that the intellectuals contributed that was unique. The intellectuals had a working knowledge of what were then wholly bourgeois institutions like city councils and parliament, and a sophisticated awareness of the historical development of these institutions that soon led to their dominance in the Socialist party and the fading of its working-class leadership. The intellectuals were probably more eager for political success, but the pressures threatening the party's autonomy and simple functioning were powerful and real. The struggle for equality and justice for workers took a back seat to fighting for prerequisite democratic rights and local reforms.

How important were international developments, particularly the rapid growth of the German Social Democratic party and emulation to the formation of the Italian Socialist party? These factors doubtless affected individual attitudes and sense of possibility, especially that of intellectuals like Antonio Labriola, Turati, and Kuliscioff. Manacorda, as this chapter noted, tries to extend this argument to those workers who attended the Paris workers' congresses in 1889, but his argument is speculative. The effect of changing patterns of economic activity and government repression of workers' institutions in Milan on both *operaista* and reformist workers' strategies was, in my view, a much more powerful influence. Analysis of these factors in Milan provides evidence that resolves the false paradox posed by Manacorda and others who perceive industrialization and the formation of class institutions as necessarily synchronic. The workers who were the major architects of the Socialist party's formation (and that of the Chamber of Labor) were workers who had experienced the technological and organizational changes of industrialization and been thwarted in their attempt to build an autonomous workers' politics based on their experience in workplace struggle. Allying politically with intellectuals in a Socialist party was a viable alternative to the failed economistic and autonomy-focused strategy of the Partito operaio and to the clientage relationship of reformist workers to the bourgeois democratic parties.

The next chapter shows that given the diversity of Italian workers, their inability during this period to act on their interests as a class on the national level, and the continuing political repression, the Socialist party's activity at the end of the century was severely limited. Nationally, the party was most concerned with the expansion or protection of democratic rights and with antimilitarism. Locally, the party's major issues were subsistence, taxation, and school reforms. On these matters, the critical question was to what extent the Socialists should work with the bourgeois democratic parties to achieve their goals.

10

The Socialist Party Redux
and the Fatti di Maggio,
1894–1898

Crispi's repression of the Socialist party and its affiliates and the prompt organization of the Lega per la difesa della liberta ended the Socialist party's formative period. These events also heralded a new period of party debate over its appropriate electoral tactics, accompanied by sporadic repression. Although many of the party's leaders lived in Milan, its business was often conducted there, and its publications were originally quartered in the city, the crisis that lasted from 1894 to 1901 must be seen in its national context. The party aspired to be a national institution (despite uneven membership and votes in the different regions) whose politics addressed the national arena as much as it did that of Milan; accordingly it founded a national newspaper published in Rome, joining the Socialist parliamentary group as the most influential institutions in the party. Its implicit claim was that it spoke for an Italian working class.

Uneven and various as was the working class of Milan, however, the national one was even more disparate, as Giuliano Procacci has shown. Using the 1901 census, he calculates that only 23.8 percent of the Italian labor force was engaged in industry (the production of goods in any type of setting, thus including artisans, small shopkeepers, salesmen, and white-collar employees but excluding waged-service workers like those in transportation, communication, and utilities), and 58.8 percent of the Italian labor force worked in agriculture. If only wage earners in large-scale (factory) production are considered, they amount to about 18 percent of the work force. Of these, about 40 percent were women, adolescents, and children because the chief large-scale industry was textiles.[1]

Only a small number—again unevenly distributed among occupations—of these various Italian workers were members of worker institutions: mutual benefit societies, improvement and resistance leagues, or chambers of labor. The membership of the Socialist party, formed by workers and intellectuals, became less representative of workers in the period after the Crispi

240

repression, as workers' institutions went their own way in order to avoid political repression because of affiliation with the party. Leo Valiani points out that Italian socialism was no longer almost completely proletarian, as it had been in the 1860s, but "something broader." According to a 1903 Socialist party membership survey (reported by Robert Michels in 1921), 42.3 percent of the respondents called themselves workers. Although this proportion was widely viewed as a sign of worker underrepresentation, it precisely reflects the proportion of production and service workers in the labor force. (If the category referred only to production workers, they would be overrepresented in the party, as they accounted for only 23.8 percent of the labor force, as Procacci shows.) Professionals and intellectuals also were overrepresented; it was the agriculturalists who were greatly underrepresented. The number of workers in the party, was only a small proportion of the total number of Italian workers, but they were not underrepresented compared to other groups. Less than half of workers' votes went to the Socialists, as Michels documented in 1908. (Italy was not alone in this respect. Although the German Social Democratic party attracted most of the workers' votes, many fewer workers voted for the Socialist and Labor parties in England, France, and the United States during this period.)[2]

The paradox of Socialist aspiration to the position of a class party in the national political arena was at least partially resolved in the crisis at the end of the nineteenth century. Several national working-class institutions—the first of them the Socialist party, soon joined by a federation of chambers of labor and federated unions—were founded in the two decades after 1890, and after 1901, they became national political actors. But what happened to the Socialist party in the political struggles between 1894 and 1901 indelibly marked and shaped Italian workers' politics in the future.

This chapter picks up the story of Milanese workers' collective action in the aftermath of Crispi's 1894 repression of the Socialist party and follows them into local politics and the national party regeneration of 1895 and 1896 when they, together with other Milanese, contributed to the downfall of the Crispi cabinet. The chapter also reviews the local political action of the Socialist party in 1897 and 1898 and its increasing electoral strength, in both Milan and Italy, to the dismay of the national government and bourgeois parties. The alarming rise of food prices in those years was accompanied by protest in many parts of Italy. In Milan, the protest about food prices was organized by the Socialists, until forbidden by prefectoral decree in January 1898. Dissent flowed in other channels, however. In May, a mass Milanese worker protest over the arrest of men distributing flyers calling for political liberties was forcibly put down. Thus began the Fatti di maggio, the three days of workers' demonstrations and violent repression that ended with the arrest of activists, parliamentary deputies, and newspaper editors; the dissolution of the Socialist party, the Chamber of Labor, and other organizations; and the closing down of the democratic opposition press. The chapter concludes with a discussion of contemporary and historical interpretations of the Fatti di maggio and an analysis of the rebellion's relationship to political processes under way in Milan and Italy in the last decades of the nineteenth century.

The Workers and the Repression of 1894

On the same day (October 23, 1894) that the Lega per la difesa della liberta published its call for the restoration of civil rights—free press, free association—"without which the nation cannot exist," the headquarters of workers' organizations at the Chamber of Labor were searched and their records seized. The chamber leadership vigorously protested. Although some workers pressed for action, the majority, headed by Filippo Turati, decided on a prudent course: joining with the broadly based Lega to protest the Crispi provocation. On October 24, forty-seven worker societies meeting at the Camera del lavoro voted to cut their bonds to the Socialist party but to continue their economic activities—cooperation, mutual benefit, improvement, employment service—within the chamber's framework. Four days later, the authorities unsealed and restored the worker societies' organizational records and funds. One of the first acts of the Lega per la difesa della liberta was a national campaign for the pardon of worker activists throughout Italy convicted by military courts under the exceptional laws passed earlier that year.[3]

Organizing quickly resumed, and the new leagues included the Federazione femminile fra arti e mestieri (women's mixed federation of arts and crafts), founded in early 1895. There was a marked step upward in the overall annual number of strikes in Milan in 1895 (as shown in Figure 6-1 and Table 6-1). Economic recovery from the long recession (described in Chapters 2 and 5), plus the resumed organizational activities of Socialist and union leaders, lay behind the increased strike activity.

The Milanese Socialists' Decisions on Electoral Tactics and the Congress of Parma

The tactical problems following the Milanese Socialists' decision to join the Lega per la difesa della liberta were brought up at its first meeting, on November 1, 1894. Democratic Radicals Felice Cavallotti and Giuseppe Mussi told two thousand supporters that the urgent task for all persons was to defend liberty. Turati was at once combative—attacking "the vampires that have reduced Italy to a conquered territory, sucking the blood of workers and now attacking them with bullets" and defensive—"affirming our union with the democrats does not mean abandoning principles." In Parliament, the Democratic Radical opposition to Crispi hardened, but no coalition of the extreme left parties that had established the Lega per la difesa della liberta emerged. Crispi's government was defeated in a vote on the Giolitti bank scandal investigation, but he obtained a royal decree adjourning the session in December. In the spring of 1895, the chamber was dissolved, and elections were called for May 26.[4]

The pressing question for the Milanese Socialists was whether their pact with the Democratic parties would extend beyond their defense of political rights to the local elections (scheduled for February 1895). At a meeting of Socialists on December 12, 1894, a minority argued that the bourgeoisie was a "single reactionary mass," thus supporting the termination of Socialist participation in the Lega per la liberta. Others suggested that the Socialists combine with the Radicals in various ways, but these proposals were rejected because they would destroy the party's autonomy. Turati and Anna Kuliscioff suggested another solution that, they argued, would give the Socialists a swing vote in any Democratic Radical victory and would serve to advance their minimum program.[5] The group agreed to support 56 of the 64 Radical candidates and present 8 of its own who would campaign separately in public meetings and through the Socialist press.

Arturo Labriola declared with heavy irony his surprise that the Milanese Socialists "from whom we have come to expect the most rigid . . . intransigent tactics" should propose their alliance policy for all of Italy. Lazzari too felt that Turati had gone too far. Leonida Bissolati pointed out that such tactics were possible only in large cities, where "mature sentiments and concepts" existed. Maurizio Punzo observes that the Milanese did not insist that others emulate them, but simply on their right to determine local electoral tactics.[6]

For Turati, Kuliscioff, Claudio Treves (1869–1933, a Turin Socialist), Camillo Prampolini, and (by the end of 1896) Bissolati, the struggle with the Democratic Radicals against Crispi was perceived as critical to the right of dissent. Because of what they saw as the "backwardness" of Italian politics, the Socialists needed to promote the development of a modern bourgeoisie. Only then could workers wage their class struggle against industrial capitalism. Their justification for surrendering their own autonomy in electoral matters was the country's historical, social, and economic diversity, which flew in the face of any single electoral policy. Their choice was for democratic rights that would provide a supportive environment for political maturation and economic development, against Crispi's authoritarianism, his alliances with southern agriculturalists and the military, and his colonial policy. In contrast, Labriola would limit any alliance to a tactical one, organized simply to win the upcoming election.[7]

The clandestine Socialist congress in Parma on January 13, 1895, rejected Turati's analysis and his proposed modification of the position on electoral tactics adopted at Reggio Emilia in 1893. The Parma congress was not very representative: Whereas there had been well over 200 delegates representing over 100,000 members (the large majority being members of the Sicilian Fasci) at Reggio Emilia, there were only 64 at Parma. The delegates were mostly from nearby regions, and they met for only one day; yet they made three important decisions: First, they changed the basis for party membership and voting for congressional delegates from membership in an affiliated economic organization to individual inscription, with members grouped in sections of the Italian Socialist party separate from those in leagues or unions. Regional congresses would elect a national party council that in turn would choose a central committee. Second, the delegates charged the central committee with drawing up minimum administrative (local) and political (national) programs. Third, they censured the Milanese electoral alliance with the Radicals and confirmed the intransigent stance, modified, however, to permit Socialists to vote for other party candidates in runoff elections.

The first decision represented, according to Michels, the "triumph of the intellectuals." Letterio Briguglio describes the decision similarly as the beginning of supremacy of political over economic action.[8] There was no support for continuing membership through labor organizations. This debate divided those who wanted local Socialist electoral circles that would agitate for the minimum program (with universal suffrage as its primary goal) and those who insisted on personal membership and local "groups" that would send representatives to regional and national congresses. The latter won the majority's support. Ironically, it was Costantino Lazzari, who had upheld the *operaista* tradition of an economic base for the party at the Reggio Emilia congress, who proposed the winning formula. Years later, at another congress, in Milan in 1910, he justified his proposal on the grounds that the Crispi dissolution had struck resistance leagues and mutual benefit societies alike. He nevertheless firmly criticized, *ex post facto,* its consequences, "small minorities deciding the course of party." Furthermore, he wrote, "minorities are not composed of proletarian delegates but of persons who consider themselves worthy of interpreting the will of the proletariat."[9]

Other former *operaisti* were even quicker than Lazzari was to express uneasiness about the increasing dominance of political over economic activity in the party. Alfredo Casati was one. When he was named a candidate for local office in Gorgonzola, he stated in an open letter to *L'Italia del Popolo* (April 10–11, 1895) that he would refuse to run if he were expected to uphold voting as the only means to power. Lazzari promptly demanded Casati's expulsion from the party; Casati objected that the party was a "new tyranny" and contested the action but ended by accepting the Parma program. Osvaldo Gnocchi-Viani also voiced his misgivings, in an article in *Critica sociale* (October 16, 1895). There he noted that although the founding congress at Genoa had respected the *operaista* concern that workers' organization and action not be overwhelmed by the political agenda of bourgeois members, now the "political–scientific point of view" had come to prevail. The Parma congress, Briguglio believes, ended the fragile consensus among the proponents of the *operaista* tradition.[10]

The question of electoral tactics was hotly contested at Parma. Gnocchi-Viani submitted his views on the problem to *Critica sociale* (January 16, 1895) and advised "coalitions, not alliances"—a retreat from his earlier unmovable position. From the point of view of the editors of *La Critica sociale,* however, he was playing the game of the reactionary parties.[11] The journal stated that the party must remain separate and distinct in administration, elections, and politics. Under particular local conditions, it continued, the Socialists could support candidates of other parties on the first ballot if their victory would lead to the restoration of civil and political liberties. The majority supported permitting Socialists to vote for candidates of other parties "who demonstrate serious concern for liberty," but only in runoff elections. The congress then unanimously voted an exception for the case of Milan in the election campaign then under way.[12]

The electoral program of the eight Milanese Socialist candidates in the February 1895 election focused on four types of reforms: administrative, economic, fiscal, and educational. The most important administrative proposals were a call for a referendum on all important issues, new sanitary services, extension of the workers' role in the operation of charitable foundations, and weekly days off for all municipal employees. The economic reforms echoed those demanded by the Partito operaio in earlier years: awarding city public works contracts to worker cooperatives and requiring that such contracts honor minimum-wage and maximum-hour standards. Fiscal and educational reforms also sounded familiar themes, some of which became increasingly politicized in later local agitation: abolition of the regressive municipal *dazio* on necessities and its replacement by a progressive tax on housing, and free provision of noon meals and recreational programs for all schoolchildren.[13]

The 1895 city electoral campaign saw the spread of the polemical phrase *lo Stato di Milano*. The concept, labeled by Fausto Fonzi as a kind of Milanese *campanilismo* (parochialism), became a theme of the Democratic

Radical candidates' meetings. The words metaphorically contrasted hard-working, democratic, rational Milan with its enemies—knee-jerk reactionaries and corrupt politicians.[14]

The moderates won enough seats to form a coherent majority in coalition with the Catholics (41 moderates and 17 Catholics versus 20 Radicals and 2 Republicans). The Catholics had lost 10 seats, the Radicals 4 (both their worker/Socialist candidates and some of their chiefs, like Giuseppe Mussi, lost), and the Republicans 7. Crispi telegraphed his colleague Giuseppe Saracco: "Italy has conquered Milan . . . Providence has led us to victory!" For them, the exclusion of the Socialists and the votes lost by the Catholics, Radicals, and Republicans were cause for exultation.[15]

The Socialists were not discouraged, however. Although none of their candidates had won, the total number of their votes had risen from 1,744 the previous year to 4,196. In the Seventh Mandamento, the local electoral district roughly equivalent to the Fifth Collegio parliamentary district, the number of Socialist votes increased by 137 percent, a harbinger of victory in the next parliamentary elections. For the first time, the increase in progressive votes could be attributed nearly exclusively to the greater number of Socialist votes, many of which had been cast for Democratic as well as for Socialist candidates. Costantino Lazzari disagreed. In an interview with *La Sera* he claimed:, "I believe that it is absolutely essential that the Socialist party preserve its independence of other parties. If the Socialists had not supported the Radicals, the clericals would not have made a deal with the moderates, and the result would have been quite different."[16] In contrast, Turati and the *Critica sociale* credited the newly flexible electoral tactics for increasing the number of Socialist votes that might influence local politics in progressive ways. The local Socialist federation stepped up its efforts to register new voters,[17] a drive that was given greater urgency by the dissolution of Parliament and the scheduling of national elections in May.

In an "extremely lively and profound debate" about electoral tactics at a meeting of the city's Socialists early in April, the new electoral committee urged the nomination of a full slate of candidates in the province of Milan. Turati, Carlo Tanzi, and others "rejected coalitions and alliances" but argued that "given the exceptional conditions of current Italian political life," the party should not nominate its own candidates in districts where democratic candidates were likely to prevail. The motion supporting the committee report was passed. Defeated in his efforts to recruit support for a united effort to replace the moderates in the city government, Turati refused to run in the First College, a seat the Socialists were unlikely to win. He also did not attend the May 14 electoral meeting in the lobby of La Scala, supporting the Socialist protest candidacy (in the Fifth College) of Nicola Barbato (a leader of the Sicilian Fasci imprisoned on charges rising from that protest movement). Ettore Ciccotti, Gnocchi-Viani, and Lazzari spoke in praise of the candidate, only to be arrested for defending his criminal behavior.[18]

Barbato was elected on the first ballot in his race with Luigi Rossi, a liberal close to the moderates. Parliament set aside the election, declaring Barbato ineligible to sit. Anti-Crispi sentiment was reflected as well in the rest of the city vote. Giuseppe Colombo, a moderate who opposed Crispi, Republicans Luigi De Andreis and Pietro Zavattari, and Democratic Radicals Malacchia De Cristoforis and Giuseppe Mussi were elected. The Socialists supported De Andreis, De Cristoforis, and Zavattari in the runoffs. (Barbato was reelected in September 1895, in an unopposed election.)

The Milanese Socialists in Local Politics

The Milanese Socialists' commitment to greater independent action on behalf of their municipal program was not limited to elections. Pursuing their fiscal reform proposals, *La Battaglia* opened its pages to discussion of the city's fiscal problems. It proposed abolition of the *dazio,* making the city "open," and substituting a direct progressive tax that would shift the burden to wealthier persons and raise more revenue for the city. It sarcastically reviewed the "achievements" of the bourgeois administration of Milan from 1885 to 1895, including heavy expenses for monuments and gardens that had been opportunities for building and bank speculation. Now the city was faced with a deficit. The Giunta's Ferrario plan (named for the assessore who had drafted it), proposed to extend the urban *dazio* to include the external *circondario* in order to cover the budget deficit. It was Mayor Negri's 1886 *dazio* expansion plan, expanded even further. In order to ease the burden for the poor in both the suburbs and the center, the proposed system would exempt bread, flour, rice, and pasta from the duty. *La Battaglia* objected vigorously that the new proposal continued the old injustice, that the *dazio* was still regressive, hitting the poor harder. The external *circondario* would have to pay far more in duty than it now paid. Would the proposal actually generate the needed increased revenue?[19]

The Giunta's proposal was a partisan issue. There was a convergence of interest among industrialists and shopkeepers—whose costs of doing business in the external *circondario* would increase—and wage earners—who would have to pay higher prices for many items: The increase would not be covered by eliminating the *dazio* on some necessities. Producers, sellers, and buyers in the external *circondario* all would be hurt. The Socialists hastened to present their position to the public before the Democratic Radicals mobilized workers behind themselves. At a protest meeting convened by the Republicans, Comrade Dino Rondani proposed a referendum on the issue in which all adult citizens, male and female, could vote, as well as a Socialist alternative to the *dazio:* a progressive direct tax.[20]

The outcome of the city council's debate on the Ferrario proposal was inconclusive. Everyone agreed that the city needed additional income, but whether expanding the *dazio* was the way to do it was not certain. Industrialist Ernesto De Angeli felt that it would be unwise, for the external *circondario* was an important source of wealth to the city, and the proposed change might drive industry away. Mayor Vigoni, however, supported the plan and argued that it was necessary in order to make Milan a truly great, unified city. The council decided to appoint a nine-person committee to examine the problem further.[21]

Hotly contested national political issues (discussed next) distracted the Milanese population early in 1896. The Committee of Nine (composed of equal numbers of supporters and opponents of the Ferrario plan plus an uncommitted chairman) proposed in September to expand the *dazio* to the entire city but to exempt many consumer products and simplify the collection process, thus cutting costs. It suggested a new graduated household tax (*tassa di famiglia*), on a limited basis at first but eventually expanded. Mayor Vigoni immediately announced that the plan was unacceptable, that the committee had gone too far in freeing the city from consumption duties. Rejecting the household tax and expanded exceptions, he offered a counterproposal, essentially a reprise of the Ferrario plan of 1895. Rancorous debate resumed. The Milanese Socialist Federation reaffirmed its minimum program position, rejecting both the Committee of Nine's project and the Giunta's counterproposal; only total abolition of the *dazio* was acceptable to the Socialists. As the council debated the issues, Ferrario fell into a deep depression and killed himself, thereby abruptly ending the discussion for that year.[22]

In January 1897, a somber city council resumed its discussion of the two plans. Meanwhile, the opposition gathered its forces to take the issue to the voters. The Socialist Federation designated Turati to speak at a Democratic Radical protest meeting on January 17.[23] With him on the platform before about four thousand persons were industrialist Guglielmo Miani, Republican Luigi De Andreis, and Democratic Radical Giuseppe Mussi. Turati's argument followed the workers' logic: The proposed *dazio* revision would hit them harder, and so a progressive tax should be substituted for the regressive one. "We demand equality of sacrifice," he declared, "not an 'improved' *dazio* that will merely put off its abolition." The anti-*dazio* coalition sponsored a petition for which signatures would be collected on one day only. The Socialist flyer regarding the petition separated the party from other opponents who objected to the proposed change but supported the current system. The Socialists opposed both.[24]

The petition collected 26,000 signatures, but those supporting the two opposition views—against extension, and for abolition—became increasingly divided. Even before the coalition broke down, however, the council approved the Giunta's plan, 42 to 29. Implementation of the plan required approval by Parliament; a bill was introduced, but suspension of the parliamentary session and new elections postponed its consideration. The issue was resolved only in the aftermath of the May events in 1898.[25]

The Socialists sponsored a similar but smaller campaign in support of free schooling and subsidies for school lunches (also part of their administrative program). In May 1896, a Socialist electoral meeting protested the Giunta's plan to charge school fees and voted that elementary schooling was not only an essential communal function but also a right of the young that should not be tied to their parents' ability to pay a fee. The city should therefore facilitate children's access to schooling through free lunches, books, and school materials.[26]

La Battaglia resumed the debate in September. Mayor Vigoni had objected that free lunches would permit parents to avoid their responsibilities. The newspaper reminded its readers that workers also have rights and that free school lunches should be one of them! This demand was granted only after 1899, when a progressive coalition council was elected when Milan returned to normal after the May rebellion.[27]

Military Defeat in Ethiopia, Milan's Reaction, and the Fall of Crispi

The divisive but eventually all-encompassing national issue of early 1896 was the Italian colonial war in Ethiopia. In late 1895, Prime Minister Francesco Crispi, hoping for a military victory in the war against Ethiopia to bolster his colonial policy, sent General Oreste Baratieri to open a new offensive. Instead of the hoped-for victory, however, news of defeat came in early December. In the Chamber of Deputies the ministerial coalition urged sending a force powerful enough to destroy Ethiopian resistance and to establish an Italian colony there. Others asked for only enough reinforcements to avoid defeat and to retreat to a secure position. The extreme left demanded an immediate withdrawal from Africa. In Milan, there was no support for the ministerial position. Even the moderate newspaper, La Perseveranza, which had approved Crispi's repression and joined him in criticizing the constant "chatter" of Parliament, was dismayed. Crispi, contending that Italy's honor must be saved, nevertheless won an overwhelming majority vote for military credits. Then, when criticism of his policies increased, he suspended Parliament at the end of December 1895.[28]

Milan's opposition to Crispi increased. Republican Luigi De Andreis proposed a protest meeting against Crispi's African policy to be held in the

Arena, but the police, on orders from Rome, canceled it. Under pressure, Crispi reluctantly set March 5 for Parliament's reopening. At the same time, he pressed Baratieri for a victory that would improve the government's position at home.

Despite a police order forbidding meetings and demonstrations regarding the government's African policy, Milan's Democratic and Republican deputies (Mussi, De Cristoforis, and Zavattari) made a "public report" to their electors. The assembled Milanese cheered the deputies' declaration that "Italy rejects war on the territory of others" and their call for an end to Italian bloodshed and wasted resources.[29] *La Battaglia* reported Turati's impassioned criticism of the moderates, who had become anti-African only as costs mounted, and his warning to democratic politicians that talk was not enough. He urged the crowd to demand that Parliament order "the immediate return of the troops from Eritrea, the abandonment of the militaristic policy of conquest, and the substitution of the nation in arms for the standing army." Although his fiery rhetoric drew cheers from the crowd, a less radical motion was passed.[30]

The news of the army's defeat at Abba Garima near Adowa reached Milan early in the morning of March 3. Telegrams from Rome warned all prefects to guard against "subversives agitating the country." Newspaper headlines screamed: "Enormous Disaster!" "Massacre!" "The Vendetta of History!" Turati's request for permission to hold a protest meeting was rejected. Crowds gathered nevertheless in the piazza del Duomo, shouting, "Abbasso Crispi! Abbasso Umberto! Morte al Re!" As the day progressed, Antonio Winspeare, Milan's prefect, "overwhelmed by panic" (according to Fausto Fonzi), telegraphed Rome for military support.

Newspaper accounts differ on the subsequent events. In the Galleria, the crowd listened on to speeches by Zavattari, De Cristoforis, Turati, and Prampolini and then crowded into the piazza della Scala, gathering outside the Palazzo Marino (the city hall). Mayor Vigoni appeared on its balcony and managed to shout, "Today as never before, I share your sentiments." Singing the "Inno dei Lavoratori," the crowd milled through the central squares and the Galleria, as police with fixed bayonets arrived to drive them out. Shouts went up: "Abbasso le armi! Via le baionette!" (down with your weapons; put away your bayonets). Carlo Osnaghi, a young printer, was injured by a bayonet and later died. By midnight, the crowd had finally been dispersed by the police and cavalry. About sixty people were arrested, some for being known Socialist and anarchist militants, others for shouting, fighting, or throwing stones. Most were released quickly.[31]

Turati later reported in *Critica sociale* and in a letter to Felice Cavallotti his own effort to calm the crowd: "I believed that . . . it behooved the Socialists to be cautious and not to play into the hands of those who are our and your enemies." He wanted the demonstration to reaffirm the right to free assembly and no more.[32]

Going beyond claims to the right to assembly, there were new demonstrations on March 4 and 5 in central Milan. On the second night the crowd also attacked the train station. *L'Italia del Popolo* had published the news that a troop train would leave the central station that night at 11:42. Young men, mostly workers, some armed with sticks and stones picked up on the way, headed for the station. The first group broke windows but was scattered. The second group attacked the station after midnight, stoning the cavalry that, firing revolvers in the air, tried to break up the effort. No one was hurt, and only six were arrested. *Il Corriere della Sera* blamed this "vandalism" on the youth of the demonstrators. The age of arrestees during the three days of demonstrations (for whom information is available) was not particularly young, however, considering the overall youth of the city and its work force. The attack seems to have been deliberately destructive of property but not to have been planned.

The arrestees whose occupations were identified ran the gamut of Milan occupations: mason, shoemaker, mechanic, carpenter, bookkeeper, notary, and clerk. All classes were involved in the demonstrations, according to the news accounts. Those arrested were predominantly workers, who also made up the large majority of the population. Both the police action at the scene and the criminal prosecution later were limited by the consensual opposition to Crispi. The city administration, all parties and officeholding politicians, and all newspapers condemned the crowd's violence. On March 6, "good sense began to prevail."[33]

Crispi's forced resignation ended further repression for the time being. Antonio Starrabba Di Rudini, scion of a noble Sicilian *latifondista* family, was appointed prime minister. A politician with a less authoritarian approach, Rudini's reputation was anticolonial and antimilitaristic.[34] The new prime minister's first acts were ambiguous, however: both moving to end the colonial war and declaring his resolve to restore Italy's national prestige. Although the democratic left preferred the ambiguity to any other alternative, the conservatives were alarmed by Rudini's few concessions, for example, amnesty for political prisoners from the Sicilian Fasci repression.

The Socialists, however, were dismayed at the conservative outcome, and the *Critica sociale* blamed the Democratic Radicals for their lack of resolve. Because of their failings, "the revolution, which would have been essentially bourgeois, had its moment and failed." The Democratic Radicals, it revealed, had shared the shopkeepers' panic over the disorder. In Turati's opinion, it was they, not the Socialists, who had—and muffed— the opportunity to achieve "the definitive uprising." Turati's reasoning was reductionist in its expectation that the Radicals would somehow launch a democratic revolution. His criticism was not only a misinterpretation of the Radicals' political potential but was also based on a mechanistic and simplistic understanding of how revolutions come about. Most revolutions start with coalitions—some very loose and informal—that bring down a regime. The ensuing struggles among members of that coalition later determine who will attain power. Bourgeois revolution thus is an outcome of, not a recipe for, action. Turati eventually accepted the Rudini government as the lesser evil and even defended the vote by Giuseppe De Felice, a Sicilian Socialist, for a government bill. *La Battaglia* (March 21, 1896) and *La Lotta de Classe* (June 6–7, 1896) both characterized the cabinet in a deprecatory tone as one of *galantuomini*—gentlemen.[35]

Socialist Congresses and the Parliamentary Election of 1896

Nicola Barbato, the Sicilian Fasci leader who had been twice elected by the Milanese to their Fifth Collegio parliamentary seat, declined the honor in mid-March. This vacated a parliamentary post in a progressive district and opened up an opportunity for a Socialist candidate.[36]

Local Socialists met to prepare for the Lombard regional congress to be held in Brescia on April 26. They favored continuing the electoral tactics agreed upon at Parma: keeping the party separate from all others and permitting support of democratic candidates in runoffs only.[37]

The Milanese view of elections was the majority position at both Brescia and the Florence national congress. Delegates at the regional congress supported the proposed administrative decentralization of the party and reaffirmed its local organization in electoral circles, while also permitting individual membership. Bissolati and the Cremonese delegates had been converted to Turati's view on tactical electoral alliances, and he offered a motion to this effect. Bissolati argued that the Socialists should be permitted to support democratic candidates in districts where there was no likelihood of their own victory. This would help promote the access to power of progressive parties that would in turn defend political liberty and permit the growth of the Socialist party. His motion was supported by Turati, Kuliscioff (who noted that intransigence for which exceptions were constantly made was useless), and Croce (who spoke of the party's "flexibility"). A majority argued that the Parma decision permitting support only in runoffs should stand; approval of the Bissolati motion, they reasoned, would diminish the Socialist party's class character. The vote was 34 to 7, in favor of the intransigent policy.[38]

Satisfied with the outcome, *La Battaglia* wrote that the "most rewarding aspect" of the congress was the role played by proletarians in its debates, which contributed greatly to the party's vitality.[39]

His own Milanese comrades opposed Turati, but the discussion did not end at Brescia. Emilio Caldara, the future mayor of Milan, published a "Defense of the Majority" in *La Battaglia* (May 16, 1896). Caldara stated that the Brescia majority was a "deep current of opinion," representing not only the largest groups but also those with strong internal majorities. The mass of workers were too often "unconscious of their manipulation by the democratic parties." Concessions would only undermine the substance of the Socialist program. Turati hoped to make the Socialist party a decisive player in the Italian political world, whereas his opponents felt a class-based suspicion of the Democratic Radicals and their shopkeeper supporters, whose interests were antithetical to those of the proletariat.[40] Nevertheless, in the June election in the Fifth Collegio, Cavallotti and *Il Secolo* helped—or at least did not hurt—the Socialist candidate, Filippo Turati.

The Socialists announced in mid-May that they would nominate Turati as their candidate. In 1892, they had received only 352 votes in the district, but in May 1895, the protest candidate Barbato received 1,820 (against Luigi Rossi's 1,556), and in September of the same year Barbato, unopposed this time, received 2,195 votes. The Republicans decided not to run a candidate of their own, because a vote for Turati would be a vote against the monarchy. The two liberal clubs nominated Luigi Rossi, who had represented the district in Parliament from 1892 to 1895. The Moderate newspaper *La Perseveranza* expressed satisfaction at a candidate who would defend the monarchical institutions against a collectivist attack. The Democratic Radicals of *Il Secolo* did not endorse Rossi but remained neutral. The Catholics, following the papal injunction of the encyclical *Non expedit*, were expected to abstain from voting unless certain conditions obtained, such as that their vote could defeat an anticlerical. Given the choice between a Socialist an an anticlerical (Rossi), most of them abstained.[41]

La Battaglia's electoral manifesto declared:

> Only the Socialist party effectively fights the political system that, via militarism, colonial adventures, indirect taxation of those who own nothing, large-scale banking and administrative robberies, and suppression of liberties guaranteed by the constitution, has trampled all sense of civil progress [and] justice . . .; only the Socialist party promises immediate improvement of the country's economic conditions.[42]

Turati's victory was unequivocal: 2,210 votes against 1,487 for Rossi. *La Battaglia* exulted that it was "the cause that makes us strong."[43] The victory was a disturbing bellwether for bourgeois moderates and liberals alike. The fact that Catholics' abstention had clinched the victory also troubled the Right, for the Catholics, like the Socialists, were considered "enemies" of the monarchy and the constitution.

La Lombardia concluded that the Socialists won not because they were a majority "but because they are strongly organized . . . the liberals lost . . . because no cohesive force binds them together." The liberal *Il Corriere della Sera* dryly remarked that "yesterday's election does not mark the ruin of Italy." For the Republican *L'Italia del Popolo,* the final vote was the expression of a "powerful political current." *Il Secolo* praised the Socialists for "the most difficult virtue: that of being temperate and serious in victory." And *La Perseveranza* published the news without comment.[44] The Socialists were serious and hardworking and were inspired by a belief in the historical inevitability of their cause. The other parties had reason to be concerned.

In his campaign, Turati had seemed at times to return to his pre-1894 position: that the democrats must be split, some finding refuge with more conservative parties and other joining the Socialists. At other times, he simply pressed the Socialists' minimum program (universal suffrage, referendum, abolition of standing army, and tax reform), which at base was democratic. Turati's own analysis of his electoral victory was that it was the result of his supporters' loyal efforts, Rossi's unpopularity, and the switch of the district's railroad workers' votes to the Socialists. He warned about increasing repression.[45] The Florence Socialist party congress (July 11–13, 1896) offered him an occasion to reiterate his call for flexible electoral tactics depending on local conditions.

The congress considered three major issues: the organization of the party, the relation of the party and the agrarian classes, and electoral tactics. Costantino Lazzari presented the executive committee's proposal for party organization: a tightly knit, disciplined network of electoral groups at the commune, parliamentary district, and local administrative level. The plan was complex, designed to avoid any strong personal influence of Socialist politicians on the masses. The debate was vigorous, for many delegates believed that the exclusively electoral sections were too limited. A compromise added economic groups to the permissible forms of organization. Members would be individually linked to the party through their payment of dues. Thus, Briguglio notes, the party was to be more than simply an electoral mechanism. The congress adopted the Partito operaio's workerist emphasis on economic action and occupational or industrial sections, but for the last time.[46]

Turati was a favored speaker in support of strikes, which were more numerous in 1896. In August, for example, he spoke on behalf of the sand pit excavators, who had struck after it became obvious that the raise their bosses had promised them was not forthcoming. Turati criticized the employers' bad faith but maintained that neither striking nor establishing a cooperative would resolve the workers' dissatisfaction. What was needed was the formation of Socialist consciousness and the conquest of public power, which were the party's goals.[47] Socialists put local strikes into the context of the Rudini government's increasingly antilabor and antiprogressive policies. Turati took bitter satisfaction in September, in writing that the Rudini cabinet "was steadily becoming more explicitly reactionary, as the Socialists . . . have predicted since its beginning."[48]

A counterdemonstration on September 20, a Risorgimento holiday celebrating the capture of Rome in 1870, offered Turati an opportunity for sharp analysis and democratic political demands.

> After twenty-six years, we are here to condemn the failure of a heroic enterprise [the unification of Italy]. . . . [W]e remain . . . the poorest people of Europe, the most illiterate after Spain . . . the most defeated in war. . . . Every year we send thirty thousand of our sons away in search of bread. . . . The constitution is ignored . . . institutions abolished. . . . We Socialists . . . respect the law, not its violation.

The angry speech ended with a demand for the recognition of popular sovereignty through universal suffrage.[49] On December 9, 1896, Turati carried his protest to the floor of Parliament, introducing a motion denouncing the "government's continuing violent and illegal policies" and calling for "respect for public liberties." It was defeated.[50]

The end of the year was marked by further government repression of worker and Socialist institutions: On December 13, the Genoa Camera del lavoro was dissolved, as were all the Socialist circles in the region; this was followed on January 6, 1897, by a similar decree against the Chamber of Labor of Rome. A new Socialist party national daily newspaper, *Avanti!*, went on sale on December 25. (*Avanti!*, edited by Leonida Bissolati, was published in Rome, thus contributing to the greater influence of the Socialist Parliamentary group.)

Support for Socialist dissent came from a surprising new source in the new year. Don Davide Albertario, the priest who edited the *Osservatore Cattolico*, a newspaper that combined firm opposition to the secular state and support for social Catholicism, wrote: "We are not Socialists, but we recognize that in a liberal system, the Socialists ought to be respected." Rudini and his supporters saw this as an alarming sign of convergence of the two extreme antiregime positions. Sidney Sonnino, a conservative member of Parliament who had been a minister and public official in many governments, anonymously published a polemical essay urging "Torniamo allo Statuto!" (Let us return to the constitution). He believed that events were pushing the nation toward extremist politics, "threatening all liberties, moral, intellectual, political, and civil." His suggestion was to give a larger role to the king, as permitted by the constitution. This antidemocratic, antiparliamentary program did not attract much support, however.[51]

National concerns crowded the agenda of Milan's socialists, and local demands also increased. A letter from Felice Anzi to *La Battaglia* (January 16, 1897) suggested that the party was too centralized and that local initiative was rare. Members felt that some officials were accumulating too many responsibilities, he reported. "Where are the speakers' bureau, the school of theory and practice, the discussion groups?" he asked, calling for a richer Socialist popular culture.

The earlier Socialist initiative establishing the Federazione femminile had increased women workers' activity, including strike participation. The federation celebrated its second anniversary at the end of January 1897 with the inauguration of its banner. Socialist teacher Linda Malnati spoke at the occasion: "Let this banner be one that illuminates the minds of the thousands and thousands still distant from us, out of ignorance, fear, or apathy." [52]

Parliamentary Elections, 1897

The Democratic Radicals hoped to press Rudini (although continuing to vote with him on some issues) to call elections, which they believed would be to their advantage. Although they rejected the Radicals' strategy, the Socialists joined them in a common defense of liberties. Antonio Maffi— sometime Socialist worker activist and now back in the Radical fold—declared in *Il Secolo* (July 16–17, 1896) that there was no split between socialism and democracy, that cooperation between the parties could be helpful to the proletariat." [53] In September 1896, Cavallotti, speaking to his constituents at Corteolona, explicitly warned Rudini that elections should be held immediately, or the Radicals would join his opposition.

Parliamentary elections began to look like a good strategy to Rudini, under pressure from both the Socialists and the Catholics. He hoped that elections would rid the Parliament of Crispi's last supporters and strengthen the constitutional parties, thus opening the way for cautious reforms like decentralization. The Milanese Moderates opposed Rudini's plan, for they feared that elections would favor the local Socialists. Rudini suspended the parliamentary session and scheduled an election for March 1897. The Milanese moderates, determined to present a united front, nominated four safe conservatives in the internal *circondario* electoral districts. These candidates were promptly supported by *La Perseveranza, Il Corriere della Sera,* the Associazione costituzionale, the Circolo popolare, and *La Sera.* Radical Giuseppe Mussi was nominated in the Sixth College (the external *circondario*) with the support of the government and *La Sera. Il Secolo* hesitated but eventually recommended him and Radical-Republican Malacchia De Cristoforis. *La Lombardia* favored the same two candidates. *Il Secolo* also supported Republicans Luigi De Andreis (First) and Pietro Zavattari (Fourth), and Socialists Osvaldo Gnocchi-Viani (Second) and Filippo Turati (Fifth).

Turati warned in the *Critica sociale* that Socialists who rejected electoral agreements with the democratic parties were making a conservative victory more likely. But the Milanese Socialists did not agree, and the intransigents prevailed once again, nominating candidates in each college. They resolved to campaign only on their own issues and to accept support only from those in full agreement with them. The two-pronged Socialist program was as follows: first, a workplace struggle by craft for immediate improvements in workers' lives—including political and wage equality between the sexes—to be achieved through the Chamber of Labor, and, second, a broader struggle to gain public power as an instrument for the economic and political expropriation of the dominant class.[54] The Republicans also chose to reject any alliances and nominated candidates in every district.

In the ensuing campaign, *La Battaglia* published dozens of breathless articles, reporting in great detail on the lives, ideas, and programs of their candidates. Among them was an unusual appeal to women, entitled "Women, Professionals, Mothers of Families," which complained about women's lack of rights, "as though we have no interests to defend." The Socialist party was the only one that supported women's economic, political, and legal equality with men. The article concluded with a request for women to work for the election of Socialist candidates who would represent their interests.[55]

On election day, March 21, the consequences of the separate Republican and Socialist candidacies became clear. Of all their candidates, only Turati won on the first ballot. Moderate Giuseppe Colombo also won outright. In three districts, runoffs between Republicans and moderates and, in one, between a Socialist and a Radical were necessary. Republicans and Socialists congratulated themselves on the substantial increase in the number of votes for each party since 1895 (35 percent for the Socialists, 31 percent for the Republicans), but their separate candidates had saved the moderates from a serious defeat. The Republicans and Socialists quickly agreed to vote for each other's candidates in the runoffs. Socialist professor Ettore Ciccotti's candidacy was challenged on the grounds that he was an employee of the state. This was cleared up, but *La Lombardia,* which supported Mussi, continued to contend that Ciccotti was ineligible, thus confusing the voters. One Republican, two moderates, and the Democratic Radical Giuseppe Mussi won the runoffs.

In Italy as a whole, Rudini's 1897 election did rid Parliament of the last Crispi supporters, but the rest of his strategy failed. The number of seats held by deputies on the extreme left rose from 63 to 75; the number of Socialist seats alone increased from 12 to 15, and the votes cast for them, from 76,000 to 135,000. Rudini had retained majority support, but his position was no stronger. In a policy of "liberal concentration," bringing together the right and left constitutional parties against "extremist" Socialists and Catholics, Rudini persuaded Giuseppe Zanardelli, spokesman for the constitutional left, to become president of the Chamber of Deputies. This maneuver won Rudini the support of some Lombard conservatives, including Giuseppe Colombo, who formed a new group dubbed the "dissident right."

Rudini reintroduced a law limiting association, and a failed assassination attempt against King Umberto led to the proposal of a law prescribing *domicilio coatto* (forced residence in exile, a kind of house arrest) for anyone who contributed to the "subversion of the social order." (The vaguely worded law was eventually passed in 1898, as protest swept the country, and was invoked well into the twentieth century.) Rudini thus reintroduced one of Crispi's exceptional laws, no longer for a fixed term, but for an indefinite period.[56]

The Bologna Socialist Congress, 1897

Writing in *Critical sociale*, Turati called again for unity of all progressive forces. "The present Rudinian reaction," he warned, "has revived the evils of the Crispi government." The Socialists and the Radicals launched a joint protest campaign against the new repressive law. (Republicans prefered to carry on their protest separately.) The Milanese Chamber of Labor added its support to the campaign, branding *domicilio coatto* a "persecutory law . . . that threatens the working class."[57]

In this climate, the Socialists met in Bologna on September 18–20 for their national congress. Turati believed that it was urgent, given the current conditions, for the Socialists to ally with parties with similar goals in order to achieve a truly democratic political system. But neither the relevant parties nor his Socialist comrades responded to his call.

On the eve of the congress, Turati published a reflective self-criticism in *Critica sociale:*

> The [Crispi] dissolutions pulled us violently apart from workers' organizations. We believed we could resolve the difficulty, adapting ourselves to these conditions, by basing the political party on individual membership. But part of the mass of workers did not follow us, and another part, which did join us, stopped acting like workers.

Turati's prescription was greater effort by the party to strengthen its worker base.[58] The party majority still opposed Turati's position on electoral tactics, but its numbers had diminished.

In a report on the party's economic action, Kuliscioff declared that politics as a principal activity "should not go forward separately from economic [activity], as the latter alone guarantees the class nature of the party and maintains its revolutionary character." Her proposal (which was accepted) specified two ways in which this could be done: first, by participating in resistance movements and, second, by pressing for protective labor legislation.[59]

At the previous year's congress, Arturo Labriola had proposed a minimum program based on the German Social Democratic party's Erfurt program, but it had been labeled inappropriate to the Italian situation, and so discussion was postponed until 1897. Now Turati's report on the minimum program, which Labriola had refused to sign, was debated. The minimum program was not a Socialist program for governing, according to Turati; that would come only after the Socialist conquest of public power. Preparing the way for "the evolution of superior forms" (the socialization of the means of production) the minimum program pushed for changes compatible with the current economic organization. The Socialist party should continue to struggle for its specific line against other parties' programs.[60]

Regarding electoral tactics, the "case of Cremona," in which Socialists had supported Radicals in order to prevent a conservative victory in the city elections, provoked a lively debate. Turati opposed sanctions against the Cremonese, but they were passed anyway, and the Florence congress's unmoving position was reaffirmed. This time, however, an amendment permitting exceptions if authorized by the party directorate was also approved. Turati, pessimistic before the congress, welcomed its outcome in *Critica sociale* (October 1, 1897).[61]

Food Prices and Popular Collective Action

Even before the Bologna congress, rising food prices had led the Milanese Socialists, working with members of bourgeois parties dissatisfied with the government's policy, to agitate for the reduction or abolition of the grain tariff.[62]

Rudini had no program for addressing Italy's problems, only a desire to retain power. He turned to *trasformismo,* a tactic frequently employed in the unstable Italian parliamentary governments under the *Statuto* of 1848. When his cabinet resigned, joining Minister of War Luigi Pelloux's protest against changes in a bill on army officers' promotion, Rudini sought an alliance with a constitutional left politician. He persuaded Giuseppe Zanardelli to join his new cabinet in December 1897 as minister of justice, thus reconstituting a viable majority in the Parliament in favor of his continued tenure as prime minister.[63]

In Milan, there were newspaper campaigns and protest meetings regarding the continued high military expenditures and consumer prices. High bread prices were supported by the grain tariff and were aggravated by fewer shipments of cheap American grain. For example, in January 1898, the Camera del lavoro denounced the high price of bread as the outcome of "speculation transformed into a system," and the Socialist speaker for a masons' league protest blamed it on the existence of private property, which led to speculation." [64]

A police inspector's report to his chief on January 20, 1898, summed up public opinion (*lo spirito pubblico*). So far, despite the pressure on poor people's food budgets, they had tolerated the price rise well. He criticized "agitation on the part of the Socialist party, which misses no opportunity to spread its unhealthy doctrines among the masses." There was a possibility of riots even in Milan. Shortly thereafter, the Rudini government reduced the tariff on wheat by one-third, effective until April 30, and at the same time recalled to active duty the class of 1874 (which had already completed its military service commitment). (In a later series of decrees, the reduction was first extended, and then the tariff was suspended altogether.) On January 25, Milan's prefect, Antonio Winspeare forbade public meetings in regard to food prices. The Mutual Benefit Society of Bakery Proprietors, proclaiming themselves sensitive to the "general distress over high bread prices, the result of the small harvest in part, but also of the high tariff and the city *dazio* on flour," voluntarily cut bread prices. The issue cooled in Milan. [65]

Popular collective action—protests against municipal taxes, objections to hated officials, and demands for jobs—had begun in January in Sicily and the south. In February and March, demonstrations, often turning violent with police repression, focused more often on food prices in Siciliy and in and around Ancona, Bari, Naples, and Florence. Starting in mid-April, the protest moved to the Po valley and the northern industrial areas, where workers and peasants made both economic (better wages) and organizational (the right to organize, strike, and demonstrate) demands.

In Milan, there were some important strikes in February and March 1898. Those by the tramline personnel and the Stigler mechanics (involving the city government, the Chamber of Labor, and the machine industrialists' association in settlement efforts) were the most publicized. A union organizing drive at Pirelli signed up 1,000 rubber workers. On March 4, a band performance of the royal march was roundly hissed at the fiftieth anniversary Festa dello Statuto. Afterwards, some 12,000 persons applauded Turati in the Arena as he demanded the political and civil rights of citizens guaranteed by the constitution. Two days later, radical leader Felice Cavallotti was killed in a duel with a conservative journalist in Rome. The funeral of the "bard of democracy" in Milan was a giant demonstration of the progressive parties. Over 100 organizations sent their banners, and wagonloads of flowers followed the hearse. Mayor Vigoni joined Cavallotti's Radical colleagues Giuseppe Mussi and Carlo Romussi and Socialist Filippo Turati in eulogizing the late deputy at the cemetery. On March 20, the anniversary of the Cinque giorni (Milan's revolution of 1848), a people's counter-demonstration claimed the heritage of the Risorgimento uprising for a more progressive cause. *L'Italia del Popolo* announced: "Milan has again demonstrated that its sentiments differ from those who govern Italy."[66]

The Fatti di Maggio, 1898

The government forbade any demonstration on May 1, a Sunday. Although orators attacked the government at several meetings, the day was calm in Milan. A police report dated May 3 noted that the Socialist party opposed illegal agitation and attempts to provoke disorder. "But," another police inspector warned, "if some incident occurs in Milan, the extremist parties are unlikely to avoid the occasion for agitation and perhaps disorders." The recall to arms on May 4 of the class of 1873 suggested a renewal of the war in Africa. On the same day, General Fiorenzo Bava Beccaris, commanding officer in Milan, ordered troops to support the police if needed, on a daily basis after 7 P.M., when the factories closed and workers flooded into the streets.[67]

On May 6, 1898, Milan was the scene of a fateful demonstration. A crowd gathered outside the police station on the via Napo Toriani in the northern factory neighborhood to demand the release of several Pirelli workers who had been arrested while distributing flyers signed "Milanese Socialists." The manifesto protested the police killing of a demonstrator (the university student son of Milan's Radical deputy Giuseppe Mussi) the day before in Pavia. The flyers called for workers to join the Socialist party to work for universal suffrage and political rights. That is, if they were to make economic gains through their own efforts, workers needed the elementary political rights of petition and assembly. Carlo Dell' Avale obtained the release of one of the men, but not the other. He then went to Pirelli himself, who—concerned about the political tone of the gathering protest—also asked the police chief to release his worker, but to no avail. The situation became calmer as the Pirelli workers went back to their work benches after the lunch break.

Late in the afternoon, Turati and Dino Rondani arrived at the Pirelli factory with the news that the city had suspended the *dazio* on bread, flour, and pasta. Hoisted onto the backs of two workers, Turati told the workers, who were filling out through the factory gates, that their still-imprisoned coworker would be freed. He urged them to act with "the reasoned prudence of citizens, of workers, conscious of your rights." "Take care, now is not the moment," he warned, "I earnestly urge you to stay calm!" But his words fell on deaf ears. At about 6:45 P.M., a march toward the center of the city began to take form, and Turati and Rodani again tried to persuade the workers to disband. For a moment, they seemed to succeed, but then some one thousand men turned on the police station on the via Napo Torriani and stoned it.[68]

Police officers emerged, firing their pistols as the troops, who had been stationed on the racetrack in anticipation of troubles, hastened to join them, also shooting. Eventually their gunfire broke up the demonstrators. One policeman was killed when he was caught in the line of fire between the two forces of repression. Several workers were wounded, some fatally. A group of their fellows started to take one of the critically wounded to the hospital of the Fatebenefratelli; when he died on the way, they continued to the piazza del Duomo, then to the Cimitero monumentale with his corpse. Police patrols prevented the demonstrators' regrouping in any numbers downtown, and a rainstorm scattered the rest.

Saturday, May 7, a turnout strike (workers went from factory to factory calling on their fellows to join them) spread in the industrial neighborhoods.[69] Demonstrating workers again skirmished with troops, who were by then stationed at the city gates. Workers who lived and worked in the suburban area outside the old walls attempted to march from the gates to the piazza del Duomo, picking up other demonstrators along the way. According to the report later submitted by Bava Beccaris, Prefect Winspeare phoned as early as 10:30 A.M. to put him in charge of restoring order. The general believed he was faced with a well-organized insurrection.[70]

In view of the readiness of the military and the police for confrontation, both the Socialist party leadership and the officials of the Chamber of Labor did as much as they could to calm the situation. Months later, Turati analyzed the days' events:

> On the morning of May 7, there was a peaceful demonstration, headed by women, the one that set out down the corso Venezia . . . if the demonstration were handled as [it would have been] in England, Belgium, America, Germany, or Austria, . . . nothing serious would have happened; everything would have evaporated in a bit of noise and some speeches. . . . But they did not want this.

The city officials, he continued, preferred to let matters degenerate so that there would be conflict. Turati was not certain if the orders for the severe repression came from Rome, but if they did not, they originated in Milan. Turati here missed the point that the repression might have been especially severe because primarily workers, and not bourgeois, were involved and because it was an issue on which ordinary soldiers (mostly peasants) agreed.[71]

There were fights between workers and soldiers in the northern half of the city that morning. Barricades were built, defended, and destroyed. Although there were frequent reports of sniping, the weapons of the battling demonstrators were stones and roof tiles to throw, boards and iron bars to break up streetcars, and heavy furniture and metal grills to use as the bases of barricades. By afternoon, violence had spread throughout the city. There was another series of barricades from the southern Ticinese Gate toward the cathedral square. Military intervention was as much a cause as a result of the violence. Mayor Vigoni telegraphed Rome for additional military aid.

From Rome came a declaration of a state of seige (which most likely preceded the receipt of Vigoni's message), and General Bava Beccaris was put in command of the city. The military established their superiority, and they spent the next two days (employing heavy artillery and under blanket orders to shoot) rooting out real and imagined troublemakers, against sporadic resistance from the workers. The bombardment of a monastery, where a group of beggars had gathered for their daily soup, was the final absurd military reaction, discrediting the army in the minds of many of the city's residents. (The officer in charge had been told that the monks and beggars were dangerous revolutionaries in disguise.) Most of the workers returned to their jobs on Tuesday, May 10.

Following two telegrams from Rudini that spoke of the necessity to punish the "subversive press," Bava Beccaris sent troops to close down *Il Secolo* and *L'Italia del Popolo* and arrest their editors. Republican deputy Luigi De Andreis was arrested at the offices of the latter. Bava Beccaris then issued a decree stating that because the Socialist party

> tends to subvert the current political and economic order of the state . . .
> [and] the Milan Chamber of Labor has served as its auxiliary . . . all
> Socialist electoral circles and the Chamber of Labor have been dissolved;
> the headquarters of these institutions will be searched, and their records
> and possessions confiscated; all offices of the circles and the Chamber of
> Labor will be closed, and all meetings forbidden.

Behind this repressive action was the belief that there was no rational economic cause behind the Milan uprising (unlike those in more poverty-stricken areas); hence it was the result of the "constant subversion and propaganda action of these circles."

On May 9, Rudini sent a telegram congratulating Bava Beccaris for having "restored Milan's tranquillity with such virility" and commanding him to keep the peace. The general promptly arrested Turati and Bissolati, who had requested an appointment to protest the arrest of Anna Kuliscioff and Andrea Costa. All the deputies were accused of incitement to revolt.[72]

Views of the Rebellion: Unanswered
Questions

Contemporary interpretations of the Fatti were quick to come. The theories that inform these interpretations may be classified crudely: the outside agitator–marginal participant theory, the repression theory, the misery theory, and the alienation–frustration theory. All of them rest on unsupported assumptions that can be tested by asking a series of simple but concrete questions about who was involved in the Fatti.

The contemporary conservative view started with the belief that there was no possible economic cause for revolt in Milan; hence outside agitators—marginal or criminal persons—were the chief perpetrators of the violence and, behind it all were scheming subversive propagandists. *Il Corriere della Sera* described "savage scenes," women shrieking unbelievable obscenities, mobs of unruly children, and the complete absence of the disciplined workers of prosperous Milan. Its editorial on May 7–8, 1898, offered the following analysis: "The movement spawned by the high price of bread began with a spontaneous outburst of the underprivileged masses but is now dominated by the subversive parties, which are inciting popular violence in hope of gain." By May 10, the *Corriere* was expressing satisfaction with the efficiency of the repression: "We know by this time that the working population is foreign to the riots and wants nothing more than to return to work."

Similar outrage, hurt disbelief, and the equation of the rioters with bums manipulated by sinister plotters appear in Mayor Vigoni's opening words at the municipal council meeting on June 2, 1898: "Our poor Milan has undergone a hard test . . . a crisis, the work of rascals whom we shall leave to the courts to judge and punish, but toward whom we cannot resist a word of moral condemnation."[73] The council joined him in voting thanks to General Bava Beccaris and his troops for restoring order, with only three members voting no.

A slightly different variation of the outside agitator–marginal participant theory, offered by conservative Pasquale Villari, emphasized uprooted migrants, recently arrived in the city, and their attraction to Socialist and anarchist agitation: "Milan had become a large, hardworking city, whose population has grown enormously and continues to grow because of the continuing in-migration of people from all parts of Italy in search of wage work. . . . Among these migrants, discontent, resentment, and class hatred from all over Italy has concentrated." The group included, according to Villari, "young women who had left their families in the country and had been indoctrinated with Socialist and anarchist ideas" and many men who had been unable to find decent jobs. All together these formed—according to Levra, who is not a conservative but accepts Villari's analysis—a *lumpenproletariat* ready to revolt.[74]

Liberal commentators were by no means alone in protesting the fierce repressiveness of the government's action, but this aspect of their interpretation is what divided them from the conservatives. The managing editor of *Il Corriere della Sera,* Torelli Viollier, announced his disagreement with the newspaper's editorial policy in a letter to *La Stampa* of Turin on June 12, 1898, and resigned in protest of the excesses of the military repression. According to this interpretation, there was no rebellion; rather, the killings were a monstrosity imposed on the populace by vengeful troops and police. In a letter to a friend, Torelli Viollier described the corso Venezia barricade, the most serious incident, but wrote that beyond this, "there was only a game of hide-and-seek between workers and troops." His concierge added, "Saturday, [May 7] was the people's revolution; today [May 9] it is the soldiers' revolution."[75]

The misery theory was not the monopoly of Socialist commentators, but it was most eloquently stated by them at that time. On June 17, 1898, the day the Rudini government was forced to resign, a speech by Deputy Nicola Badaloni described the protest as an outgrowth of high bread prices and misery unheeded by the government. He then attacked the government's claim that the protest was political in the north, specifically in Milan, and the result of a plot of "enemies of our institutions." Although there was unemployment and misery in Milan, most Milanese did not lack bread. They had to pay high prices for it, however, and for this they blamed the government. Badaloni concluded that the suppression of workers' rights was a central factor; he, too, placed the onus on Rudini.[76]

Napoleone Colajanni, a positivist sociologist, popular lecturer, journalist, and member of Parliament, attacked Socialist analysis like that of Badaloni for being too simple: "Aren't the Socialists reducing the social question to absurdity when they make it a simple 'question of the stomach'?" The main thrust of Colajanni's interpretation is that the Fatti were essentially political, a "moral" protest against dishonest and unjust government policies. He also brought up a more subtle misery thesis, emphasizing relative deprivation, a theory of rising expectations: "The influence of unfavorable economic setbacks is suffered more quickly and more intensely where prosperity is greater." Colajanni went on to contend also that an important number of the rioters were uprooted, alienated, and newly urbanized peasants, "elements without work and stable residence." Here the psychological alienation–frustration theory meets the conservative sociological marginal-participant theory.[77]

Much of the evidence that could support or challenge the theories discussed here—namely, the outside agitator–marginal participant, repression, misery, and alienation–frustration theories—lies in careful identification of the participants in the Fatti de maggio.

Were there indeed many women and children among casualties and arrestees, which there should have been if *Il Corriere della Sera* were correct in identifying them as the chief instigators and participants? Or was the impression an artifact of the youthfulness of the industrial work force and the heavy employment of women in the industries that struck? Did other marginal persons—criminals, hoodlums, bums, the unemployed, and subversive Socialists—play an important role?

If repression were the principal shaper of events, as liberal witnesses argued, the people who were arrested or wounded should reflect a random sample of persons likely to be in the streets of Milan on a spring day. Does an analysis of those involved bear this out? The argument that emphasizes repression also belittles the participants (Torelli Viollier called them a *canaglia*, a rabble) and postulates a lack of consciousness of group identification, felt injustice, or program for change. Were the participants merely swept along by a movement out of control, or were their actions linked to peaceful political action and organization?

To judge the adequacy of the misery theory, those who actually were involved in the Fatti must be compared with those who were the most poorly paid, the least secure economically. What is the fit between the occupations of the arrested and the relative wages of these occupations? If the persons with the lowest wages, those most likely to suffer from increased food prices, were not overrepresented among those involved in the Fatti, the value of the misery theory as an explanation is vitiated.

Uprootedness and alienation as causative factors would be supported if a larger proportion of country-born peasants than native Milanese were among those involved. A majority of Milan residents were not native born, however, and if the nativity of the rebels did not differ substantially from that of the population at large, the alienation argument is invalid because it cannot differentiate participants from nonparticipants.

We turn now to the variously labeled rabble, hoodlums, alienated peasants, and hungry workers, whose involvement in the Fatti di maggio has left historical traces. We shall use the description in Chapters 2 and 3 of the demographic, social, and economic characteristics of the Milanese working class, and the enumeration of the proportions of the population who shared these characteristics, to establish the population at risk to participate in the Fatti. Comparisons between the two populations will test the assumptions of observers and historians and answer the questions that grew out of their interpretations.

The question of involvement is not as precise as it could be, given the nature of the protest and the zealous official reaction. Some people were doubtless caught on the street unaware of the deadly nature of the conflict. Once wounded, arrest was likely; a good deal of the police's effort went into tracing down wounded persons and investigating their role in the disorders. Because of these uncertainties, I have designated my combined list of casualties, arrested, and tried as "persons involved," rather than "rebels" or "rioters," as the authorities labeled them.

Names and biographical information were gathered from two newspapers: the conservative–liberal *Il Corriere della Sera* and the liberal *La Lombardia*. As far as I know, no official list of casualties was ever issued, although the newspapers carried allegedly complete lists of the dead. I collected from newspaper accounts of casualties 236 male names, 27 female names, and 1 of unknown sex (this excludes police and military casualties). For 94 men, 11 women and the 1 unknown, name was the only information found. In regard to arrests (which came close to 1,700, according to *Il Corriere della Sera*), I identified 602 men and 25 women who appeared to be residents of Milan and were involved in incidents in the city between May 6 and May 10, 1898. There was more information about the arrestees because most of those arrested who were reported in the press eventually came to trial. In the trial reports, age, occupation, and sometimes place of birth and paternity were mentioned. I included with the arrestees 30 people who were tried *in absentia*, most of them convicted. These were Socialist and anarchist activists not arrested at the time because they had fled Milan and Italy. Among those arrested, the tiny middle-class contingent was largely from the newspaper editors and publicists representing the entire range of the antiregime press, from Socialist to clerical, and 3 members of the Chamber of Deputies.[78]

Looking at age and sex, the simplest demographic characteristics of those involved, we can see immediately that the number of women was insignificant, 8.5 percent of casualties and 4 percent of arrests. The median age of male and female casualties was 24, that of women arrested, 22, and that of men arrested, 25. These numbers place doubt on the heavy involvement of women that contemporaries reported but confirm in a general way the relative youth of those were involved.

Most of the persons participating in the Fatti di maggio were employed workers. Only 4 percent of those whose occupation was given were listed as unemployed, which is close to the prevailing unemployment rates for Milan in this period, discussed in Chapter 4. Forty-one of the men brought to trial were identified as having previous convictions (some for political offenses), a rate of 6.8 percent. After publishing the names of all those tried and the identifying information on which these estimates are based, *La Lombardia* summed up the results as follows: "The population of Milan is dedicated to productive work and turns away from rash agitation. This was proved by the general abstention of the working class from any part in the riots, which were the deed of the lowest and most loathsome dregs of society."[79] The newspaper erred, as did most contemporary analysts, who could not imagine such widely based participation in an affair of this kind. This and the following evidence also cast doubt on Levra's belief that a large *lumpenproletarian* element took part in the events.

Industrial workers as a group were the most involved, but this category covers a range of patterns of life and work. There was no correspondence, however, between economic disadvantage and rate of involvement. (Rates of involvement by industry were constructed by standardizing the numbers involved by sex in each industrial category and by the total population of that category in the 1901 census.)[80] The most actively involved men and women were employed in occupations with wages in the middle range. The most poorly paid workers, that is, the unskilled peasants, servants, and garment and textile workers, seldom took part. Nor did the best-paid workers. Women chemical workers were one exception to this conclusion, as they were both poorly paid and active in the Fatti. So too were the brick makers who set up a barricade in the southern part of the city. (These 23 men could be the origin of the tattered peasant hypothesis discussed next, as they were rural and poor.) Any chain of causation in the order of low wages–misery–rebellion would be an oversimplification.

What conclusions can we make about the roles of women and the young, from participation rates standardized by age and sex? The role of women looks unimpressive, except for the chemical workers, primarily from Pirelli. It is possible that women were not as likely to be arrested as men were, although no contemporary accounts suggest this. Furthermore, even if they were less often arrested, if women were in the frontlines of the demonstrations, as reported, they should have been included more often in the casualties. The same was true regarding the gunfire aimed at the rooftops from which stones and tiles were thrown. Women made up well under 10 percent of the casualties, even though all casualties reported (including some who were clearly accidentally hurt) were taken into account in this calculation. Women were surely in the street during the first day's demonstration, for over 800 of them worked at the Pirelli plant. They also took part in the May 7 demonstrations, as they were, after all, an important part of the city's work force (and both the Pirelli and the tobacco monopoly factories, in which hundreds more women worked, were near the neighborhoods in which barricades were built that morning). Overall, however, the women's presence was overestimated and publicized out of proportion to its importance.

What about the nativity argument, that alienated newcomers fed the fires of rebellion? (We can discount in advance the agitated report of the conservative *La Perseveranza* that "tattered, barefoot peasant, hatless and with twisted faces" came from the countryside to join the troubles. Colajanni gave this "news" report as fact to support his interpretation.)[81] A pamphlet that gave the place of birth of the 80 dead reported that 28, or 35 percent, were native born, a rate remarkably like that of the population as a whole.[82] Simply comparing the occupations of the involved individuals that I identified with the average nativity in those industries (in Table 3-2) reveals no discernible patterns.

I therefore also compared the nativity of individuals among those involved with the respective working-class occupational groups at risk. Birthplace was available for only 283 arrested persons. Overall, 96 men were born in Milan, or 33.9 percent, and only 27.3 percent of the male population over age 6 was born in the city. For women, almost half the arrestees whose birthplace was given were native Milanese. Industry by industry, arrestees were more likely to be native born than was the population at risk in the metal and machine, construction, chemical (including rubber), wood and straw, and food industries. In the case of stone, sand and clay, textile, and garment industrial workers, arrestees were less often native born. In the first two categories the number of men whose birthplace was known is tiny, and so no conclusion is possible. The garment workers included 12 shoemakers, all born outside Milan. Shoemakers were the third largest occupational category of in-migrants in these years. Although they were considered a rough and undisciplined group, there had been efforts to organize them into a craft union and to promote self-help, and at least 2 of the arrested shoemakers were accused political activists.[83] The shoemakers are the chief exception to the absence of relationship between birth outside Milan and activism in the Fatti, but the numbers involved are too small for any generalization.

Despite the overall failure of contemporary explanations, it is clear that the persons participating in the Fatti di maggio were not a random sample of the population of Milan. Although metal and machine workers were the most involved in absolute numbers, in proportionate terms the stone, sand, and clay group heads the list. This conclusion is based on one important incident already mentioned—the brick makers' derailing a tram and knocking down telegraph poles at Gratosoglio, in the southern rural area of the city. Because of the comparatively small population at risk, the 23 men who were tried for this episode alone had great proportionate importance. Another overrepresented industry was sales, which included mainly street vendors and small shopkeepers, types who bring to mind their participation in other protests like the micca and anti–African war demonstrations. These cases serve as a reminder that there was a much broader base to the rebellion than the workers in factory-based industry represented by the Pirelli and machine shop workers who triggered the first day's demonstrations.

Comparing those persons taking part in the Fatti di maggio with the population at risk demonstrates that none of the explanatory theories, namely, the outside agitator–marginal participant theory, the repression theory, the misery theory, or the alienation–frustration theory, is adequate. To make sense of the Fatti, we must return to the political arena in which workers were a group with limited opportunities for collective action. Behind the Fatti lay earlier street protests with wide participation by various occupational groups, the intense organizational activity of the period after 1894 that centered on industrial workers like the machine and metal and Pirelli workers but reached into many occupational categories, and Socialist action in local and national political arenas. As the flyer distributed on May 6 claimed, the right to demonstrate, the right of workers to organize, and the right to participate in the political process were the issues. If they were to make economic gains by means of their own efforts through building associations based on occupation and industry, workers needed the elementary political rights of petition and assembly.

These had long been matters of intense political contention. This time, the city and the prefect responded to challenge partly in their usual way, by banning subversive propaganda and forbidding meetings; then by greatly increasing police surveillance and military presence in the city, and finally by sending out troops with orders to shoot. As both Turati and Torelli Viollier recognized, the government interpreted the demands of workers as revolutionary and so treated the demonstration as a revolution, thus transforming it into something that was threatening in reality. The demands that the progressive press, union organizing and mobilization, and Socialist propaganda had set into motion led to the popular involvement and violence that the Socialist leaders feared. Turati warned the demonstrating workers outside the Pirelli factory not to let the authorities ''choose the day of revolt.'' But the violence was not a new idea to working-class politics. It grew out of the clash between the workers' claims to the right to demonstrate and the guns that the state assembled to reject those claims.

Conclusion

By 1894, Milanese working-class institutions like the Chamber of Labor and national ones like the Italian Socialist party had existed for several years, running candidates in elections, organizing and mobilizing workers, and promoting collective action. A national working class could be said to exist, but it was one with a particular physiognomy—combining peasants, landless agricultural laborers, and urban craft, home, and industrial workers—in an autocratic political system. An alliance of workers' groups with bourgeois intellectuals had founded the party, and the injection of peasants in the Sicilian Fasci had greatly increased its membership. Crispi's repression of the Fasci, Socialist institutions, and organizations in the chambers of labor affiliated with the party led to two important changes: first, the elimination from the party of the most numerous group of members—peasants in the Sicilian Fasci—and, second, the rapprochement of democratic and Socialist parties in defense of political rights. These profound changes undermined the base of the coalition that founded the Socialist party. Repression posed for the Socialists the question of how far and under what conditions such association should continue. Thus no sooner had the working class arrived on the national and local scenes in institutionalized form than political repression greatly reduced party membership, limited party autonomy, and pushed some of its leaders to develop strategies of alliance with democratic parties. In effect, the party that claimed to speak for the working class was forced back to the Milanese and Lombard arena, there to work politically for rights and to continue to support the workplace struggle.

Until 1898 a majority among Socialists opposed any alliance strategy in debates in their press and in regional and national congresses. At the same time, however, the Socialists were building their local organization in Milan and campaigning, sometimes with the democratic parties, for local reform. Milanese workers continued to organize, mobilize, and strike. On both the local level, where the agrarian-based *consorteria moderata* dominated politics, and the national level, where it was the historic leftist ministry of the Sicilian Rudini that governed with the support of other conservatives, the increasing Socialist and worker activity was viewed with dismay. Given their experience in 1894, leaders like Turati tried to keep the agitation in 1898 focused on narrowly formulated issues. They encouraged protest against bread prices until it was forbidden at the end of January. They supported labor organizing and strikes. They joined demonstrations demanding civil and political rights. Socialist slogans like class struggle, change in the organization of production, and political and social revolution were muted.

The violence that swept the country, like that in Milan, was the result of the interaction between protesters (issues varied: often bread prices, sometimes workplace demands or political rights) and the authorities. Once the wave of protest started, a process of contagion carried it along, even though the government severely repressed most collective action from February 1898 onward. The fact that Socialists in Milan preached prudence on May 6 was less significant than the fact that they had earlier organized protest against food prices, against government policies, and for civil and political rights and had supported a mobilized working class itself acting collectively in the workplace.[84] Milan's ruling elite and the national government both saw the Milanese workers' protest as a threat to social and political order, and they acted accordingly.

The ferocity of the repression led some members of the Lombard business elite and democratic parties to join with the Socialists to unseat the moderate Milanese *consorteria*. On the national level a liberal coalition was constructed following a period of parliamentary obstructionism mounted against the government's continued illiberal policy. The acceptance by the government of the workers' economic and political rights then set the scene for a broader-based working class to form in the ensuing period, as sketched in the next, and final, chapter.

11

Conclusion

With the state of siege firmly in place and much of his Milanese opposition in jail, Rudini designated a military court for the trial of those arrested. He was ready to use the army even further in domestic matters and requested King Umberto's permission to legislate appropriations by royal decree. But this was too much for the king, who refused to dissolve Parliament, and so Rudini resigned. He was succeeded as prime minister on June 24, 1898, by Luigi Pelloux, who had successfully managed protest in Apulia earlier in the year without extreme measures. Pelloux was conservative but recognized the economic distress behind many of the disorders and was ready—for a while—to include liberals in his cabinet and minimize the repressive measures.

The Milan trials were held in Castello Sforzesco and were reported in great detail by the newspapers still publishing, of which only *La Lombardia* was even cautiously Democratic. Edoardo Sonzogno, publisher of the silenced *Il Secolo,* issued a daily bulletin on the trials called *I Tribunali* (the courts). The first of the trials, starting on May 23, were of little-known persons implicated in the popular collective action. Accused of inciting to class hatred and civil war, instigating criminal acts, resisting or insulting the authorities, the accused were regularly questioned about their political activities and any previous criminal record. According to a contemporary report, about 56 (8 percent) of those tried were identified as Socialists, many of them leaders who had played no proven active role.[1] Two group trials of the "journalists" (newspaper editors, publicists, writers) and the "deputies" attracted the most attention. Familiar names among the convicted include Angiolo and Maria Cabrini, Emilio Caldara, Silvio Cattaneo, Carlo Dell' Avale, Anna Kuliscioff, Costantino Lazzari, Dino Rondani, Giuseppe Scaramuccia, Carlo Tanzi, Filippo Turati, and Paolo Valera. Their sentences ranged from 6 months to 16 years (for Rondani); Turati received 12 years; Cattaneo, also 12 years; Kuliscioff, 2 years and a heavy fine; Lazzari, 1 year and a fine; and Valera, 1.5 years and a fine.

On June 3, the city council convened to hear Mayor Vigoni state that "the unfortunate events had been the doing of the enemies of order," un-

checked by a permissive government. He praised the army and General
Bava Baccaris for saving Milan, and then former Mayor Gaetano Negri
introduced a motion expressing Milan's gratitude to Bava Baccaris.[2] The
city began to return to normal, although the state of siege was not lifted
until September 6, 1898.

In the same month, *Il Secolo* resumed publication, blaming local authorities for using the protest as an excuse to suppress, illegally, their elected
opposition.[3] Political debate resumed, with Socialists, Democratic Radicals, and Republicans circling warily around the question of some kind of
electoral alliance. There were, as usual, deep disagreements among the Socialists regarding this question. Enrico Ferri believed that such an alliance
would be a big mistake, a return to "dangerous political obfuscation."[4]
Claudio Treves, who had begun publishing a newspaper called *La Lotta* of
Milan which continued the volume number of *La Lotta di Classe,* supported
the idea of an electoral coalition. Gaetano Salvemini was uncompromising
on the issue, believing that for any alliance to work, the Socialists should
be sought out as allies rather than being the seekers of allies.[5]

Socialists and other political progressives began to meet and organize
again, with the immediate goals of raising funds to aid political prisoners
and their dependents and petitioning the government to pardon the convicted. The Milan's city council (caught off guard by a motion from a
minority member during the public part of one of its meetings) overrode
Vigoni's procedural hesitation and voted unanimously in late November
1898 in favor of a general pardon. *Il Secolo* published a Democratic Radical
council member's reflections on a possible alliance of Socialists, Republicans, and clericals.[6] At the end of the year, Pelloux commuted sentences
of up to two years, affecting about 3,000 persons, including Anna Kuliscioff and Paolo Valera, but this partial pardon satisfied no one.

The prime minister refused to proclaim a general pardon or to free Filippo Turati and Luigi De Andreis, both deputies who had received heavy
sentences, as the extreme left in the Chamber of Deputies proposed. the
constitutional left, including Giovanni Giolitti, supported the government
even while criticizing its political repression. In February 1899, Pelloux
introduced three laws limiting individual civil rights and permitting severe
censorship in the interest of public security. Some of the more liberal members of his cabinet objected to the limitations on the press; they were replaced with more reliable conservatives. *Il Secolo* declared that such laws
went against the principles fought for in the Risorgimento, and other Lombard newspapers (including *L'Osservatore Cattolico*) and groups joined their
objections. The laws were passed, however, with the support of Lombard
moderates.[7]

Partial elections for the city council were scheduled for June. The March
reelection of the imprisoned Turati to the Fifth College's parliamentary seat

(which Pelloux had declared vacant because of his conviction) was a harbinger of things to come. Together with the reelection of the likewise-
imprisoned Luigi De Andreis at Ravenna, it was, writes Alfredo Canavero,
"the ultimate condemnation by public opinion of the policies of Pelloux—
and in Milan—of the Moderates." [8]

In May, Pelloux and his cabinet resigned over an issue of foreign policy. He was thereupon invited by King Umberto to form a new government,
and he did so with additional conservative Lombard members, thus
strengthening his support among the Milanese Moderates. Pelloux also declared his intention to govern by royal decree, without parliamentary ap-
provals of bills. As their ability to criticize and oppose his policies was
undermined by the draconian censorship, the extreme left decided to ob-
struct passage of bills in the Chamber of Deputies and began a filibuster.

The "popular parties" of Milan—the Democratic Radicals, Republicans, and Socialists—agreed on a common list for the city council election,
excluding any member who had voted the previous June for the motion
congratulating General Bava Beccaris. (This was especially serious for the
Radicals, as several of their council members had so voted.) Turati, finally
pardoned on June 6, 1899, had returned to Milan by election day. Many
democrats feared the Socialist alliance and refused to vote for them. Voter
turnout was exceptionally high, however, and the result was a triumph for
the progressive parties. According to *La Lotta,* "The *consorteria,* which
for 40 years has controlled Milan, has ignominiously fallen." The moderate–
clerical coalition, however, had a one-vote majority, and progressive council members refused three times to vote for mayor. Hence, a royal commis-
sioner was named as a temporary mayor, and new council elections were
set for December 1899. Turati was reelected in August to the Fifth College
parliamentary seat against publisher Pietro Vallardi, running as an anti-So-
cialist for the Moderates (with, however, a liberal position on freedom of
the press). [9]

In the Chamber of Deputies, meanwhile, Socialist, Democratic Radical,
and Republican deputies took turns speaking in the filibuster. When the
speaker attempted to call for a vote on the rules, they turned the filibuster
tactic on that motion. At one point the speaker declared the debate closed
and brought the question for the vote. Socialists Leonida Bissolati and Giuseppe De Felice overturned the ballot boxes; both were arrested, and Parliament was suspended for three months.

Socialists began to look further for help in the political battle, to debate
their position toward politicians of the constitutional left like Giolitti and
Zanardelli. Claudio Treves had already stated his willingness to support
them to the extent that they worked with the extreme left to achieve its
goals: "The Socialist party substitutes itself in defending and conquering
bourgeois liberalism; there is no contradiction in this." [10] Socialists like

Ferri and Salvemini continued to refuse to go so far. Salvemini objected that the *Statuto* (constitution) itself was at fault, such as the numerous powers it reserved for the king and its neglect of political and civil rights; its problems were manifold.[11]

The Milanese Socialists parted ways over the upcoming repetition of the local elections. Former *operaisti,* headed by Lazzari, and their allies distrusted any alliance except as a transitional tactic, "to wash away the blood of May."[12] Turati and Treves were more convinced than ever: "This is the moment for democracy," Turati announced, "a brave enlightened democracy." It was necessary, he believed, to support the democrats in the struggle against Pelloux's reactionary policies and then to fight them, in turn, to establish Socialism. Emilio Caldara again supported Turati: In the Lombard regional Socialist congress he led the fight to approve a Socialist alliance with the Radicals and Republicans in the December local elections.[13] Both nationally and in Milan, Socialist leaders put too much faith in Giolitti's and Zanardelli's liberalism. They rushed to support these well-established politicians who cautiously expressed liberal principles but neglected to consult their own constituents.

In Milan's December 1899 local election, the victory of the progressive parties was even more decisive, and Giuseppe Mussi was elected mayor. In the aftermath of the successful union of the democratic parties, Turati and Anna Kuliscioff defined revolution "as emerging from the nature of things . . . schools, better health, higher standard of living for the poor, protective labor laws, and so on." Their advice was to "increase these latent forces and to work each day, making revolution your daily task."[14]

In February 1900, the high court of appeals declared that executive arbitrariness through decree law had no standing, that parliamentary approval was necessary for all legislation. Pelloux again attempted to bring his bill to a vote in April 1900, against the filibuster. This drove the extreme left to leave the chamber in protest. At this point, Zanardelli threatened to associate himself, along with his constitutional left colleagues, with the walkout. Believing that he could rebuild a majority, Pelloux dissolved the Parliament and called a general election. But he was wrong: The extreme left and Zanardelli's constitutional left only increased their representation. Pelloux still had a majority, but nonetheless he resigned in June 1900. King Umberto, not bound by the constitution to choose a member of the Chamber of Deputies as prime minister, invited the elderly Senator Giuseppe Saracco to form a government. Several days later, the king was assassinated by an Italian anarchist, an immigrant to the United States whose voyage back to Italy had been sponsored by other migrants who shared his politics. His deed, the assassin maintained, was revenge for the deaths inflicted in the Fatti di maggio.

Saracco served only briefly. He restored constitutionalism but continued censorship of the press and repressive measures against strikes. The new king, Vittorio Emanuele III, was more willing to take an active role in politics than his father had been. He called on Zanardelli and the constitutional left to form a government in 1901, with Giolitti as minister of the interior. Parliamentary government was restored; debate and criticism once again were permitted; and a new coalition was built. The Socialists supported the Zanardelli government because of Giolitti's willingness to permit labor organization and strikes as legitimate expressions of workers' interests. Turati wrote in a "declaration of principles" (*Critical sociale* 11 [16 July 1901]) that change would be "slow and gradual" but inevitable. He believed that the Socialists could divide the bourgeoisie through parliamentary action in the interests of the working class. This became the reformist credo.

After Giolitti himself became prime minister in 1903, such direct support became more difficult, however. There remained plenty for progressives to object to—continued repression, especially of agricultural strikes, and high military expenditures, for example. In the Socialist party itself, revolutionary syndicalists and other radicals severely criticized the reformists' excessive concern with the organized northern workers and their neglect of the problems of the south, the less well organized or nonorganized workers, and the poor. Socialists disagreed on the tactics and strategy to be used on both the national and local levels. Overall, although some democratic political rights were won, some social legislation enacted, and Socialist votes became more numerous in the Giolitti years up to 1914, in general the Socialists were so divided that decisive stands on their own agenda were the exception rather than the rule.

Organized labor benefited to a greater extent. The freedom to organize and the legalization of strikes granted in 1900 by Saracco (discussed in Chapters 5 and 6) opened the way on a national level as well as in Milan for mass worker mobilization and a strike wave in 1901 and 1902. Both the metal and machine workers and agricultural workers established national federations in 1901. The Federazione dei lavoratori della terra (Federterra) represented primarily the landless *braccianti* of the Po valley. New chambers of labor proliferated. By the end of 1902, half of the organized workers in Italy were in agriculture. In 1902 the Central Secretariat of Resistance, a loose committee coordinating labor organizations, was founded. There was an interval from 1903 to 1905 in which the annual number of strikes declined, and a mass general strike in 1904 (which began in Milan under the leadership of the revolutionary syndicalists) failed to achieve its goals or demonstrate the effectiveness of such tactics. The General Federation of Labor (C.G.L.) was founded by the chambers of labor and the federated unions in 1906, and a similar (if weaker) syndicalist federation was established in 1907. Yearly congresses of the C.G.L. debated the relationship of workers' organizations with capital and the relationships with politics and the Socialist party. New national peaks in the number of strikes and strikers were reached in 1906–8, which was at least partly the result of the spread of syndicalism in some areas.

Four features marked the labor movement in the period before World War I: (1) the industrialization of new sectors and new geographical areas (here the rise of the automobile industry in Piedmont and Turin is most striking); (2) the high levels of strikes, which after 1906 never fell again to pre-1901 levels except at the height of the war; (3) the nationalization and greater centralization of workers' institutions; and (4) the much greater involvement of agricultural workers in both organizations and strikes. In general, the national coordination of the labor movement was never strong; rather, local leagues and chambers of labor were its vital core. Nevertheless, these changes greatly reduced the significance of Milan in labor struggles.

In their 1900 national congress in Rome (September 8–11), a majority of Socialists voted for a minimum program as a step toward a maximum program. The atmosphere was very different from that in 1892, the year the party was founded. The Rome congress was a gathering of "prison veterans, men returned from exile" who warmly pledged unity and papered over their differences. Gone was the notion of the state as a class institution; new on the scene was democratic politics, with reformism carrying the day.[15] As Luigi Cortesi and Giuseppe Mammarella point out, Italian Socialist debates took place at the same time as the German revisionist debates and French reformist debates were held. It was reformism that was the more salient in Italy. Turati and his colleagues in 1899 "pointed openly at democracy—not the specifications and 'tactics' of 1894–95 but a tight political alliance." The 1900 congress was especially concerned with tactics, economic organization, and the minimum program. Full autonomy for local electoral committees in regard to a political alliance with the extreme left parties was voted by a large majority. This vote also distinguished two groups: the reformists (Turati, Treves, Bissolati, Camillo Prampolini, and others) and the future revolutionary syndicalists (Enrico Ferri and a group of Piedmontese, Tuscans, and southerners). The debate between those who insisted on full autonomy for the labor movement and those who preferred political direction for labor was not resolved. The minimum program of democratic reforms was approved, but the problem of how the party's revolutionary character was to be preserved over a period of such reforms was sidestepped.[16]

During the years up to 1914, the balance of power of the factions in the Socialist party shifted several times, often without a clear victory for any single group. The reformists dominated the congress of Imola in 1902. By 1903, however, the revolutionaries and revolutionary syndicalists had become stronger in the party, and in 1904, the syndicalists began a brief hegemony. But then the 1904 general strike discredited the syndicalists, and in the parliamentary election called that year by Giolitti, the Socialists lost seats (although the number of votes cast for the party rose). By 1906, syndicalism was fading (although it was still important in some regions). Despite continuing opposition to their views, the reformists prevailed in the Rome Congress in 1907.

The Socialists supported social legislation in Parliament in this period but did not push for the expansion of suffrage, which seemed to the reformists to open up the possibility of a mass (peasant) conservative vote. By the Florence congress in 1908, the reformists had regained control of the party, and the predominance of political over economic organizations was reaffirmed. The Socialists increased their representation in Parliament in 1909, with most of the deputies belonging to the reformist wing. Giolitti pursued his program of social reform, stressing universal male suffrage in 1910; the Socialists, forced to take a stand on the issue, supported his program. During the Milan congress in 1910, divisions within the party were more contentious than ever. Salvemini again appealed for universal suffrage as a means to improve the situation in the south, and Bissolati expanded his concept of collaboration and reform. Turati and the left reformists avoided a break among factions, although they increasingly disagreed, not only with revolutionaries like Benito Mussolini and Lazzari on their left, but also with the right reformists, led by Bissolati. Turati saw reformism as a way to achieve socialism over the long term (the problem, however, was that Giolitti was unwilling to deal with the P.S.I. as a political equal), whereas Bissolati, Salvemini, and Bonomi supported social reform as an end in itself.[17]

When Giolitti returned to the prime ministry in 1911 after an interval out of power, he invited Bissolati to become minister of Agriculture, Industry, and Commerce. Bissolati refused, but he did agree to consult with the king on the government's program—a betrayal of Socialist principle. The revolutionary wing, now led by men like ex-*operaista* Costantino Lazzari and Giovanni Lerda, launched a new weekly newspaper, *La Soffita* *(The Attic)*, calling for a return to the principles of the party's founding congress. Young Socialists like Mussolini, at times joined by disenchanted older ones, like Salvemini, expressed their dissent in *L'Avanguardia* (the youth federation's newspaper).

A special congress was convened, but in the interval between the call to the meeting and its opening, Giolitti, in response to the growing nationalism, declared war with Turkey over Tripoli, thus returning Italy to the competition for colonies. The long-festering questions of collaboration with bourgeois parties, anticolonialism, and pacifism were brought into the open. A general strike, called by the Socialists, failed. At the special congress in

Modena, there was a clear break between the left and right reformists. The congress rejected ministerialism and condemned the war, thus rejecting the right reformist project. (As the sole antiwar party in Italy, the Socialists were roundly condemned by the press and other parties.)

Schism finally came nine months later at the regular congress in Reggio Emilia (July 1912), at which Leonida Bissolati, Ivanoe Bonomi, and Angiolo Cabrini were expelled; Lazzari took the party's helm; and Benito Mussolini became editor of *Avanti!*. Although the expelled reformists founded a new party, the others, along with the labor leaders, stayed in the Socialist party—and the minority—until 1922. The maximalist majority moved away from the P.S.I.'s old leaders, its parliamentary delegation, and its roots among workers despite the presence of the old *operaista,* Lazzari. His influence waned in contrast with that of younger extremists like Mussolini, whose role as editor of *Avanti!* gave him an effective platform from which to preach. It was this revolutionary majority that split once again (Mussolini left the party to become an interventionist) over entering World War I: The majority of the party refused to support intervention.

The Socialist alliance with the progressive parties in Milan lasted until 1906, although it had begun to crumble in 1903. At that point, the Socialists were divided into a Milanese reformist minority and a national revolutionary syndicalist, or revolutionary, majority, with an outpost in Milan led by Arturo Labriola (discussed in Chapter 5). The Socialists in the city council worked with their allies, especially in the first three years of the coalition, on issues like school lunches and inexpensive housing, but municipal reform was modest during this period. The Socialists then dropped out of the coalition in 1906, and a clerical–moderate alliance won the Palazzo Marino in 1909. The reformists and maximalists split over whom to nominate for the Sixth College parliamentary seat in 1913: The old Garibaldino Amilcare Cipriani was chosen with the support of Paolo Valera, Mussolini, and Lazzari against the opposition of Treves, Turati, and Alessandro Schiavi. In 1914 a Socialist majority was returned, and Emilio Caldara (a long-standing advocate of municipal autonomy) was elected mayor, as a brief "golden age" of reformist Socialism began in Milan.

In 1900, Milan—its workers, its intellectuals, its organizational models, and its ideas—could still claim a central place in both the national institutions and the struggles of workplace and Socialist politics; by 1914, both labor and party institutions were more centralized and fully national. The period of this study, then, is that of the greatest intersection of Milanese class politics with the process of national working-class formation.

Economic structural change, proletarianization, and industrial reorganization gathered force in Lombardy and Milan in the 1880s. Flows of capital and the ensuing economic change brought people to the city and altered the conditions of work for city-born workers and migrants alike. Milan's labor force was characterized by great diversity. The different groups of workers had varying histories, and experienced different patterns of change. Many tried to do something about maintaining or improving their workplace situation, but some were more successful than others. The political process of workers' efforts to respond in workplace and electoral politics found workers on different sides of many issues, sometimes pursuing quite diverse strategies and alliances. The outcome, which cannot be read as a straight-line evolutionary process inherent in economic change, was an outcome as well of a political process in which workers, employers, and city and national officials were critical actors (but none were undifferentiated groups).

Those who succeeded in the workplace in building institutions and protecting or forwarding their interests—printers and construction workers, for example—were instrumental in breaking earlier patron–client relationships with Democratic Radical politicians in the Consolato operaio. New kinds of autonomous worker institutions supplemented or replaced mutual benefit societies in many occupations or industries. After suffrage was expanded in 1881, the Partito operaio continued this process in the workplace and the local political arena. It both encouraged and supported worker organization and helped develop an independent workers' politics. The Partito operaio's resources and approach were inadequate to the task it set itself, however. Despite the reluctance of the *operaisti* to become involved in politics and run candidates for office, much of the party's program required some kind of electoral activity, at least on a municipal level. Indeed, the Chamber of Labor project got off the ground and came to fruition only once there was an ally in the city council to push it. The chamber, established in 1891 and staffed by former Partito operaio members, worked energetically on behalf of workers' interests and in support of their organizations. The political problems that the Partito operaio identified were adopted as its own by the Socialist party, which was established by union leaders, Partito operaio members, and Socialist intellectuals from Milan and elsewhere. By 1892, specialized institutions had taken over the Partito operaio's functions in both workplace and local politics and had linked up nationally with allies elsewhere.

The Socialist party had no hesitation regarding electoral involvement or taking positions on national political issues like imperialism and its wars. The worker–intellectual alliance and the entry of the party into national politics made it an active challenger of the state. The circumstances that the party faced were different from those that the Partito operaio had faced in the 1880s. Earlier, the state had facilitated worker organization and political involvement through expanded suffrage. When the Partito operaio became too demanding in the economic and political arena, the state was able to stop and eventually crush it as a local institution through heavy-handed repression.

By 1894, the Socialist party was a national challenger, because of its support for the Sicilian Fasci (members of the Fasci provided the majority of party members). It took strong positions against war and colonial adventures. The party was closed down because of its growing national force, but once its Sicilian membership was lost, it retreated to its Milan nucleus. Turati and others considered possible tactical alliances with democratic parties but could not persuade the majority of their party to support this position. The party was stifled even more severely in 1898, as it was once again becoming integrated into national politics and contributing to the growing number of labor challenges in the industrial north where industry was recovering and prospering after the long recession. These two repressions shaped the Milanese Socialists' priorities and strategies. Gone was the optimism of the party's founding period, significant was the authoritarian state and the limits it posed for Socialist politics. Milanese Socialists were willing to ally with the bourgeois parties in hopes of breaking the impasse. Although the national party included many who would not go along with that view, the groups did agree to disagree. The result was years of shifting majorities, an absence of clearly delineated policies, and little coordination of the party's constituent parts: leadership, press, parliamentary delegation, labor organizations, and members. Labor organized the agricultural proletarians and centralized their associations. It was able to maintain and increase its action in workers' interests. At the same time, however, employers adopted collective strategies of their own, establishing federations of both industrial and agricultural capitalists, and so they were able to stop the forward motion of the workers' organizations.

Let us return to the propositions with which I concluded Chapter 1. First is the demonstration of the importance of political interaction and individual actors in the highly contingent process of working-class formation and transformation. Second is the evidence that although there was no mechanistic playing out of the forces of economic structural change, there is no doubt that such changes were a prerequisite to class formation. The experiences of industrialization and proletarianization shaped the men and women workers who were actors in the process. Third, the formation of a working class in Italy was highly regional, and the case study of Lombardy and Milan illuminates its early period because that region and city were the home of the chief actors and institutions participating in the process. Fourth, although the Italian Socialist party was founded "early" in terms of national indicators of industrialization, it was founded primarily by intellectuals and workers from Lombardy and Milan, where structural change had produced new conditions and new interests that contributed to the breaking of patron–client ties with the Democratic Radicals, the formation of autonomous institutions—leagues, the Partito operaio, the Chamber of Labor—and the development of new dispositions.

Milan had achieved a mature and organized working class by the early 1890s, but the establishment in 1892 of a national Socialist party that claimed to speak for a national working class was ambiguous at best. After the repression of 1894, its membership was scarcely national. The brief summary of events after 1901 given in this chapter suggests that an equally critical period of national working-class formation followed the decline of Milan's influence, with new industrialization and labor organization elsewhere in the north (especially Turin), and the rise of agricultural proletarians' organization, mobilization, and collective action to advance their interests.[18] The political process that shaped Milan's workers and intellectuals, their ideas, their organizations, and their strategies was regional and local. It contributed importantly to Italian working-class formation, offering models of organization, ideas, and leadership. It was only after Milan's influence declined, partly as a consequence of state repression, that a more distinctively national set of institutions and diverse actors (shielded from repression by the loose democratic alliance of Socialists, extreme left and liberals) transformed the political process after 1900. Milan's contribution, then, was vital, but it was just that—a contribution.

Stepping back for a moment from the Milanese–Lombard–Italian formation of a working class, let us look once again at Aristide Zolberg's generalizations and comparisons in his conclusions regarding class formation in France, Germany, and the United States.[19]

Zolberg first shows that working-class formation followed different paths in the three cases discussed in the Katznelson–Zolberg volume (France, Germany, and the United States), plus England, along several broad variables related to the form of the state and structure of the economy. Institutionally, nevertheless, before World War I, the three European working classes seemed to be converging on social democratic parties and linked national federations of labor unions pursuing economic class interests within the framework of capitalism and of pluralist parliamentary national states. The war drastically changed the picture, with the result being much more differentiated for many decades. To what extent was this true for Italy?

There is indeed some "fit." Certainly most of the Italian local leagues and the union federations—including those of landless agrarian workers, the *braccianti* of the Po valley—were working for their interests within the framework of capitalism. There was a relatively high sense of class in the northern industrial rank and file and their leadership. The labor movement was relatively autonomous vis-à-vis the Socialist party in practice, partly because of the continuing diversity in productive relations and types of workers. Further, a large proportion (if only some of the time a majority) of the Socialist party's membership and leadership was reformist and willing to work within the bourgeois state and the parliamentary system for reforms that would establish the conditions for socialism in some future period.

Nevertheless, I would argue that although the war exacerbated problems in Italy and contributed to the rise of Fascism, there were long-standing institutional obstacles in the Italian working class's path to social democracy.

What and why?

The highly regional pattern of class formation (Lombardy in the early decades, Piedmont and the Po valley in the later period) meant that there were not only large peasant and independent producer and worker populations outside the movement but also politically powerful economic elites who were committed to neither liberal economics nor democratic politics. Italy's late and slow industrialization and its uneven structural change also meant that the urban industrial capitalists themselves were from a different mold than were those to the north in Europe. In short, urban industrial capitalists were not a progressive, unified, or powerful political group. The economic transformation of the north affected the whole country, but large groups of both workers and elites persisted outside the industrial system. Italy's constitution and centralized state had not produced the electoral parties common in England and the United States, the political rights and parliamentary system of France, or the large class party committed to democracy of Germany. In Italy, the party system was much less developed, and the constitution permitted highly authoritarian rule. The formation of working-class autonomous institutions was a more contested process.

Some historians would contend that Germany was in the same situation. Using the argument that parliamentary control over taxation was more important than antimilitarism alone, Zolberg makes the case that the German unions were becoming increasingly reformist, noting that the Social Democrats voted for new taxes for military credits.

The dynamic of working-class formation in Italy—already present in the period examined in this book and still salient in the early twentieth century—echoed the process in other countries: early clientelism; the growth of increasingly autonomous labor organizations, federations, efforts to guarantee democratic rights, and workers' rights to organize and strike; national-level class parties; and class interest–based political and economic action. The political and economic conditions in Italy were extremely unpromising compared with those in other countries, however. On the one hand was the continuation of a constitutional and state structure in which authoritarian rule was embedded and the potential for repression was undiminished. On the other hand were social and economic structures based on very uneven development, with important sectors outside industrial capitalism and broadly based interests that were not concerned with, or were opposed to, those of the industrial or agrarian proletarians. The case of Italy suggests that although Zolberg's critical variables—economic and state structures—are determining, there may be extremes on the scale measuring those variables in which convergence toward a social democratic model was not likely before World War I.

In the years before 1914, nationalism grew as a political force in Italy. The country hesitated but entered World War I on the side of the Allies, rather than with the Central Powers, with which it had treaty ties. The Socialists split over supporting the war, with the majority opposing it. Divisions among the Socialists and between workers and the state continued through the war years. After the armistice, membership in unions and Socialist institutions increased enormously. The Socialist party elected 156 deputies (131 from the north, the Po valley, and Tuscany, and only 10 from the south) in the 1919 election. The Socialists and the Catholic Popular party together formed a parliamentary majority outside the old constitutional parties, but no coalition between the two was ever achieved.

Popular demands—some postponed over the war years, others new—multiplied. The social Catholics mobilized small landowners and landless laborers to demand the redistribution of the land. Conservative veterans and nationalists countermobilized. The Fascist movement (at first simply another conservative nationalist group with a mildly progressive social and political program) was launched at a rally in Milan in 1919. Within a month after their formation, the Fascists physically attacked Socialist institutions in the city. During the same year, they offered their services to the Nitti government to "preserve order" during a general strike. Although the government prohibited their autonomous action, it accepted their "cooperation." Mobilized urban and rural workers struck in a huge wave, culminating in the occupation of the factories in September 1920; land occupations roiled rural areas. Many of the strikes failed, as landowners and industrialists stiffened their resistance to popular demands. The Fascists then gathered counterforces and made a more sustained attack on Socialist institutions, violently destroying or paralyzing many of them. Early in 1921, the Communists broke with the Socialists to set up their own party, which promptly became the target of Fascist violence. Most of northern and central Italy's local governments and workers' progressive institutions were eventually forcibly seized and controlled by Fascist squads. Although Socialist party disagreements over tactics and strategy contributed to the debacle (most importantly their mistaken belief that Italy was in a revolutionary situation in 1919 and their simultaneous lack of planning or organization for such an outcome), I emphasize here institutional and external factors, which seem to me to have been greatly underestimated. Threatening to "march on Rome" with his armed followers, Mussolini was invited by the king (who had prevented any strong governmental actions from being taken against the Fascists) to take office as prime minister in 1922. Within several more years, the working class was essentially demobilized by the Fascists, claiming to act in defense of law and order.

Regional differences in economic and class structures, and the limitations of the Italian parliamentary system, which prevented progressive governments from taking strong stands and made it all too easy for conservative and reactionary governments to repress any opposition, had profoundly undermined the conditions making the working class an autonomous collective actor. A class had been made and unmade.

NOTES

Chapter 1

1. Costa's speech of July 2, 1886, abstracted from the *Atti del Parlamento italiano, Camera dei deputati, Discussioni,* vol. 419, p. 426, is reprinted in Gastone Manacorda, ed., *Il socialismo nella storia d'Italia* (Bari: Laterza, 1966), pp. 164–65. (All translations are mine unless otherwise noted.)

2. Karl Marx, *The Poverty of Philosophy* (New York: International Publishers, 1963 [1847]), p. 173.

3. Gastone Manacorda, *Il Movimento operaio italiano attraverso i suoi congressi: Dalle Origini alla formazione del Partito socialista (1853–1892)* (Rome: Rinascita, 1953). A second edition with a new introduction was published in Rome in 1963 by Editori Riuniti. Manacorda's study and the contemporaneous publication of Ernesto Ragionieri's study, *Un Comune socialista: Sesto fiorentina* (Rome: Rinascita, 1953), sparked a debate among labor historians in Italy about the appropriate focus of labor and working-class history. Founded in 1948, the journal *Movimento operaio* and its editor, Armando Saitta, had published what might be called compensatory history: little-known documents, annotated bibliographies of pamphlets and newspapers, and articles on forgotten figures, organizations, or events in the history of the working class. Saitta argued in *Movimento operaio* (vol. 7, 1955–56, in a column entitled "Pro e contra") that working-class history should not be "corporatist" and local but, instead, national. Indeed, the journal ceased publication, and approaches like that of Manacorda—that is, efforts to provide a working-class perspective on national history—became dominant.

4. Alexander Gerschenkron, "Notes on the Rate of Industrial Growth in Italy, 1881–1913," pp. 72–80, in Alexander Gerschenkron, *Economic Backwardness in Historical Perspective* (Cambridge, MA: Harvard University Press, 1962), first published in the *Journal of Economic History* 15 (1955); Luciano Cafagna, "L'industrializzazione italiana: La Formazione di una 'base industriale' fra il 1896 e il 1914," *Studi storici* 2 (July–December 1961): 690–724; Giuliano Procacci, "La Classe operaia agli inizi del secolo XX," *Studi storici* 3 (1962): 3–62, also published, with updated references, in Giuliano Procacci, *La Lotta di classe in Italia agli inizi del secolo XX* (Rome: Riuniti, 1970), pp. 3–75. See also Stefano Fenoaltea, "Riflessioni sull'esperienza industriale italiana dal risorgimento alla prima guerra mondiale," pp. 121–156 in Gianni Toniolo, ed., *Lo Sviluppo economico italiano, 1861–1940* (Bari: Laterza, 1973); Luciano Cafagna, "Intorno alle origini del dualismo economico in Italia," pp. 103–150, in Alberto Caracciolo, ed., *Problemi storici dell'industrializzazione e dello sviluppo* (Urbino: Argalia, 1965); and Valerio Castronovo, "La Storia economica," pp. 5–50 in *Storia d'Italia,* vol. 4: *Dall' Unità d'Oggi* (Turin: Giulio Einaudi, 1975).

5. Giuliano Procacci, *The Italian Working Class from the Risorgimento to Fascism* (Cambridge, MA: Harvard University Center for European Studies, 1979), p. 5. Manacorda adopted a similar view in the 1963 edition of his *Movimento operaio italiano*. The introduction to the second edition was also published as an article entitled "Formazione e primo sviluppo del partito socialista in Italia: Il problema storico e i piu recenti orientamenti storiografi," *Studi storici* 4 (1963): 23–50 and, without the subtitle, pp. 165–92 in Giuseppe Manacorda, *Rivoluzione borghese e socialismo: Studi e saggi* (Rome: Editori riuniti, 1975). Idomoneo Barbadoro shares the Manacorda–Procacci revised view in his *Storia del sindacalismo italiano della nascita al fascismo* (Florence: La Nuova Italia, 1973), vol. 2, p. 18: "At the beginning of the twentieth century, the Italian working class possessed small numbers and a rather fragile structure; it was fragmented and lacked homogeneity. Hence it ought to have been weaker and backward compared with those [working classes] of more industrially developed countries."

6. Stefano Merli, "La Grande fabbrica in Italia e la formazione del proletariato industriale di massa," *Classe* 1 (1968): entire issue. Page citations here are from the revised version of this study, *Proletariato di fabbrica e capitalismo industriale*, 2 vols. (Florence: La Nuova Italia, 1972), vol. 1, p. 4.

7. Ibid., p. 16.

8. See my review of Merli's book in the *Journal of Modern History* 48 (June 1976): 339–42. Other reviews include those by Giuseppe Barone, "La Nascita del proletariato industriale in Italia," *Quaderni storici* 24 (May–August 1973): 699–709; Aldo Monti, "Alle Origine della classe operaia italiana: Un tentativo di revisione," *Quaderni storici* 24 (September–December 1973): 1040–1048; Giorgio Mori, "Un Infanzia lunga 150 anni: Formazione ed evoluzione dell'industria italiana dagli esordi alla fine del secolo XIX," *Passato e presente* 1 (January–June 1982): 91–114; and Andreina De Clementi, "Appunti sulla formazione della classe operaia in Italia," *Quaderni storici* 32 (May–August 1976): 685–88. Alessandra Pescarolo, "From Gramsci to 'Workerism': Notes on Italian Working-Class History," pp. 273–78 in Raphael Samuel, ed., *People's History and Socialist Theory* (London: Routledge & Kegan Paul, 1981), uses the categories *Gramscian* for Manacorda and Procacci and *Workerist* for Merli and his followers, contrasting both with E. P. Thompson, *The Making of the English Working Class* (London: Victor Gollancz, 1963); and with other English and American historians who focus on neither political institutions nor workplace collective action but instead on working-class life and popular culture.

9. De Clementi, "Appunti," pp. 685–88. De Clementi's essay is now included in a volume of papers that she edited entitled *La Società inafferrabile: Protoindustria, città e classi sociali nell'Italia liberale* (Rome: Edizioni lavoro, 1986).

10. De Clementi, "Appunti," pp. 688, 713 (citations here are from the journal version of the article).

11. Ibid., pp. 714–15.

12. Ibid., p. 719.

13. Giuseppe Berta, "Dalla Manifattura al sistéma di fabbrica: Razionalizzazione e conflitti di lavoro," in *Storia d'Italia, vol. 1: Dal feudalismo al capitalismo* (Turin: Giulio Einaudi, 1978), p. 1095.

14. Giuseppe Berta, "La Formazione del movimento operaio regionale: Il Caso dei tessili (1860–1900)," pp. 297–327, in Aldo Agosti and Gian Mario Bravo,

eds., *Storia del movimento operaio, del socialismo e delle lotte sociali in Piemonte* (Bari: De Donato, 1979), vol. 1.

15. Berta, "La Formazione," pp. 300–1, 312, and pass.; E. P. Thompson, *The Making of the English Working Class.* See also Franco Ramella, *Terra e telai: Sistemi di parentela e manifattura nel biellese dell'ottocento* (Turin: Einaudi, 1984). Ramella delves more deeply into the situation of the Biella handloom weavers, showing the importance of kin and community networks and small property holdings to the weavers' long fight against proletarianization. He also incorporates gender much more fully into his analysis. Once the male weavers had been demobilized, the new women workers were highly vulnerable to dismissal and replacement. Industrialists sometimes relocated the mills where new women workers with less knowledge of the earlier history and culture of the industry and very different material conditions and organizational possibilities were less likely to protest.

Some of the earlier local studies are those by Paolo Spriano, *Storia di Torino operaia e socialista: Da De Amicis a Gramsci* (Turin: Einaudi, 1972) (this is the reprinting of four volumes published in 1958 and 1960, with new introductory, concluding, and connecting sections); and Ottavio Cavalleri, *Il Movimento operaio e contadino nel bresciano* (Rome: Edizioni Cinque Luni, 1972). These studies examine local economy and demography, institutions, leaders and followers, the press, strikes, and other collective action. Spriano focuses especially on the Socialist context and Cavalleri, on Catholic worker associations. Local studies that support Merli's interpretation include that by Stefano Musso, *Gli operai di Torino* (Milan: Feltrinelli economica, 1980), a strongly empirical study of occupation and civil status by industry, sex, age, and the like, along with a more speculative argument about the position and strength of labor and reformist Socialists, and other Torinese class actors vis-à-vis the developing Fascist menace. (See also Volker Hunecke, and the articles in *Classe* 5, cited and discussed in notes 28 and 30).

Other excellent local studies, more influenced by anthropology, were done in the last decade. In addition to Ramella, *Terra e telai*, these include Donald Bell, *Sesto San Giovanni: Workers, Culture, and Politics in an Italian Town, 1880–1922* (New Brunswick, NJ: Rutgers University Press, 1986). Focusing primarily on this Milan suburb in the twentieth century, Bell argues for a broader view of the formation of an industrial working-class consciousness through earlier worker culture and community life, and he concludes that the defense of craft prerogatives and opposition to Catholic counterorganization were factors critical to the process. Also Maurizio Gribaudi, *Mondo operaio e mito operaio: Spazi e percorsi sociali a Torino nel primo novecento* (Turin: Einaudi, 1987). Gribaudi's study of off-work class activity, based on demographic and spatial analysis, demonstrates that pre-World War I working-class life was precarious, even in its family and household aspects and that the Socialists contributed practical support through urban politics and educational and recreational activities.

16. Franco Andreucci and Gabriele Turi, "La Classe operaia: Una storia nel ghetto," *Passato e presente* 10 (January–April 1986): 3–7. See also Giovanni Gozzini "Lavoro e classe. Le tendenze della storiografia," *Passato e presente* 24 (September–December 1990): 97–111.

17. Ira Katznelson, "Working-Class Formation: Constructing Cases and Comparisons," pp. 3–41 in Ira Katznelson and Aristide R. Zolberg, eds., *Working-Class Formation: Nineteenth-Century Patterns in Western Europe and the United*

States (Princeton, NJ: Princeton University Press, 1986), pp. 6–7. The reference here is to E. P. Thompson, "The Poverty of Theory," in E. P. Thompson, *The Poverty of Theory and Other Essays* (London: Merlin Press, 1978).

18. Katznelson, "Working-Class Formation," pp. 10–11.

19. Ibid., pp. 30–31. See also the excellent review of Katznelson's and Zolberg's book by Margaret Ramsay Somers, "Workers of the World, Compare!" *Contemporary Sociology* 18 (May 1989): 325–29. Victoria E. Bonnell, *Roots of Rebellion: Workers' Politics and Organizations in St. Petersburg and Moscow, 1900–1914* (Berkeley: University of California Press, 1983) offers a different but in some ways similar model for looking at economic and demographic structural, political, and cultural factors in working-class formation.

20. Aristide Zolberg, "How Many Exceptionalisms?" pp. 397–455 in Katznelson and Zolberg, eds., *Working-Class Formation.*

21. Ibid., p. 450.

22. Adriana Lay and Maria Luisa Pesante, *Produttori senza democrazia: Lotte operaie, ideologie corporative e sviluppo economico da giolitti al Fascismo* (Bologna: Il Mulino, 1981), likewise analyze patterns of economic development and strikes in cross-national comparative perspective, to address a different but connected problem (the extent to which Fascism was the outcome of exceptionally violent worker collective action and lagging Italian industrialization, which led industrialists and government to accept—or welcome—the new regime). Although the period that interests Lay and Pesante (1900–22) is beyond the period of greatest concern to me, I shall examine further their provocative argument and valuable compilation of evidence in Chapters 2 and 6.

23. E. P. Thompson, "Eighteenth-Century English Society: Class Struggle Without Class?" *Social History* 3 (May 1978): 149.

24. Yves Lequin, *Les Ouvriers de la region lyonnaise (1848–1914),* 2 vols. (Lyons: Presses universitaires de Lyon, 1977), pp. 186, 158.

25. Philip Abrams; *Historical Sociology* (Ithaca, NY: Cornell University Press, 1982), p. 16. In this study, Abrams refers to Marx's well-known statement that "Men make their own history, but they do not make it just as they please; they do not make it under circumstances chosen by themselves, but under circumstances directly encountered, given and transmitted from the the the past." Karl Marx, "The Eighteenth Brumaire of Louis Bonaparte," in Karl Marx and Friedrich Engels, *Selected Works* (Moscow: Foreign Languages Publications, 1958 [1850]), vol. 1, p. 247.

26. Charles Tilly, "How—and What—Are Historians Doing?" Working Paper 58 (January 1988), Center for Studies of Social Change, New School for Social Research, p. 1.

27. Charles Tilly, "Flows of Capital and Forms of Industry in Europe, 1500–1900," *Theory and Society* 12 (March 1983): 123–43. For the Italian case, see Luciano Cafagna, "Italy 1830–1914," pp. 279–328 in Carlo Cipolla, ed., *The Fontana Economic History of Europe: The Emergence of Industrial Societies* (London: Collins/Fontana, 1973), pt. 1, esp. pp. 323–25; and Luciano Cafagna, "Protoindustria o transizione in bilico? (A proposito della prima onda della industrializzazione italiana)," *Quaderni storici* 54 (December 1983): 971–84. Some of the material in this section and some in Chapter 2 as well are taken from Louise A.

Tilly, "Lyonnais, Lombardy and Labor in Industrialization," pp. 127–47 in Michael Hanagan and Charles Stepenson, eds., *Proletarians and Protest: The Roots of Class Formation in an Industrializing World* (Westport, CT: Greenwood Press, 1986).

28. Economic historians and geographers like Gino Luzzatto and Ferdinando Milone have accepted regional political boundaries in Italy as defining meaningful economic units: Gino Luzzatto, "L'Evoluzione economica della Lombardia dal 1860 al 1922," pp. 449–526 in *La Cassa di Risparmio delle Provincie Lombarde nella evoluzione economica della regione* (Milan: Alfieri and Lacroix, 1923); Ferdinando Milone, *L'Italia nell' economia delle sue regioni* (Turin: Einaudi, 1955). In practice, I differ from Aldo Agosti and Gian Mario Bravo, who write in the editors' introduction to their *Storia del movimento operaio*, p. 9: "The panorama delineated here is 'Piedmontese' in intentions, often Turinese in results." The essays in that volume focus on various cities or subregions, whereas I look at regional patterns of economic and population change and how Milan was affected by them. See also the thoughtful discussions about the meaning and value of regional studies by Giovanni Levi, "Regioni e cultura delle classi popolari," pp. 720–31, and Raffaele Romanelli, "Il Sonno delle regioni," pp. 778–81, both in *Quaderni storici* 41 (May–August 1979). Peter Hertner, Louis Bergeron, and Giorgio Mori, "La Geografia dell'industrializzazione, *Passato e presente* 2 (1982): 9–26; and Istituto Ernesto Ragionieri, *Storia regionale e storia del movimento operaio*, Conference, Florence, May 29, 1981, esp. Franco Andreucci, "Introduzione," pp. 1–6.

29. Volker Hunecke, *Classe operaia e rivoluzione industriale a Milano, 1859–1892* (Bologna: Il Mulino, 1982) (originally published in Germany, 1978). See also Adalberto Nascimbene, *Il Movimento operaio in Italia: La Questione sociale a Milano dal 1890 al 1900* (Milan: Cisalpina–Goliardica, 1972). This study's misleading title suggests that what happened in Milan is the history of the Italian working class. Nascimbene focuses almost exclusively on national politics, from the point of view of Rome and is careless about economic history. Further, he totally ignores both social process and differences among either workers or capitalists in his crudely teleological argument.

30. Berta, "La Formazione," p. 320.

31. Ada Gigli Marchetti, "Gli operai tipografi milanesi all' avanguardia della organizzazione di classe in Italia," pp. 1–82, and Maria Teresa Mereu, "Origini e primi sviluppi dell'organizzazione di classe dei muratori milanesi," pp. 243–331, and Luciano Davite, "I Lavoratori meccanici e metallurgici in Lombardia dall' unita alla prima guerra mondiale," pp. 333–433, all in *Classe* 5 (1972); Duccio Bigazzi, " 'Fierezza del mestiere' e organizzazione di classe: Gli operai meccanici milanesi (1880–1900)," *Società e storia* 1 (1978): 87–108; and Luisa Osnaghi Dodi, "Sfruttamento del lavoro nell' industria tessile comasca e prime esperienze di organizzazione operaia," pp. 83–151, and Andre Cocucci Deretta, "I Cappellai monzesi dall' avvento della grande industria meccanica alla costituzione della Federazione nazionale," pp. 153–242, both in *Classe* 5 (1972).

32. Giorgio Mori, " 'Dimensione stato' e storiografia economica e sociale: Lo Stato e la rivoluzione industriale," pp. 84–90 in Franco Andreucci and Alessandra Pescarolo, eds., *Gli spazi del potere: Aree, regioni, stati: Le Coordinate territoriali della storia contemporanea* (Florence: La Casa Usher, 1989).

Chapter 2

1. Giuseppe Colombo (1836–1921), one of the energetic promoters, entrepreneurs, and capitalists who changed the city's face in the last third of the nineteenth century, was the son of a goldsmith, born in Milan. He received a classical education and then went on to teach at the Società d'incoraggiamento d'arti e mestieri, a technical school. When Milan's technical university, the Politecnico, opened in 1863 with a grant from the newly unified Italian state, Colombo played an important role in the development of its engineering department, committed to training business leaders "devoted to the scientific culture that is the base of all economic progress." In the 1870s, Colombo joined the cotton manufacturer, Eugenio Cantoni, in establishing a company to import textile machines, and in the 1880s he was among the promoters of the Società Edison, the first electric company in Italy. Active also in politics, Colombo was elected to the Milan City Council and the Chamber of Deputies; he served in several cabinets and was appointed a senator. See Edoardo Borruso, "La Formazione dell'imprenditorialità in Lombardia nel XIX secolo: Il giovane Colombo (economia, istruzione, scienza, tecnica)," Tesi di laurea, Università degli studi di Milano, Facoltà di scienze politiche, 1975, pp. 203–4, 178, 196–97.

Colombo's 1881 speech is also discussed in Comune di Milano, *Esposizione Nazionale di Milano 1881: Documenti e immagini 100 anni doppo* (Milan: Comune di Milano), 1981, p. 35. One of the judges of the exposition offered a less optimistic but instructive estimate of the machine industry's progress: The engineering plants "built machines of the sort that are individually constructed, not in series, those that present few technical problems, that do not require a large range of machine tools, and that are more suitable for small or medium industries than for large." J. Benetti, "Meccanica speciale," in *Esposizione industriale italiana del 1881 in Milano, relazione dei giurati,* vol. 10: "Le Industrie meccaniche" (Milan, 1884), p. 4, quoted in Duccio Bigazzi, " 'Fierezza del mestiere' e l'organizzazione di classe: Gli operai meccanici (1880–1900)," *Società e storia* 1 (1978): 93.

2. Carlo M. Cipolla, "The Decline of Italy: The Case of a Fully Matured Economy," *Economic History Review* (2nd series) 2 (1952): 178.

3. Quoted in Salvatore Francesco Romano, *Le Classi sociali in Italia dal medioevo all'eta contemporanea* (Turin: Piccola biblioteca Einaudi, 1952), pp. 158–59. Alain Dewerpe, *L'Industrie aux champs: Essai sur la protoindustrialisation en Italie du nord (1800–1880)* (Rome: Ecole française de Rome, 1985), p. 14 and elsewhere, also discusses Sacchi's cautious attitude toward industrialization.

4. Quoted in Valerio Castronovo, "La Storia economica," in *Storia d'Italia,* vol. 4: *Dall'Unità a Oggi* (Turin: Giulio Einaudi, 1975), p. 61. Stefano Jacini (1826–1891), a member of the Catholic Right, served in many cabinets of unified Italy, starting with Cavour's in 1861, and chaired the Parliamentary Committee of Inquiry on Agriculture from 1877 to 1885.

5. Dewerpe, *L'Industrie aux champs,* pp. 38–39; see also Luciano Cafagna, "Proto-Industria o transizione in bilico? (A proposito della prima onda della industrializzazione italiana)," *Quaderni storici* 54 (December 1983): 971–84. Cafagna's concept, "balanced transition," is similar to Dewerpe's use of the word *protoin-*

dustrization. Two substantive (and critical) reviews of Dewerpe are by Salvatore Ciriacono, *Società e storia* 38 (1987): 1066–68; and Silvana Patriarca, "Tra vecchio e nuovo: Un libro sulla protoindustria in Italia," *Quaderni storici* 68 (1988): 629–35. Other recent works treating the subject include those by Paul Corner, "Manodopera agricola e industria manifattura nella Lombardia postunitaria," *Studi storici* 25 (October–December 1984): 1019–27 (who points out the inefficiencies of protoindustrial silk production and argues that it slowed specialization, the division of labor, and the expansion of markets); Giuseppe Berta, "Dalla Manifattura al sistema di fabbrica: Razionalizzazione e conflitti di lavoro," in *Storia d'Italia*, vol. 1: *Dal Feudalismo al capitalismo* (Turin: Giulio Einaudi, 1978); Franco Ramella, *Terra e telai: Sistemi di parentella e manifattura nel biellese dell'ottocento* (Turin: Giulio Einaudi, 1984); and Anna Cento Bull, "Proto-Industrialization, Small-Scale Capital Accumulation and Diffused Entrepreneurship: The Case of the Brianza in Lombardy (1860–1950)," *Social History* 14 (May 1989): 177–200. Guido Quazza, *L'Industria laniera e cotoniera in Piemonte dal 1831 al 1860* (Turin: Museo nazionale del risorgimento, 1961); and Kent Roberts Greenfield, *Economics and Liberalism in the Risorgimento: A Study of Nationalism in Lombardy, 1814–1848* (Baltimore: Johns Hopkins University Press, 1965) discussed the phenomenon but not the concept of protoindustry. The germinal statements on protoindustrialization are by Franklin Mendels, "Protoindustrialization: The First Phase of the Industrialization Process," *Journal of Economic History* 32 (March 1972): 241–61; and Peter Kriedte, Hans Medick, and Jurgen Schlumbohm, *Industrialization Before Industrialization: Rural Industry in the Genesis of Capitalism* (Cambridge: Cambridge University Press, 1981).

6. See Alessandra Pescarolo and Gian Bruno Ravenni, *Il Proletariato invisibile: La Manifattura del capello di paglia nella Toscana mezzadrile, 1820–1950*, forthcoming.

7. Alain Dewerpe, "Genèse protoindustrielle d'une region developée: L'Italie septentrionale (1800–1880)," *Annales: Economies, sociétés, civilisations* 39 (September–October, 1984): 896–914, sums up his continuity argument. On cotton, see also Roberto Romano, "Le Basi sociali di una localizzazione industriale: L'Industria cotoniera lombarda nell'ottocento," *Storia urbana* 11 (January–April 1978): 3–19 (reprinted on pp. 51–72 in Andreina De Clementi, ed., *La Società inafferabile: Protoindustria, città e Class sociali nell'Italia liberale* (Rome: Edizioni lavoro, 1986).

8. Giorgio Mori, "Industrie senza industrializzazione: La Penisola italiana dalla fine della dominazione francese all'unità nazionale (1815–1861)," *Studi storici* 30 (July–September 1989): 603–35. See also Giorgio Mori, "Il Tempo della protoindustrializzazione," in Giorgio Mori, *L'Industrializzazione in Italia (1861–1900)* (Bologna: Il Mulino, 1977), and Giorgio Mori, "La Genesi dell'industria," *Studi storici* 24 (July–December 1983): 397–420.

9. Dewerpe, *L'Industrie aux champs*, p. 148.

10. Domencio De Marco, "L'Economia degli stati italiani prima dell'unità," *Rassegna storica del risorgimento* 54 (1957): 257; Marta Petrusewicz, *Latifondo: Economia morale e vita materiale in una periferia dell'ottocento* (Venice: Marsilio, 1989).

11. Frank J. Coppa, "The Italian Tariff and the Conflict Between Agriculture and Industry: The Commercial Policy of Liberal Italy, 1860–1922," *Journal of Economic History* 30 (December 1970): 742–69.

12. The debate usually compares the promising developments of 1880 to 1888, which did not promote self-sustaining growth, with the period after 1896, which did.

Under the title *Risorgimento e capitalismo* (Bari: Laterza, 1959), Rosario Romeo republished two of his essays, the more important of which—"Lo Sviluppo del capitalismo in Italia"—made a case against Gramsci's *rivoluzione mancata* (the failed revolution). Contrary to Gramsci's thesis that the nature of Italian unification—in particular what he saw as its failure to achieve an agrarian revolution and expropriate large landholders—had undermined the development of the Italian bourgeoisie, Romeo contended that agriculture had accumulated capital for the later expansion of industry and the development of Italian capitalism. Alexander Gerschenkron attacked Romeo's analysis and claimed that his use of the concept of original accumulation was inadequate and that the available data did not support it. In fact, Italian savings increased at a very modest rate after unification and peaked in a period of agricultural depression, after 1880. Gerschenkron argues instead that the 1880s, which Romeo characterized as the birth of large-scale industry—*Breve storia della grande industria, 1861–1961* (Bologna: Il Mulino, 1961), chap. 3—failed to achieve such an outcome, largely because of the lack of state intervention and energetic bank support. See Alexander Gerschenkron, "Rosario Romeo and the Original Accumulation of Capital," in Alexander Gerschenkron, *Economic Backwardness in Historical Perspective* (Cambridge, MA: Harvard University Press, 1962) (first published in *Rivista storica italiana* 72 [1960]), 117.

Gerschenkron based his analysis on combined industrial output rates for six industries that he had calculated for an earlier article. See "Notes on the Rate of Industrial Growth in Italy, 1881–1913," first published in 1955 in the *Journal of Economic History*, and reprinted in his *Economic Backwardness*. In that study, Gerschenkron hypothesized that late industrializing countries that began from a position of "considerable economic backwardness" were likely to launch industrialization with a big initial push (p. 73). Although both Romeo (in these essays) and Gerschenkron held anti-Marxist positions, Gerschenkron's alternative, like most Marxist interpretations, saw industrialization as a sharp discontinuity in economic activity. Most Italian historians have granted Gerschenkron some points in showing that Romeo's notion that the 1880s saw an Italian industrial revolution is untenable, but they disagree with him about how applicable his notion of a big push is to Italy's economic history between 1896 and 1913 and about the extent to which its tariff policy was misguided. Their own interpretations tend to be more complicated, seldom as stimulating, and not always more convincing than Gerschenkron's. Recent Italian studies of the question include those by Luciano Cafagna, "L'Industrializzazione italiana: La Formazione di una 'base industriale' fra il 1896 e il 1914," *Studi storici* 2 (July–December 1961): 690–724, and "The Industrial Revolution in Italy, 1830–1914," pp. 279–325 in Carlo Cipolla, ed., *The Fontana Economic History of Europe: The Emergence of Industrial Societies* (London: Collins/Fontana, 1973), pt. 1 (first published separately in 1971); Valerio Castronovo, "La Storia economica" and "Lo Sviluppo economico nella storia dell'Italia unita,"

Rivista storica italiana 91 (1979): 107–43; and Giorgio Mori, "Storiografia dell'industria e storiografia dell'imprese in Italia," *Studi storici* 24 (January–June 1983): 127–35. Stefano Fenoaltea expands and elaborates Gerschenkron's view of the shortcomings of Italian state intervention in "Decollo, ciclo e intervento dello stato," pp. 95–113 in Alberto Caracciolo, ed., *La Formazione dell'Italia industriale* (Bari: Laterza, 1969), and "Riflessioni sull'esperienza industriale italiana dal Risorgimento alla Prima guerra mondiale," pp. 93–103 in Gianni Toniolo, ed., *L'Economia italiana, 1861–1914* (Bari: Laterza, 1978).

13. Alexander Gerschenkron, "Notes on the Rate of Industrial Growth in Italy, 1881–1913," in Gerschenkron, *Economic Backwardness*, p. 78; *Formazione, distribuzione, e impiego del reddito dal 1861*, vol. 1 of Giorgio Fuà, *Lo Sviluppo economico in Italia*, 3rd ed. (Milan: Franco Angeli, 1978).

14. Gerschenkron, "The Rate of Industrial Growth in Italy," p. 78; Cafagna, "The Industrial Revolution in Italy," pp. 321, 297; Alberto Caracciolo, "La Crescita e la trasformazione della grande industria durante la prima guerra mondiale," in Fuà, ed., *Lo Sviluppo economico in Italia* vol. 3, p. 197. Valerio Castronovo provides a similar interpretation in "Il 'Caso' italiano di rivoluzione industriale," pp. 99–129 in *Storia d'Italia* (Turin: Einaudi, 1975), vol. 4; Luigi De Rosa, "Urbanization and Industrialization in Italy (1861–1921)," *Journal of European Economic History* 17 (Winter 1988): 477, argues that the rapid expansion between 1896 and 1914 "did not amount to an industrial revolution—in Italy this would be completed much later, after the second World War." For De Rosa, industrial revolution requires substantial structural change, that is, the involvement of a high proportion (majority?) of the labor force in large-scale industry. Hartmut Kaelble, "Was Prometheus Most Unbound in Europe? The Labour Force in Europe During the Late Nineteenth and Twentieth Centuries," *Journal of European Economic History* 18 (Spring 1989): 65–104, shows that only in Europe did the industrial sector outweigh the service sector for a long period and that Italy was in the company of countries like France and Sweden in the continuing importance of agricultural employment until World War II or after. By most definitions, Italy was actively industrializing between 1896 and 1914, and this seems a more useful concept than seeking out the singular moment of an industrial revolution.

15. Cafagna, "The Industrial Revolution in Italy," pp. 298–99; Shepard B. Clough, *The Economic History of Modern Italy* (Cambridge: Cambridge University Press, 1964), p. 97. Giulio Sapelli, "Modelli della crescita e progresso tecnico: Reflessioni dall' Italia," *Società e storia* 45 (1989): 619–59, is largely concerned with later developments but makes some interesting points about the period considered here as well (pp. 621–25). See Aristide Zolberg, "How Many Exceptionalisms?" in Ira Katznelson and Aristide Zolberg, eds., *Working-Class Formation: Nineteenth-Century Patterns in Western Europe and the United States* (Princeton, NJ: Princeton University Press, 1986), pp. 437–39, on differences in "industrialness" in industrialized countries. He argues that the more relevant factor for the timing and form of working-class formation is not the pace of economic growth but the structure of the economy, especially the proportion of the labor force in industry and its capital intensiveness.

16. Cafagna, "The Industrial Revolution in Italy," pp. 300–3.

17. Ibid., pp. 311–16.

18. Ibid., pp. 323–25. Giorgio Mortara, *Le Popolazioni delle grandi città ital-iane: Studio demografico* (Turin: Unione tipografico–Editrice torinese, 1908), lays the north's differentiation (and its economic development) from the center and south to a higher level of capital accumulation from the protoindustrial period, and better access (via the ports of Genoa and Venice) to imports and export outlets.

19. Domenico Sella, *Crisis and Continuity: The Economy of Spanish Lombardy in the Seventeenth Century* (Cambridge, MA: Harvard University Press, 1979), p. 136. See also Bruno Caizzi, "La Crisi economica del Lombardo–Veneto nel decennio 1850–1859," *Nuova Rivista storica* 42 (1958): 205–22.

20. Bruno Caizzi, *Storia del'industria italiana del XVIII secolo ai giorni nostri* (Turin: Unione tipografico–Editrice torinese, 1965), pass.; Mario Romani, "L'E-conomia milanese nel settecento," in *Storia di Milano,* (Milan: Fondazione Trec-cani degli Alfieri), vol. 12, pp. 677, 679; Ira Glazier, "Il Commercio estero del Regno Lombardo–Veneto dal 1815 al 1865," *Archivio economico dell'unificazione italiana* (1st series) 15 (1965): 1–47, discusses the Austrian tariff regime and trade regulations.

21. Greenfield, *Economics and Liberalism;* Romani, "L'Economia milanese," pp. 689–91; Raffaele Ciasca, "L'Evoluzione economica della Lombardia dagli inizi del secolo XIX al 1860," in *La Cassa di risparmio delle provincie lombarde nella evoluzione economica della regione, 1823–1923* (Milan: Alfieri e Lacroix 1923), pp. 362–65; Ferdinando Milone, *L'Italia nell' economica delle sue regioni* (Turin: Einaudi, 1955), pp. 83–115.

22. Bruno Caizzi, *L'Economia lombarda durante la restaurazione (1814–1859)* (Milan: Banca commerciale italiana, 1972), pp. 15–40; Romani, "L'Economia milanese," p. 693. Luciano Cafagna, "L'Avventura industriale di Giovanni Ag-nelli e la storia imprenditoriale italiana," *Quaderni storici* 22 (January–April 1973), p. 157, chides those who would rigidly counterpose agrarian and industrial "men-talities" and insists that agriculture was commercial. Further, the production and export of raw silk and other agricultural products were closely connected through the individuals, the banks, and the commercial institutions involved.

23. Greenfield, *Economics and Liberalism,* pp. 90–92; Caizzi, *L'Economia lombarda,* pp. 151–82.

24. Greenfield, *Economics and Liberalism,* pp. 82–83. Compare Dewerpe, *L'Industrie aux champs.*

25. Greenfield, *Economics and Liberalism,* pp. 84–86; Romani, "L'Economia milanese," p. 702; Caizzi, *Storia dell'industria italiana, pp. 230–39; Caizzi, L'E-conomia lombarda,* pp. 15–40; Ciasca, "L'Evoluzione economica," pp. 372–73. See Glazier, "Il Commercio estero," for evidence gleaned from Austrian govern-ment export records.

26. Romani, "L'Economia milanese," p. 694; Caizzi, *Storia dell'industria italiana,* pp. 210–16; Greenfield, *Economics and Liberalism,* pp. 86–96; Ciasca, *L'Evoluzione economica,* p. 375. See also Romano, "Le Basi sociali di una loca-lizzazione industriale," pp. 51–52.

27. Caizzi, *L'Economia lombarda,* pp. 111–33.

28. Caizzi, *Storia dell'industria italiana,* pp. 247–54; Caizzi, *L'Economia lom-barda,* p. 159.

29. Greenfield, *Economics and Liberalism*, pp. 92–93; Caizzi, *L'Economia lombarda*, p. 46, notes also that wool "spinning and weaving were intertwined in the form of domestic industry or small shops in the eastern parts [of Lombardy]"; Edoardo Borruso, "Evoluzione economica della Lombardia negli anni dell'unificazione italiana," *Quaderni storici* 32 (1976): 520, although insisting that this expansion of domestic industry was "nonrevolutionary," acknowledges that it contributed to the diffusion of technology and the development of a purely industrial labor force.

30. Cafagna, "L'Industrializzazione italiana," p. 707, notes the transfer of available workers into cotton spinning as silk declined. The cotton industry, he continues, then provided a model of wage labor for later industrializing sectors. Caizzi, *L'Economia lombarda*, pp. 98–100, discusses the problem of disciplining rural silk workers that preoccupied silk merchant capitalists.

31. Romani, "L'Economia milanese," pp. 733–34; A. M. Galli, "Il Comasco nella 'grande crisi' bachiola (1854–1874)," *Economia e storia* 14 (1967): 185–229; Caizzi, *L'Economia lombarda*, pp. 46–53; Borruso, "Evoluzione economica," pp. 520–23; Dewerpe, "Genèse protoindustrielle," p. 906. See also Glazier, "Commercio estero," for the collapse of export trade.

32. Borruso, "Evoluzione economica," pp. 522–23; Gino Luzzatto, "L'Evoluzione economica della Lombardia dal 1860 al 1922," in *La Cassa di Risparmio delle Provincie Lombarde nella evoluzione economica della regione* (Milan: Alfieri e Lacroix 1923), pp. 458, 467. Cafagna, "L'Industrializzazione italiana," p. 707, observes that the Banca di Busto Arsizio was the nucleus of later bank consolidations—the Società bancaria italiana and the Banca italiana di sconto.

33. Borruso, "Evoluzione economica," pp. 523–24.

34. G. Scagnetti, *La Siderurgia in Italia* (Rome: Industria tipografia romana, 1923), p. 167, quoted in Luciano Davite, "I Lavoratori meccanici e metallurgici in Lombardia dall'unità alla prima guerra mondiale," *Classe* 5 (1972): 337.

35. Davite, "I Lavoratori," pp. 338–39.

36. Luzzatto, "L'Evoluzione economica," pp. 468, 465.

37. The description of canals and railroads in Etienne Dalmasso, *Milan, capitale économique de l'Italie: Etude géographique* (Gap: Editions Ophrys, 1971), pp. 69–72, 45–47; Ferdinando Reggiori, *Milano, 1800–1943: Itinerario urbanistico-edilizio* (Milan: Edizioni dei milione, 1947), pp. 425–27, 469–78. See also *Milano nel 1906*, Edizione fuori commercio, 1906, pp. 175–76, 178–80.

38. Dalmasso, *Milan, capitale économique*, p. 152; see also Comune di Milano, *Esposizione Nazionale di Milano*.

39. Giulio Belinzaghi (1818–1892) (mayor also from 1889 to 1892) was an orphan without means who made his fortune as a banker and investor in railroads; he was also president of the Milan Chamber of Commerce. Quoted in Franco Nasi, "1860–1899, Da Beretta a Vigoni: Quarant'anni di amministrazione comunale," *Città di Milano* 5 (1968): 46.

40. C. Saldini, "L'Industria," in *Milano 1881* (Milan: Giuseppe Ottino, 1881), pp. 365–66.

41. "Milano industriale," in *Mediolanum* (Milan, 1881), vol. 3, p. 40.

42. Ibid., p. 51.

43. Ibid., p. 42. Ambrogio Binda (1811–1874), born in Milan and orphaned as a child, started a button business with two machines at age 18. With his profits from this business, he established the paper mill in 1855.

44. Ibid., p. 43. Volker Hunecke, *Classe operaia e rivoluzione industriale a Milano: 1859–1892* (Bologna: Il Mulino, 1982), pp. 187–88, citing Camera di Commercio di Milano, *Statistica al giugno 1891 delle caldaie a vapore, dei motori a vapore, a gas, elettrici ed idraulici nel distretto camerale di Milano* (Milan, 1891), pp. xiii, xviiii, points out that Alberto Riva, the author of this account, gives a closely reasoned criticism of earlier estimates, including Colombo's, concluding that they are too small by 1,000 horsepower.

45. *Annuario dell'industria e degli industriali di Milano* (Milan, 1890), pp. 49–53; Società italiana Ernesto Breda per costruzioni meccaniche—Milano, *Per la millesima locomotiva* (Milan, 1908), p. 7; the number of workers in the various engineering plants is cited by Bigazzi, " 'Fierezza del mestiere'," p. 93, based on report of the exposition committee on the machine industry.

46. *Annuario*, pp. 33–37.

47. Ibid., pp. 127–132.

48. Giovanni Battista Pirelli (1848–1932) was, like Colombo (who was one of his teachers), an innovative entrepreneur. He had humble origins and was a migrant to Milan from Lake Maggiore. Paolo Valera, the sardonic socialist chronicler of Milanese mores and politics, wrote that Pirelli was the son of a doorman. The official company history notes that Pirelli, after receiving an engineering degree as a scholarship student, sought a challenging new industry with growth potential. Rubber was his choice. He traveled in Europe to study the organization of factories elsewhere as a model for his new enterprise. With the support of Colombo, he then persuaded wealthy Lombard investors to back him. Paolo Valera, *Le Terribili giornate del maggio '98*, ed. Enrico Giudetti (Bari: De Donato, 1973) (originally published in 1913), p. 20. Valera (1850–1926), born in Como to a match seller and a cook, ran away to join a Garibaldian unit in 1867 and then went to Milan. Friendly with leftists and intellectuals, he fashioned a career for himself as a writer of realistic fiction and muckraking journalism. See Pirelli e C., *Nel suo cinquantenario, 1872–1922* (Milan: Alfieri and La Croix, 1922), pp. 5–6.

49. Pirelli e C., *Nel suo cinquantenario*, notes 256 workers, whereas *Annuario*, pp. 21–22, lists 250.

50. *Annuario*, pp. 105–6, 75–81; Bigazzi, " 'Fierezza del mestiere'," p. 93.

51. *Annuario*, pp. 81–86, 110–13, 129–30, 268–70, 152–55.

52. Ibid., pp. 71–72; Martino Pozzobon, "L'industria tessile nel Milanese, 1900–1930," in Maria Cristina Cristofoli and Martino Pozzobon, eds., *I Tessili milanesi: Le Fabbriche, gli industriali, i lavoratori, gli sindacati dall'ottocento agli anni '30* (Milan: Franco Angeli, 1981), pp. 21–22.

53. *Annuario*, pp. 72–74.

54. Giuseppe Colombo, "Milano industriale," in *Mediolanum* (Milan, 1881), vol. 3, pp. 51–52.

55. Ibid., pp. 54–55; Direzione generale della statistica, "Notizie sulle condizioni industriali della Provincia di Milano," ed. Leopoldo Sabbatini, *Annali di statistica*, 4th series, vol. 65 (Milan, 1893), pp. 403–9.

56. Ibid., pp. 208–17, 120–26, 176–78.

57. Ada Gigli Marchetti, "Gli operai tipografi milanesi all'avanguardia della organizzazione di classe in Italia," *Classe* 5 (1972): 4.

58. Ibid., pp. 13, 16.

59. Direzione generale della statistica, "Notizie," pp. 91–97, 115–20, 273, 276; Marchetti, "Gli operai tipografi milanesi," p. 14.

60. Giuseppe Sacchi, "La Vita intima," in *Mediolanum* (Milan, 1881), vol. 2, pp. 84–87.

61. Colombo, "Milano industriale," pp. 60–62. Such ambivalent statements were not uncommon in other industrializing countries. For abundant French examples, see Peter Stearns, *Paths to Authority: The Middle Class and the Industrial Labor Force in France* (Urbana: University of Illinois Press, 1978).

62. Hunecke, *Classe operaia*, p. 113.

63. See Volker Hunecke, "Cultura liberale e industrialismo nell'Italia dell' ottocento," *Studi storici* 18 (October–December 1977): 30–31, for the suggestion that Colombo promoted electricity for home industry, such as garment making, as much as for large-scale industry. Hunecke notes that in 1867, after his trip to the Paris Exposition of that year, Colombo had praised the prospects of a resurgent dispersed domestic industry with small efficient motors. Perhaps, Hunecke continues, Colombo's promotion of electricity was conceived as a step in that direction, with the ultimate goal of a manageable and passive labor force. The first concrete information about the use of electric motors in the garment industry comes from the industrial census of 1911, cited in the next note. By then, most electric motors were in garment shops with more than ten workers, but not in smaller units. To what extent this census takes account of individual workers' machines is not clear. However, the pedal sewing machine, not the electrically powered one, was typically used for homework; so although Hunecke may be correct about Colombo's motives, the outcome was not what he may have anticipated, at least not for many years.

64. Direzione generale della statistica e del lavoro, *Censimento degli opificie delle imprese industriale al 10 Giugno, 1911*, Rome, vol. 2 (establishments with fewer than ten workers), pp. 76–85, vol. 3 (establishments with ten or more workers other than owner), pp. 44–55; Direzione generale della statistica, "Notizie," pp. 403–9. See also Piero Angiolini, "La Svolta industriale italiana negli ultimi anni del secolo scorso e le reazioni dei contemporanei," *Nuova rivista storica* 56 (January–April 1972): 56–60, for a discussion of the development of the electrical industry in Italy as a case of Gerschenkron's theory that new technologies are seized and elaborated, especially in late-developing countries, thus giving them some advantage over already industrialized countries with older technologies.

65. Borruso, "Evoluzione economica," pp. 530–31.

66. Luzzatto, "L'Evoluzione economica," pp. 471–72.

67. Ernesto Breda (1852–1918), a native of the Veneto, learned the business from a cousin who established the Terni Steelworks in Rome. Breda benefited, during the company's first years, from government munitions orders, and later, he was able to specialize in building locomotives. Società italiana Ernesto Breda, *Per la millesima locomotiva*, pp. 8–9, 15; *La Società italiana Ernesto Breda per cos-*

truzioni meccaniche dalle sue origini ad oggi, 1886–1936 (Verona: Mondadori, 1936), pp. ix–xi; Bigazzi, " 'Fierezza del mestiere'," pp. 93; 68, Bull, "Protoindustrialization, Small-Scale Capital Accumulation," pass.

69. Quoted in Giancarlo Galli, *Il Movimento operaio milanese alla fine dell'ottocento,* Pubblicazione edita dall'ufficio stampa del comune di Milano (Milan: IGIS, 1971), p. 15. Gaetano Negri, a writer and geologist, was a multiterm municipal councillor in Milan, first elected in 1873. He was mayor from 1884 to 1889.

70. Galli, *Il Movimento operaio milanese,* pp. 14–15. See Chapter 7 for a discussion of the 1891 machinists' strike, which was an effort to resist these changes.

71. Armando Sapori, "L'Economia milanese dal 1860 al 1915," *Storia di Milano* (Milan: Treccani degli Alfieri, 1962), vol. 15, p. 869.

72. Luzzato, "L'Evoluzione economica," pp. 474, 480; Cafagna, "L'Industrializzazione," p. 710. Following Gerschenkron's periodization of Italian industrialization and his emphasis on industrial credit banks, Jon S. Cohen discusses the Banca commerciale in "Financing Industrialization in Italy, 1894–1914: The Partial Transformation of a Late-Comer," *Journal of Economic History* 27 (September 1967): 363–82, and "Italy, 1861–1914," pp. 58–90, in Rondo Cameron, ed., *Banking and Economic Development: Some Lessons of History* (New York: Oxford University Press, 1972). See also, however, the much fuller and more critical study by Antonio Confalonieri, *Banca e industria in Italia, 1894–1906,* vol. 2: *L'Esperienza della Banca commerciale italiana* (Milan: Banca commerciale italiana, 1976).

73. *Annuario,* p. 53.

74. *Il Corriere della Sera,* February 9–10 and 10–11, 1890.

75. Ibid., March 16–17, 1890, editorial.

76. Luzzatto, "L'Evoluzione economica," p. 480; Pozzobon, "L'Industria tessile," p. 41.

77. Aldo De Maddalena, "Rilievi sull'esperienza demografica ed economica milanese dal 1861 al 1915," in *L'Economia italiana dal 1861 al 1961* (Milan: Giuffrè, 1961).

78. Gerschenkron, *Economic Backwardness,* pp. 75, 77. See also Idomeneo Barbadoro, *Storia del sindacalismo italiano, dalla nascita al fascismo,* vol. 2: *La Confederazione generale del lavoro* (Florence: La Nuova Italia, 1973), p. 12; and Stefano Fenoaltea, "Decollo, ciclo e intervento dello stato."

79. Largest firms in *Annuario,* as selected by Adalberto Nascimbene, *Il Movimento operaio in Italia: La Questione sociale a Milano dal 1890 al 1900* (Milan: Cisalpino Goliardica, 1972), p. 36.

80. Direzione generale della statistica, "Notizie," pp. 403–9, Table I regarding the commune of Milan. All numerical statements referring to 1893 are based on this table. See also Louise A. Tilly, "The Working Class of Milan, 1881–1911" (Ph.D. diss., University of Toronto, 1974). Procacci showed that the word *opificio,* as used in contemporary statistics, referred to a permanent unit of production with a common source of motor power employing waged labor (in this case, with common motor power, there is no limit on the number of workers), or if there is no motor power, the *opificio* must have 10 or more workers. See Giuliano Procacci, "La Classe operaia italiana agli inizi del secolo XX," *Studi storici* 3 (1962): 7–8. An arithmetic check shows that for all large categories of industry, except food production, the average size of the units in the Notizie, is 10 or more workers.

In the food industry, the average number of workers per unit is 9.8. However, in the subcategories, units are included that employed fewer than 10.

81. See L. A. Tilly, "The Working Class," pp. 102–6, 417–43.

82. Giuseppe Colombo, "Le Industrie meccaniche italiane all' esposizione di Torino," *Nuova Antologia* (4th series) 77 (1898): 388–95.

83. Raffaele Calzini, *Milano "fin de siècle," 1890–1900* (Milan: Hoepli, 1946), p. 159.

84. Giancarlo Consonni and Graziella Tonon, "Casa e lavoro nell'area milanese: Dalla fine dell'ottocento all'avvento del fascismo," *Classe* 14 (1977): 183.

85. Giuseppe Dematteis, Gino Lusso, and Giovanna DiMeglio, "La Distribuzione territoriale dell'industria nell'Italia nord-occidentale, 1887–1927," *Storia urbana* 3 (1979): 117–56.

86. Luzzatto, "L'Evoluzione economica," pp. 485–89; see also Pozzobon, "L'Industriali tessile," pp. 13–20.

87. *Milano nel 1906* (Milan: Edizione fuori commercio, 1906), p. 194. I have not used Direzione generale della statistica, *Statistica industriale: Riassunto delle notizie sulle condizioni industriale del regno* (Rome, 1906), which is based on the Sabbatini survey, updated unsystematically.

88. *Milano nel 1906*, pp. 194–204, pass.

89. Ibid., p. 184.

90. Direzione generale della statistica e del lavoro, *Censimento degli opifici e delle imprese industriali al 10 Giugno 1911* (Rome: Tipografia nazionale G. Bertero, 1913–15), vol. 2, pp. 84–85, for firms with 10 or fewer employees, vol. 2, pp. 54–55 for firms with over 10 employees.

91. Maria Christina Cristofoli, "Le Lotte e l'organizzazione dei lavoratori e delle lavoratrici tessili, 1900–1930," in Cristofoli and Pozzobon, eds., *I Tessili milanesi*, p. 116.

92. Tessie Pei-Yuan Liu, "From Protoindustry to Sweated Work: Household Producers, Small Scale Manufacturing, and Rural Development in Southern Anjou, 1780 to 1914" (Ph.D. diss., University of Michigan, 1987); Anne McKernan, "The Dynamics of the 'Linen Triangle': Factors, Family and Farm in Rural Ulster, 1740–1825" (Ph.D. diss., University of Michigan, 1991). See also Louise A. Tilly, "Linen Was Their Life: Family Survival Strategies and Parent–Child Relations in Nineteenth-Century France," in Hans Medick and David W. Sabean, eds., *Interest and Emotion: Essays on the Study of Family and Kinship* (Cambridge: Cambridge University Press, 1984), for the case of a community of hand weavers persisting in the highly industrialized department of the Nord in early twentieth-century France. See also Charles Sabel and Jonathan Zeitlin, "Historical Alternatives to Mass Production," *Past and Present* 108 (August 1985): 133–76.

Chapter 3

1. *La Plebe* (Milan) 16, (March 1883).

2. Antonio Golini, *Distribuzione della popolazione, migrazioni interne e urbanizzazione in Italia* (Rome: University of Rome Istituto di demografia, 1974, pubblicazione 27, p. 138. On the northern rural-to-urban migration, see Alain De-

werpe, *L'Industrie aux champs: Essai sur la protoindustrialisation en Italie du nord (1800–1880)* (Rome: Ecole française de Rome, 1985), pp. 388–93; and Valerio Castronovo, "La Storia economica," in *Storia d'Italia* (Turin: Einaudi, 1975), vol. 4, pt. 1, pp. 112–13, 146–47. Rural-to-urban migration was not a new phenomenon; population mobility had long existed, with children not needed on the farm sent to urban jobs in service or menial occupations and young men and women without an inheritance or dowry seeking their fortunes in the cities. This migration was likely to be a relatively short distance and often for short periods only.

3. Carlo Carozzi and Alberto Mioni, "Il Processo di urbanizzazione," in Carlo Carozzi and Alberto Mioni, eds., *L'Italia in formazione: Ricerche e saggi sullo sviluppo urbanistico del territorio nazionale* (Bari: De Donato, 1970), pp. 26–27; Giovanni Aliberti, "Sviluppo urbano e industrializzazione nell'Italia liberale: Note su un modello d'interdipendenza," pts. 1 and 2, *Storia contemporanea* 6 (March and September 1970), pp. 218, 226; and Giancarlo Consonni and Graziella Tonon, "Casa e lavoro nell'area milanese: Dalla fine dell'ottocento all'avvento del fascismo, *Classe* 14 (1977): 196.

4. Paul Bairoch, "Urbanization and Economic Development in the Western World: Some Provisional Conclusions of an Empirical Study," in H. Schmal, ed., *Patterns of European Urbanization Since 1500* (London: Croom Helm, 1981), pp. 64–65.

5. Alberto Caracciolo, "Some Examples of Analyzing the Process of Urbanization: Northern Italy (Eighteenth to Twentieth Century)," in Schmal, ed., *Patterns of European Urbanization*, pp. 131–41; Aliberti, "Sviluppo urbano," pp. 226–27, 236. See also Luigi De Rosa, "Urbanization and Industrialization in Italy (1861– 1921)," *Journal of European Economic History* 17 (Winter 1988): 467–90; Carlo Olmo and Roberto Curto, "La Città tra mercato e industrializzazione: Il Caso di Torino," *Passato e presente* 5 (January–June 1984): 27–56; Giorgio Mortara, *Le Popolazioni delle grandi città italiane: Studio demografico* (Turin: Unione tipografico– editrice torinese, 1908).

6. See Golini, *Distribuzione della popolazione*. Maurizio Gribaudi, *Mondo operaio e mito operaio: Spazio e percorsi sociale a Torino nel primo novecento* (Turin: Einaudi, 1987), takes a much closer look at workers' migration to Borgo San Paolo, a Turin neighborhood, in the first decades of this century, through oral histories and manuscript census (the latter was unfortunately not available for Milan).

Sources for the quantitative material in this chapter, unless otherwise specified, are the city publications regarding the population censuses of 1881, 1901, and 1911, reclassified into comparable occupational and industrial categories, as described in Louise A. Tilly, "The Working Class of Milan, 1881–1911" (Ph.D. diss., University of Toronto, 1974), app. A. Complete citations for the census volumes are as follows: Milan, Giunta communale di statistica, *La Popolazione di Milano secondo il censimento 31 dicembre 1881* (Milan, 1883); Comune di Milano, *La Popolazione di Milano secondo il censimento del 1901* (Milan, 1903); and Comune di Milano, *La Popolazione di Milano secondo il censimento eseguito il 10 giugno, 1911* (Milan: Stucchi, Ceretti, 1919).

7. Figure based on Etienne Dalmasso, *Milan, capitale économique d'Italie: Etude géographique* (Gap: Editions Ophrys, 1971), p. 450.

Proportions were calculated by dividing net immigration by total population increase per 100 inhabitants. The yearly population figures (births, deaths, and registered migration) were collected from Comune di Milano, *Dati statistici,* the city's annual statistical report. Retrospective figures to 1881 are provided in the first volume in the series, that for 1884, which was published in 1885. The *Anagraphe,* or population register, recorded movement in and out of the city through a system that required (by a national law) that individuals and heads of household register with the municipal authorities themselves and household members, as well as all changes in household composition and residence. Landlords were expected to inform the city when their apartments changed hands. Although there were many people who did not register in a city as large as Milan, the city did maintain a systematic and relatively efficient register. More people were regularly reported as residents than the census counted. A well-run population register is considered by most demographers to be a more accurate indicator of population levels then censuses are. For a discussion of the use of Milan's *Anagrafe,* for purposes of social and political control in the first half of the nineteenth century, see Olivier Faron, "L'ordre statistique: Sur l'usage politique d'un registre démographique à Milan au XIX siècle," *Revue d'histoire moderne et contemporaine,* 36 (1989): 586–604. Comparisons with other cities are based on Adna Ferrin Weber, *The Growth of Cities in the Nineteenth Century: A Study in Statistics* (Ithaca, NY: Cornell University Press, 1963), pp. 238–39 (first published in 1899).

8. Aldo De Maddalena, "Rilievi sull'esperienza demografica ed economica milanese dal 1861 al 1915," in *L'Economia milanese dal 1861 al 1961* (Milan: Giuffre, 1961), p. 88. See also Maria Teresa Mereu, "Origini e primi sviluppi dell'organizzazione di classe dei muratori milanesi," *Classe* 5 (1972): pp. 245–46.

9. Comune di Milano, *La Popolazione di Milano secondo il censimento del 1901,* pp. 22–24.

10. Francesco Coletti, "Zone grigie nella popolazione di Milano," in Carlo Carozzi and Alberto Mioni, eds., *L'Italia in formazione: Ricerche e saggi sullo sviluppo urbanistico del territorio nazionale* (Bari: De Donato, 1970), pp. 107–9 (article was originally published in *Il Corriere economico,* June 1917).

11. Comune di Milano, *Dati statistici, 1884* (Milan, 1885), p. 122. For comparisons with labor force patterns aggregated at the level of the national state, see Ornello Vitali, *Aspetti dello sviluppo economico italiano alla luce della ricostruzione della popolazione attiva* (Rome: University of Rome Istituto di demografia, 1970), pubblicazione 20.

12. Direzionale generale della statistica, "Notizie sulle condizioni industriali della provincia di Milano," ed. Leopoldo Sabbatini, *Annali di statistica,* fasc. 44, Milan, 1893.

13. Weber, *The Growth of Cities,* p. 281.

14. Società Umanitaria, *Le Condizioni generali della classe operaia in Milano* (Milan, 1907), p. 87. This survey, conducted in 1903, tried to interview every Milanese working-class household. It collected information on 75,321 families and 68,255 isolated individuals. Its total population count was 280,519 or 57 percent of the 491,460 persons enumerated by the 1901 census (p. 12). The other communities sending masons were in the Cremonese, Piacentino, Lodigiano, and Brianza, or even as far away as Tuscany (p. 34). Also see Giuseppe Paletta, "Strategia rivendicativa di fabbrica e rapporto di delega nelle organizzazioni operaie milanesi

(1900–1906)," in Alceo Riosa, ed., *Il Socialismo riformista a Milano agli inizi del secolo* (Milan: Franco Angeli, 1981), pp. 130–33.

15. Lucio Villari, "I Fatti di Milano del 1898. La Testimonianza di Eugenio Torelli Viollier," *Studi storici* 8 (July–September 1967): 540. Torelli wrote that Pirelli had told him in 1895 that his workers were "impervious to socialism, because they were peasants, young women, all crude and exceedingly ignorant."

16. Ada Gigli Marchetti, "Gli operai tipografi milanesi all'avanguardia della organizzazione di classe in Italia," *Classe* 5 (1972): 1–82, shows that printers no longer enjoyed guild protections or the opportunity to become masters. Too much capital was required for the large-scale printing establishments that were necessary to compete in Milan for an individual worker to aspire to own his own shop. See also Duccio Bigazzi, " 'Fierezza del mestiere' e organizzazione di classe: Gli operai meccanici milanesi (1880–1900)," *Società e storia* 1 (1978): 87–108.

17. Compare Leslie Page Moch, *Paths to the City: Regional Migration in Nineteenth-Century France* (Beverly Hills, CA: Sage, 1983); and David Kertzer and Dennis Hogan, *Family, Political Economy, and Demographic Change: The Transformation of Life in Casalecchio, Italy, 1861–1921* (Madison: University of Wisconsin Press, 1989). Kertzer and Hogan examine changing patterns of rural-to-urban migration from the perspective of a rural community. Moch and Kertzer and Hogan all show that there was no urban–rural dichotomy in late nineteenth-century Europe. Proletarians frequently moved from country to city, and vice versa.

18. This measure of the "femaleness" of an industry is derived from Edward Gross, "Plus ça change . . .? The Sexual Structure of Occupations over Time," *Social Problems* 16 (Fall 1968): 200–21. The phenomenon of sex segregation in industries, occupations, and jobs is a very general one, visible in nineteenth-century Europe and today's industrialized West. Some of the material in this section is taken from Louise A. Tilly, "Urban Growth, Industrialization and Women's Employment in Milan, Italy, 1881–1911," *Journal of Urban History*, 3 (1977): 467–84.

19. This is confirmed by detail from the 1871 census, in which sex, marital status, and age are available by industry. In 1871, most seamstresses were single and very young. Domestic servants, laundresses, and the like were overwhelmingly single, but more than half of them were over 30. I estimate that in 1871 80 percent of single women aged 15 and over, 52 percent of widows, and 40 percent of married women worked. All together, 49.6 percent of the city's female population listed occupations. In order to arrive at this estimate, because age, marital status, and occupation were cross-tabulated one by one, I assumed that all females under 15 (the youngest age break) were single, and then I subtracted the number of females 0 to 15 (21,609) from the total of single females (53,930) to use as a divisor for the proportion of single women working. I reduced the number of single working women by 10 percent to allow for working females under age 15. I then divided the estimated number of employed single women over 15 by the estimated total number of single women over 15 to produce the 80 percent employed figure. Comune di Milano, *Censimento della popolazione della città di Milano, 1871* (Milan, 1872), pp. 26–77.

20. Yearly returns of the population register were reported in informative cross-tabulated form in the city statistical yearbook, *Dati statistici*. All annual data in this

chapter are based on this source unless otherwise stated. Further graphic details and a fuller analysis may be found in L. A. Tilly, "The Working Class," chap. 4.

21. De Maddalena, "Rilievi sull'esperienza," p. 80.

22. Renzo del Carria, *Proletari senza rivoluzione* (Milan: Edizioni oriente), 1966, vol. 1, pp. 194, 220–21.

23. Quoted in Mario Romani, *Un Secolo di vita agricola in Lombardia (1861–1961)* (Milan: Giuffrè, 1963), pp. 135–36.

24. Comune di Milano, *Dati statistici, 1884* (Milan, 1885), p. 122.

25. Società Umanitaria, *Condizioni generali,* p. 30.

26. Coletti, "Zone grigie," p. 111.

27. Società Umanitaria, *Condizioni generali,* p. 27.

28. Weber, *The Growth of Cities,* pp. 257–59; Moch, *Paths to the City;* and Kertzer and Hogan, *Family, Political Economy, and Demographic Change.*

Chapter 4

1. Alessandro Schiavi, "Il Partito socialista e le elezioni amministrative in Milano," *Critica sociale* 17 (1908): 155. Schiavi (1872–1965) worked for regional publications and became a member of the first editorial board of *Avanti!.* A close collaborator of Filippo Turati, he edited Turati's correspondence and wrote biographies of him and other early Socialist leaders in the post-World War II period.

2. Antonio Gramsci, writing in 1920, quoted in Paolo Spriano, *Storia di Torino operaia e socialista: Da De Amicis a Gramsci* (Turin: Giulio Einaudi, 1972), p. 241.

3. Franco Della Peruta, *Milano: Lavoro e fabbrica, 1814–1915* (Milan: Franco Angelli, 1987), presents extensive evidence on these matters.

4. Società Umanitaria (hereafter, S.U.), *Le Condizioni generali della classe operaia in Milano: Salari, giornate di lavoro, reddito, ecc.* (Milan: Società Umanitaria, 1905), pp. 101–3. See also S.U., *Contro la disoccupazione* (Milan: Società Umanitaria, 1905), pp. 30–32, based on the 1903 survey.

5. S.U., *Condizioni generali,* p. 105. Unstable employment was probably the normal condition for many workers in this period (for the United States, see Alexander Keyssar, *Out of Work: The First Century of Unemployment in Massachusetts* (Cambridge: Cambridge University Press, 1986). According to Mark Granovetter and Charles Tilly, "Inequality and Labor Processes," pp. 175–221 in Neil J. Smelser, ed., *Handbook of Sociology* (Newbury Park, CA: Sage, 1988), the larger capitalists were trying to build more regular work forces and to reduce turnover.

6. Camera del lavoro, *Relazione, 1905,* p. 2, and *Relazione, 1906,* p. 6, quoted in Della Peruta, *Lavoro e fabbrica,* p. 150.

7. S.U., *Contro la disoccupazione,* p. 12.

8. S.U., *Origini, vicende e conquiste delle organizzazioni operaie aderenti alla Camera del lavoro in Milano* (Milan: Società Umanitaria, 1907), p. 347.

9. Ibid., pp. 401–2.

10. Giancarlo Consonni and Graziela Tonon, "Casa e lavoro nell'area milanese: Dalla fine dell' ottocento all' avvento del fascismo," *Classe* 14 (October 1977): 167. See also Alberto Cova, *L'Occupazione e i salari: Contributi per una*

storia del movimento sindacale in Italia (Milan: Franco Angeli, 1977), who quotes (p. 11) Gino Luzzatto, "Il problema della disoccupazione in Italia nei primi settant' anni dell' Unita," in *Atti della Commissione parlamentare d'inchiesta sulla disoccupazione: Studi speciali,* vol. 4, p. 4. Luzzatto argues that "it is difficult to speak of unemployment [in the modern sense]" when agricultural, craft, and household employment were typical and independent wage earning was rare. Luzzatto also notes that long-term unemployment was attenuated by migration.

11. *Bollettino dell' Ufficio del lavoro* (hereafter *B.U.L.*) 14 (1910): 2–3. See also S.U., *Colonia agricola per operai disoccupati* (Milan: Società Umanitaria, 1907), regarding a project to place unemployed workers from Milan on a model farm.

12. S.U., *Condizioni generali,* pp. 105–6.

13. Ibid., pp. 21–22.

14. Camera del lavoro, *Relazione, 1904,* p. 5, quoted in Della Peruta, *Lavoro e fabbrica,* p. 151.

15. "Una proposta di Ferdinando Fontana per ben impiegare i milioni del lascito Loria," in *L'Italia del Popolo,* January 14–15, 1893, quoted in Della Peruta, *Lavoro e fabbrica,* p. 150.

16. Renza Casero, "La Camera del lavoro di Milano dalle origini alla repressione del maggio 1898," in Marina Bonaccini and Renza Casero, *La Camera del lavoro di Milano dalle origini al 1904* (Milan: SugarCo, 1975), pp. 18–19. The founding of the Chamber of Labor is discussed in greater detail in Chapter 5.

17. Ibid., pp. 83–84, 77. Giuseppe Croce (1853–1915), a self-taught socialist, was one of the founders of the Partito operaio and editor of its newspaper, *Fascio Operaio,* in the 1880s. He also served on the committee that drafted the constitution of the Socialist party and supported unification in 1892 of the *operaisti* with socialist intellectuals in the party.

18. S.U., *Condizioni generali,* p. 116.

19. S.U., *Origini, vicende,* LVI.

20. Alberto Cova, "La Lega del lavoro del Milano dalle origini al 1914," in *Bollettino dell' Archivio per la storia del movimento sociale cattolico in Italia,* 1979, p. 54, cited in Della Peruta, *Lavoro e fabbrica,* p. 156.

21. *B.U.L.* 9 (1908): 883.

22. Ibid., pp. 883–84. See also Francesco Cafassi, "La Disoccupazione nell'industria tipografica," in *La Disoccupazione* (Milan: Società Umanitaria, 1906).

23. Speech to the 1907 Congress of the FIOM (Federation of metal workers), quoted in Della Peruta, *Lavoro e fabbrica,* p. 157.

24. S.U., *Condizioni generali,* pp. 250–53.

25. Alberto Geisser and Effren Magrini, "Contribuzione alla storia e statistica dei salari industriali in Italia nella seconda metà del secolo XIX," *La Riforma sociale* 14 (1904): 866.

26. S.U., *Condizioni generali,* pp. 250–53.

27. Ibid., p. 129.

28. Ibid., p. 133.

29. Della Peruta, *Lavoro e fabbrica,* p. 163.

30. F. Pagliari, "Le Condizioni della classe operaia milanese," *Critica sociale* 17 (1907): 173.

31. S.U., *Origini, vicende,* pp. 29–30.

32. Giovanni Montemartini, "L'Evoluzione dei salari industriali nella seconda metà del secolo XIX in Italia," *Critica sociale* 15 (1905): 11–12; Geisser and Magrini, "Contribuzione."

33. Angelo Pugliese, *Il Bilancio alimentare di 51 famiglie operaie milanesi* (Milan: Società Umanitaria, 1914), p. 29. See Casero, "Camera del lavoro," p. 248, for other estimates of the importance of food in workers' family budgets.

34. *B.U.L.* 1 (1904): 135.

35. S.U., *Origini, vicende,* p. 81.

36. The data discussed here are from Louise A. Tilly, "The Working Class of Milan, 1881–1911" (Ph.D. diss., University of Toronto, 1974), pp. 447–50.

37. Alexander Gerschenkron, "Notes on the Rate of Industrial Growth in Italy," in Alexander Gerschenkron, *Economic Backwardness in Historical Perspective* (Cambridge, MA: Harvard University Press, 1962), p. 86. See also Samuel Surace, *Ideology, Economic Change and the Working Classes: The Case of Italy* (Berkeley and Los Angeles: University of California Press, 1966).

38. Schiavi, "Il Partito socialista," p. 155.

39. Guglielmo Tagliacarne, "Il Progresso economico di Milano negli ultimi cinquant' anni," in *Nel cinquantenario della Società Edison, 1884–1934* (Milan, 1934), vol. 4, p. 22.

40. See L. A. Tilly, "The Working Class," pp. 253–55.

41. Comune di Milano, *Dati statistici,* 1897, p. 173.

42. *B.U.L.* 17 (1912): 647.

43. Pugliese, *Il Bilancio alimentare;* Stefano Somogyi, "Cento anni di bilanci famigliari in Italia (1857–1956)," in Istituto Giangiacomo Feltrinelli, *Annali* 2 (1959): 146–48. Calories are calculated from soldiers' ration found in Geisser and Magrini, "Contribuzione," p. 803.

44. Pugliese, *Il Bilancio alimentare,* p. 25.

45. Ibid., p. 29.

46. Ibid., p. 28.

47. Ibid., p. 17.

48. Ibid., p. 27.

49. Ibid., pp. 30–31. Pugliese also published a guide to nutrition for workers, which included severe restrictions on the use of wine, "which ruins health, makes one brutish, leads to degenerate offspring, and, often, insanity." S.U., *Nozioni d'alimentazione popolare,* ed., Angelo Pugliese (Milan: 1916), p. 21.

50. Quoted in Della Peruta, *Lavoro e fabbrica,* p. 170.

51. Carlo Zambelli, "Studi statistici sul movimento economica e sociale della città di Milano raccolti del Municipio," in *Mediolanum* (Milan: Vallardi, 1881), vol. 4, pp. 133–36; *Milano 1906* (Milan: Edizione fuori commercio, 1906), pp. 106–7.

52. *Milano 1906,* p. 107; Volker Hunecke, *Classe operaia e rivoluzione industriale a Milano: 1859–1892* (Bologna: Il Mulino, 1982), pp. 277–78.

53. Stefano Allochio, *La Nuova Milano* (Milan: Hoepli, 1884). See also Franco Nasi, "1860–1899: Da Beretta a Vigoni: Quarant'anni di aministrazione comunale," *Città di Milano* 5 (1968): 50–55.

54. Allochio, *Nuova Milano,* p. 47.

55. Nasi, "1860–1899," p. 60.

56. *Atti del Municipio di Milano, 1890–1891* (Milan, 1891), pp. 148–53, 259–86, esp. 269–70; Hunecke, *Classe operaia,* p. 283. Osvaldo Gnocchi-Viani (1837–1917), a native of Mantua who was active in student patriotic politics in the Risorgimento, received a law degree, but instead became a journalist. He was associated with many worker and socialist organizations, in particular the Partito operaio italiano.

57. Giovanni Montemartini, *La Questione delle case operaie in Milano* (Milan: Società Umanitaria, 1903).

58. Ibid., p. 20; Mario Punzo, "riformisti e politica comunale," in Alceo Riosa, ed., *Il Socialismo riformista a Milano agli inizi del secolo* (Milan: Franco Angeli, 1981), p. 219. See also Valeria Rossetti, "Edilizia popolare e cooperazione," pp. 275–378 in Riosa, ed., *Il Socialismo riformista.*

59. Montemartini, *La Questione,* p. 25. Montemartini argued that the city residents' demand for new housing, because of marriage and household formation was quite small; rather, increased density was more likely to result from these events. However, even though newlywed couples might double up with parents or others at first, eventually they would seek housing on their own. Hence, his estimates were probably low.

60. S.U., *Condizioni generali,* pp. 54–58.

61. Consonni and Tonon, "Casa e lavoro," p. 176.

62. Comune di Milano, *Relazione della Commissione municipale d'inchiesta sulle abitazioni popolari,* extract from *Atti del municipio* (Milan: Reggiani, 1905), p. 22.

63. Comune di Milano, *Relazione,* p. 68 and Table 17, pp. 62–67.

64. Quoted in Della Peruta, *Lavoro e fabbrica,* p. 112.

65. *Milano 1906,* p. 108; S.U., *Le Case popolari dell' Umanitaria in via Solari,* extract from *Il Politecnico* (Milan, 1906); S.U., *L'Opera della Società Umanitaria dalla sua fondazione ad oggi* (Milan, 1906), quoted in Ornella Selvafolta, "La Società Umanitaria e le case popolari a Milano, 1900–1910," *Storia urbana* 11 (April–June 1980): 43. See also Luisella Pizzetta, "La Questione delle abitazioni popolari a Milano, 1859–1908," *Storia urbana* 11 (April–June 1980): 3–27.

66. *B.U.L.* 9 (1908): 360–63.

67. Ibid., p. 737.

68. See *B.U.L.* 9 (1908): 1047; 10 (1908): 1028–29; 11 (1909): 692–93, 1175–76; and 12 (1909): 479.

69. *B.U.L.* 14 (1911): 154.

70. *Ibid.,* pp. 765, 905.

71. "L'Azione dei comuni nei riguardi delle case popolari," *B.U.L.* 12 (1914): 146–59. Tagliacarne, "Il Progresso economico," p. 48, shows that the situation was exactly the same in 1921 and almost as bad in 1931. See also Alessandro Buzzi-Donato, "Note sullo sviluppo di Milano negli ultimi cento anni," *Quaderni di documentazione e studio* (Milan: Servizio statistica, 1970), pp. 10–12.

72. Mary Nolan, "Economic Crisis, State Policy, and Working-Class Formation in Germany, 1870–1900," in Ira Katznelson and Aristide Zolberg, eds., *Working-Class Formation: Nineteenth-Century Patterns in Western Europe and the United States* (Princeton, NJ: Princeton University Press, 1986), pp. 360–61; Douglas E.

Ashford, *The Emergence of the Welfare States* (New York: Basil Blackwell, 1986), p. 78; Shepard B. Clough, *The Economic History of Modern Italy* (New York: Columbia University Press, 1964) pp. 162–63.

73. *B.U.L.* 9 (1909): 692; and 21 (1914): 53.

Chapter 5

1. Alessandro Schiavi, "Introduzione" to Società Umanitaria, *Origini, vicende e conquiste delle organizzazioni operaie aderenti alla Camera del lavoro in Milano* (Milan, 1909), pp. 11–12.

2. Ira Katznelson, "Working-Class Formation: Constructing Cases and Comparisons," in Ira Katznelson and Aristide R. Zolberg, eds., *Working-Class Formation: Nineteenth-Century Patterns in Western Europe and the United States* (Princeton, N.J.: Princeton University Press, 1986), pp. 17–19.

3. Volker Hunecke, *Classe operaia e rivoluzione industriale a Milano, 1859–1892* (Bologna: Il Mulino, 1982), pp. 340–56.

4. Guido Neppi Modona, *Sciopero, potere politico e magistratura, 1870–1922* (Bari: Laterza, 1969), p. 3. Giuseppe Zanardelli (1826–1903) was born in Brescia, served as a volunteer in the war of 1848, taught law, and was first elected to Parliament in 1859; he served various Left governments after 1876.

5. Hunecke, *Classe operaia*, pp. 359–60.

6. See Neppi Modona, *Sciopero,* p. 81, who describes the decade between 1890 and 1900 as one in which abuses of police power "were accompanied by a general involution" that attacked freedom of association, assembly, and the press, even in Parliament.

7. Ibid., p. 93. Giuseppe Saracco (1821–1907), a lawyer who began his public service in Piedmont before Italy's unification, served in several Depretis and Crispi ministries. Giovanni Giolitti (1842–1928) was a liberal politician from Piedmont, who launched the reform programs of the first decade of this century. Filippo Turati (1857–1932) was a progressive Milanese lawyer, one of the founders of the Italian Socialist party, and the architect of its reformist tendency.

8. This section is based largely on Giancarlo Galli, *Le Origini del movimento operaio milanese,* Pubblicazione edita dall' Ufficio stampa del Commune di Milano (Milan: IGIS, 1970), pp. 31–35; and S.U. (Società Umanitaria), *Origini, vicende.*

9. Quoted in Galli, *Le Origini,* p. 37.

10. Antonio Maffi, *Il Consolato operaio milanese e i suoi trent' anni di vita: Relazione alle società consociate in occasione del loro congresso tenutosi il 16 Novembre, 1890* (Milan: Tipografia degli operai [Società cooperativa], 1891), p. 5. Maffi (1845–1912), self-taught typecaster who was secretary of the typecasters' mutual benefit society and a teacher in the Consolato workers' school, was elected to Parliament in 1882 on the Democratic Radical list.

11. Felice Cavalotti (1842–98) fought with Garibaldi as a young man, later became a journalist for democratic journals and newspapers, and was elected to Parliament where he was the spokesman for Italy's (and Lombardy's) Democratic Radicals.

12. "Il Programma amministrativo del Fascio dei lavoratori," in Stefano Merli,

Proletariato di fabbrica e capitalismo industriale: Il Caso italiano (Florence: La Nuova Italia, 1972), vol. 2, pp. 11, 19–31.

13. *Il Muratore,* November 7, 1889, quoted in Maria Teresa Mereu, "Origini e primi sviluppi dell' organizzazione di classe dei muratori milanesi," *Classe* 5 (1972): 298. Silvio Cattaneo (1861–1928), born in the Alto Milanese to a humble family, started work at 10 as a mason. He later joined the Partito operaio and was one of the founders of the masons' cooperative society and their mutual benefit association, as well as editor of *Il Muratore.*

14. The disagreement between Casati and Romussi is discussed in *Il Corriere della Sera,* April 22–23, 1890; the second meeting is reported in *Il Corriere della Sera,* April 24–25, 1890. The text of police order is in *La Lombardia,* April 26, 1890, and is quoted in Adalberto Nascimbene, *Il Movimento operaio in Italia: La Questione sociale a Milano dal 1890 al 1900* (Milan: Cisalpino–Goliardica, 1972), p. 83. Alfredo Casati (1857–1920) was born in Milan and became a skilled worker; he began his political career as a Mazzinian republican, briefly supported the Democratic Radicals, was one of the founders of the Partito operaio, and fought for the principle of worker collective action in the economic rather than the political arena throughout his life. Carlo Dell' Avale (1861–1917) was born in Milan, was self-taught, moved toward socialism in the early 1890s, and became one of the founders of the Italian Socialist party, in which he served on its executive committee and on the editorial board of *Lotta di Classe.*

15. Nascimbene, *Il Movimento operaio,* pp. 87–88, citing telegrams from the prefect in Milan to the minister of interior, in the Crispi Papers, ACS (Archivio centrale dello stato), Box 316; see also ASM (Archivio dello Stato, Milano) Q 58, police documents concerning May Day 1890, and *Corriere della Sera,* April 30–May 1, 1890.

16. *L'Italia del Popolo,* June 19–20, 1890. Costantino Lazzari (1857–1927) was born in Cremona, attended technical school in Milan, joined the Circolo operaio, in which he first met Gnocchi-Viani, and later the Lega dei figli del lavoro. He was one of the founders of the newspaper *Fascio operaio* and the Partito operaio and a member of the constitutional committee for the Socialist party, in which he frequently differed from Filippo Turati and the reformists. See Costantino Lazzari, "Memorie," ed. Alessandro Schiavi, *Movimento operaio* 4 (July 1952): 598–633, and 5 (September 1952): 789–837.

17. *L'Italia del Popolo,* June 22–23, 1890.

18. Maria Grazia Meriggi, *Il Partito operaio italiano: Attivita rivendicativa, formazione e cultura dei militanti in Lombardia* (Milan: Franco Angeli, 1985), p. 55; and Guido Cervo, "Le Origini della Federazione socialista milanese," in Alceo Riosa, ed., *Il Socialismo riformista a Milano agli inizi del secolo* (Milan: Franco Angeli, 1981), p. 56.

19. *Il Muratore,* November 1, 1890, quoted in Mereu, "Muratori milanesi," p. 299.

20. Cervo, "Le Origini della Federazione socialista milanese," p. 62, citing the account of the conference in *Il Secolo,* November 17–18, 1891; Nascimbene, *Il Movimento operaio,* pp. 116–18.

21. Maffi, *Il Consolato operaio,* pp. 8, 21.

22. Cervo, "Le Origini della Federazione socialista milanese," pp. 63–66, citing *L'Italia del Popolo*, March 9–10 and 16–17, 1891, and *Il Secolo*, April 10–11, 1891.

23. Cervo, "Le Origini della Federazione socialista milanese," pp. 66–67.

24. Nascimbene, *Il Movimento operaio*, pp. 163–64, quoting *La Lotta di Classe*, September 17–18, 1892, *Il Secolo*, October 21–22, 1892, and *La Lombardia*, November 17, 1892.

25. Merli, *Proletariato di fabbrica*, vol. 2, p. 64.

26. Ibid., vol. 2, p. 93.

27. Renza Casero, "La Camera del lavoro di Milano dalle origini alla repressione del maggio 1898," pt. 1 of Marina Bonaccini and Renza Casero, *La Camera del lavoro di Milano dalle origini al 1904* (Milan: SugarCo, 1975), pp. 15–17.

28. S.U., *Origini, vicende*, p. XIVIII.

29. Osvaldo Gnocchi-Viani, *Le Borse del lavoro*, a cura del Comitato centrale del P.O.I. (Alessandria: G. Panizza, 1889), pp. 37–38.

30. "Il Comitato centrale dell' associazione fra gli operai tipografi italiani alle società operaie tipografiche per la istituzione delle borse del lavoro," in Merli, *Proletariato di fabbrica*, vol. 2, pp. 34–36.

31. Casero, "La Camera del lavoro," pp. 20–24; cf. Letterio Briguglio, *Il Partito operaio italiano e gli anarchici* (Rome: Edizioni di storia e letteratura, 1969), p. 108.

32. Casero, "La Camera del lavoro," pp. 17–18.

33. *Il Corriere della Sera*, February 9–10, 1890.

34. *Il Corriere della Sera*, February 14–15, 1890. Gaspare Finali (1829–1914), a patriot in the liberal movement of the Romagna in 1859–60, was elected to the Italian Parliament in 1860, and served as a member of the Crispi and Saracco cabinets. In his interview with the Milanese workers, Finali disingenuously scolded them for joining their bosses, who, he claimed, were simply trying to get something out of the government and were not interested in promoting the workers' interests. He was probably correct in his analysis but was trying to reduce the pressure by splitting the forces demanding that he act. For other workers' meetings—in which Brando was heckled by anarchists—see *Il Corriere della Sera*, February 17–18, 19–20, 26–27, and March 9–10, 1890. See also ASM Q 47, pass.

35. *L'Italia*, August 29–30, 1890; *L'Italia del Popolo*, September 13–14, 16–17, 1890; and *La Lombardia*, November 23, 1890.

36. ASM Q 47 letter from inspector to police chief dated March 3, 1890; *Il Corriere della Sera*, March 4–5, 1890.

37. *Il Corriere della Sera*, March 5–6, 1890.

38. *Il Corriere della Sera*, March 10–11, 11–12, 12–13, 13–14, 14–15, 16–17, 1890.

39. Although the report title referred to *borsa del lavoro* (labor exchange), the usage in the text was *camera del lavoro* (chamber of labor). The latter title was formally substituted (because it was thought to be more appropriate to its actual functions) in the March 30, 1890, meeting that approved its constitution. See Merli, *Proletariato di fabbrica*, vol. 1, pp. 634–36.

40. *La Borsa del lavoro in Milano: Suoi scopi, benefici e modo di funzionare*

(Milan: Tipografia degli operai, 1890), discussed in Casero, "La Camera del lavoro," pp. 28–32; quotation from the abstract of the report in Merli, *Proletariato di fabbrica,* vol. 2, pp. 45–46.

41. Casero, "La Camera del lavoro," p. 29.

42. Ibid., pp. 30–33; the constitution as finally ratified, in Merli, *Proletariato di fabbrica,* vol. 2, pp. 47–50. Almost twenty years later, Samuel Gompers, an American labor leader, remarked on the Milanese workers' tendency (which he found distinctive) to organize in separate societies by occupational specialty, rather than in broader categories, like "metal workers." See Samuel Gompers, *Labour in Europe and America* (New York and London, 1910), p. 162.

43. Casero, "La Camera del laboro," pp. 38–40.

44. *Atti del Municipio di Milano: Annata 1890–91* (Milan: Tipografia Luigi di Giacomo Pirola, 1891), pp. 148–53.

45. Casero, "La Camera del lavoro," pp. 49–50; cf. *L'Italia del Popolo,* February 1–2, 17–18, 1891. See also Giuseppe Paletta, "Dinamiche occupazionali e sindacali nell' industria a Milano tra i censimenti del 1901 e del 1911; *Economia e lavoro* 16 (1982): 91–104.

46. Casero, "La Camera del lavoro," pp. 54–55. The report is published in full in *Atti del Municipio di Milano: Annata 1890–91,* pp. 259–85.

47. Cervo, "Le Origini della Federazione socialista milanese," p. 85.

48. S.U., *Origini, vicende,* pp. 8–9. In 1897, the National Federation of Chambers of Labor was launched.

49. Casero, "La Camera del lavoro," p. 165, quoting *"La Lotta di Classe,* June 10, 1899.

50. Ibid., pp. 165–67.

51. Ibid., p. 173. Born in Milan in 1859, Giuseppe Scaramuccia was active in the Federazione dei lavoratori del libro, and led the Chamber until 1906, when he was forced to resign because of accusations of fraud. He fled to the United States where he died in New York in 1929.

52. Ibid., pp. 171–72, 179.

53. Giuseppe Paletta, "Strategia rivendicativa di fabbrica e rapporto di delega nelle organizzazioni operaie milanesi (1900–1906)," in Riosa, *Il Socialismo Riformista a Milano,* p. 164.

54. Alessandro Schiavi, "Gli scioperi e la produzione," *Critica sociale,* March 1 and 16, 1902, p. 73; and Filippo Turati, "Variazioni sul tema dell'articolo precedente," *Critica sociale,* December 1, 1901, p. 356, both quoted in Paletta, "Strategia rivendicativa," p. 166.

55. Marina Bonaccini, "Ricostituzione e sviluppo della Camera del lavoro di Milano (1898–1904)," in Marina Bonaccini and Renza Casero, *La Camera del lavoro di Milano dalle original 1904,* Milan: SugarCo, 1975, 262. Arturo Labriola (1873–1959), born in Naples, was active in republican and socialist politics at the University of Naples and as an organizer of the Sicilian Fasci. A writer and lecturer, he was accused and convicted in 1898 of incitement to revolt.

56. Paletta, "Strategia rivendicativa," p. 90, quoting an article by "Girondino" in *Avanguardia Socialista,* January 17, 1904.

57. S.U., *Origini, vicende,* p. 52. See also Paletta, "Strategia rivendicativa," pp. 180–99.

58. Antonio Gramsci, "Il Problema di Milano," from *L'Unità*, February 21, 1924, unsigned; reprinted in Antonio Gramsci, *Sul Fascismo*, ed. Enzo Santarelli (Rome: Editori riuniti, 1973), quotations on pp. 196–97.

Chapter 6

1. Francesco Crispi (1819–1901), an ex-Garibaldino and politician of the Left served as prime minister from 1887 to 1890 and again from 1893 to 1896.

2. Orley Ashenfelter and George E. Johnson, "Bargaining Theory, Trade Unions, and Industrial Strike Activity," *American Economic Review* 59 (1969): 35–48.

3. Emile Durkheim, *Suicide* (New York: Free Press, 1951), pp. 241–76; Neil J. Smelser, *Social Change and the Industrial Revolution: An Application of Theory to the British Cotton Industry, 1770–1840* (Chicago: University of Chicago Press, 1959); Clark Kerr, John T. Dunlop, Frederick H. Harbison, and Charles A. Myers, *Industrialism and Industrial Man* (Cambridge, MA: Harvard University Press, 1960).

4. Antonio Di San Giuliano in the Chamber of Deputies, April 23, 1884, quoted in Adrianna Lay, Dora Marucco, and Maria Luisa Pesante, "Classe operaia e scioperi: Ipotesi per il periodo 1880–1923," *Quaderni storici* 22 (January–April 1973): 90–91. Lay, Marucco, and Pesante critically examine Italian published strike statistics and discuss the conditions under which the data were collected, the quality of the information, and its form. San Giuliano (1852–1914), scion of a Sicilian noble family, became mayor of Catania at 27. He was a deputy at 30 and served in the cabinets of Giolitti and Pelloux.

5. Charles Tilly, *From Mobilization to Revolution* (Reading, MA: Addison-Wesley, 1978), p. 7. See also (specifically on strikes) Edward Shorter and Charles Tilly, *Strikes in France, 1830–1968* (Cambridge: Cambridge University Press, 1974).

6. Cf. Stefano Merli, *Proletariato di fabbrica e capitalismo industriale: Il Caso italiano, 1880–1890* (Florence: La Nuova Italia, 1972), vol. 1, p. 563:

> The numerical leap that occurs in strikes in 1890 and the years following is certainly to be understood as a qualitative change determined by the presence of workplace and political organization that promised moral and material support, by a greater degree of shared experience and agreement among workers, and by the growth and maturation of their oppositional consciousness.

7. For example, Maurice Neufeld, *Italy, School for Awakening Countries* (Ithaca, NY: New York State School of Industrial and Labor Relations, Cornell University Press, 1960), pp. 219–20, writes: "Lombard and Venetian peasants, in their search for factory work, had shifted from quiet village life to an urban atmosphere of political turmoil. These countrymen, without willing it, upset the social equilibrium in Milan."

8. Lorenzo Bordogna, Gian Primo Cella, and Giancarlo Provasi, 1881–1923: A Quantitative Analysis," pp. 217–46 in Leopold H. Haimson and Charles Tilly, eds., *Strikes, Wars, and Revolutions in International Perspective: Strike Waves in the Late Nineteeenth and Early Twentieth Centuries* (Cambridge and Paris: Cambridge University Press and Editions de la Maison des Sciences de l'Homme, 1989).

9. David Snyder, "Institutional Setting and Industrial Conflict: Comparative

Analyses of France, Italy and the United States," *American Sociological Review* 40 (June 1975): 259–78.

10. Bordogna, Cella, and Provasi, "Labor Conflicts in Italy," pp. 240–41.

11. Adriana Lay and Maria Luisa Pesante, *Prodottori senza democrazia: Lotte operaie, ideologie corporative e sviluppo economico da Giolitti al Fascismo* (Bologna: Il Mulino, 1981).

12. Charles Tilly, "Theories and Realities" and "Introduction to Part IV," in Haimson and Tilly, eds., *Strikes, Wars, and Revolutions,* esp. pp. 6–7 and 435–39. For Russian similarities to Italian patterns see also Victoria E. Bonnell, *Roots of Rebellion: Workers' Politics and Organizations in St. Petersburg and Moscow, 1900–1914* (Berkeley: University of California Press, 1983).

13. Annalucia Forti Messina, "Agitazioni e scioperi operai a Milano all'indomani dell'Unita," *Nuova rivista storica* 52 (January–April 1968): 74–110; Volker Hunecke, *Classe operaia e rivoluzione industriale e Milano, 1859–1892* (Bologna: Il Mulino, 1982), p. 372, adds two other strikes to Messina's total. See also Leo Valiani, "Le Prime grandi agitazioni operaie a Milano e a Torino," *Movimento operaio,* October–November 1950, pp. 362–67.

14. Messina, "Agitazioni," pp. 73–74, 110–11.

15. Hunecke, *Classe operaia,* p. 374; Eva Civolani, "Scioperi e agitazioni operaie dell'estate 1872 nei comparti manifatturieri di Milano e di Torino," *Movimento operaio e socialista* 22 (1977): 427–55.

16. The quantitative criteria defining a strike wave suggested by Charles Tilly cannot be verified for 1860, as there has been no compilation of strikes before 1860. Hunecke's compilation of strikes between 1868 and 1872 (*Classe operaia,* p. 374) averages 1.4 per year; hence the eight strikes in 1872 qualify as a strike wave.

17. Ministero di Agricoltura, industria, e commercio, Direzione generale della statistica, *Statistica degli scioperi avvenuti nell' industria e nell' agricoltura* (Rome: Tipografia nazionale, 1878–94, 1895–97, 1896, 1898–1903); Società Umanitaria, *Scioperi, serrate, e vertenze fra capitale e lavoro in Milano nel 1903* (Milan, 1904), p. 10; and Hunecke, *Classe operaia,* pp. 369–76. David Snyder coded the information in the official strike statistics in machine-readable form and reported his analysis in "Industrialization and Industrial Conflict in Italy, 1878–1903," unpublished paper, Department of Sociology, University of Michigan, 1970. I recoded his files to fit my industrial categories for analysis. The data he coded recorded all the information available in the documents.

18. Ada Gigli Marchetti, "Gli operai litografi milanese all'avanguardia della organizzazione di classe in Italia," *Classe* 5 (1972): 3.

19. Quoted in ibid., p. 54.

20. Ibid., pp. 35, 53–55.

21. Ibid., p. 56.

22. *Fascio Operaio,* June 6–7, 1885.

23. ASM (Archivio di Stato di Milano) Q (Fondo Questura) 45. Vallardi strike, May 7, 1886: printers' petition to police chief and printed broadside of the Sede compositori di Milano, operai tipografi italiani per introduzione ed osservanza della tariffa; Marchetti, "Gli operai tipografi milanese," p. 61, dates this strike in 1885.

24. Società Umanitaria (S.U.), *Origini, vicende e conquiste delle organizzazioni operaie aderenti alla Camera del lavoro in Milano* (Milan, 1909), p. 141;

Giuseppe Paletta, "Strategia rivendicativa di fabbrica e rapporto di delega nelle organizzazioni operaie milanesi (1900–1906)," in Alceo Riosa, ed., *Il Socialismo riformista a Milano agli inizi del secolo* (Milan: Franco Angeli, 1981), p. 136, citing a lecture "attributed to" Turati in *Il Litografo,* November 1, 1901.

25. Merli, *Proletariato di fabbrica,* vol. 2, p. 5.

26. Marchetti, "Gli operai litografi milanesi," pp. 61–62; ASM Q 46, "Sciopero alla Tipografia Stefani di Circlei," March 1889.

27. Marchetti, "Gli operai litografi milanesi," pp. 61–62; Renza Casero, "La Camera del lavoro di Milano, delle origini alla repressione del maggio 1898," in Marina Bonaccini and Renza Casero, eds., *La Camera del lavoro di Milano dalle origini al 1904* (Milan: SugarCo, 1975), p. 93; *La Lotta di Classe,* October 29–30, and November 12–13, 1892; printers' "Appeal for Solidarity," in Merli, *Proletariato di fabbrica,* vol. 2, p. 127.

28. ASM Q 50, police reports dated December 6 and 8, 1892. Vallardi continued his opposition to the union; see the uniform factory regulations he helped draw up for Milanese publishers in 1900, in Merli, *Proletariato di fabbrica,* vol. 2, pp. 773–77.

29. Casero, "La Camera del lavoro," pp. 129–30; ASM Q 51, a series of police reports and news clippings from *La Lombardia* dated from September to November 1895.

30. ASM Q 52, police reports: January 1897, re strike at the Tipografia Pizzi Giovanni, and April 1897, re dispute at the Tipografia Bonetti Enrico. The strikes in 1898 were in February (Tensi) and April (Belloni). See also Casero, "La Camera del lavoro," pp. 140–42.

31. S.U., *Origini, vicendi,* p. 133; Casero, "La Camera del lavoro," p. 164; Paletta, "Strategia rivendicativa," p. 141.

32. Ada Gigli Marchetti, *I Tre anelli: Mutualita, resistenza, cooperazione dei tipografia milanesi (1860–1925* (Milan: Franco Angeli, 1983), pp. 138–40, 150.

33. Paletta, "Strategia rivendicativa," p. 138; S.U., *Scioperi, serrate e vertenze,* p. 15.

34. S.U., *Scioperi, serrate,* p. 15.

35. Quoted in Hunecke, *Classe operaia,* p. 405.

36. ASM Q 50, inspector's report dated May 9, 1892.

37. Messina, "Agitazioni," pp. 101–5; Giancarlo Galli, *Le Origini del movimento operaio milanese,* Publicazione edita dall' Ufficio Stampa del Commune di Milano (Milan: IGIS, 1970), p. 43; Hunecke, *Classe operaia,* Table 6.4, p. 382.

38. Maria Teresa Mereu, "Origini e primi sviluppi dell'organizzazione di classe dei muratori milanesi," *Classe* 5 (1972): 282–84; Civolani, "Scioperi," Table 5, p. 451.

39.
Mereu, "Origini e primi sviluppi," pp. 284–86, 291.

40. Ibid., pp. 305–13.

41. Ibid., 314; quotations from *Il Corriere della Sera,* October 4–5, 1887; *Fascio Operaio,* October 15–16, 1887; and *Il Muratore,* June 23, 1889. There is a notarized copy of the convention of September 30, 1887, in ASM Q 48 (the box contains primarily documents from 1891).

42. Mereu, "Origini e primi sviluppi," pp. 287–89.

43. Ibid., pp. 315–17; *Corriere della Sera*, March 10–11, 11–12, 12–13, and 16–17; 1890.

44. *L'Italia del popolo*, April 12–13, 1891.

45. September 1891, documents in ASM Q 48; Chamber of Labor invitation in *Il Muratore*, October 31, 1891, quoted by Casero, "La Camera del lavoro," p. 72; motion of November 15, 1891, in *Il Muratore*, November 29, 1891, reproduced in Merli, *Proletariato di fabbrica*, vol. 2, p. 98. The flyer calling masons and associated workers to a meeting on November 15, 1891, at the Chamber of Labor to discuss abolishing *il ponte* was found in ASM Q 50.

46. Casero, "La Camera de lavoro," pp. 73–74; ASM Q 50.

47. *La Lotta di Classe*, July 30–31, 1892.

48. Casero, "La Camera del lavoro," p. 99; Paletta, "Strategia rivendicativa," p. 144; see Merli, *Proletariato di fabbrica*, vol. p. 264, for the program of the resistance league.

49. *La Battaglia*, September 8, 1894.

50. *La Battaglia*, October 20, 1894; Mereu, "Origini e primi sviluppi," p. 296.

51. S.U., *Origini, vicende*, p. 82.

52. Ibid., p. 86.

53. Paletta, "Strategia rivendicativa," p. 139.

54. S.U., *Origini, vicende*, p. 84; see also Paletta, "Strategia rivendicativa," p. 140.

55. Paletta, "Strategia rivendicativa," p. 142.

Chapter 7

1. Volker Hunecke, *Classe operaia e rivoluzione industriale a Milano, 1859–1892* (Bologna: Il Mulino, 1982), p. 207; Giuseppe Colombo, "Milano industriale," in *Mediolanum* (Milan, 1881), vol. 3, pp. 51–52; Direzione generale della statistica, "Notizie sulle condizioni industriali della Provincia di Milano," ed. Leopoldo Sabbatini, *Annali di statistica*, 4th series, vol. 65 (Milan, 1893), pp. 266, 282.

2. Archivio di Stato di Milano (ASM) Q 45: Strike in Ditta Osnago Ambrogio.

3. Società Umanitaria (S.U.), *Origini, vicende e conquiste delle organizzazioni operaie aderente alla Camera del lavoro in Milano* (Milan: Società Umanitaria, 1909), pp. 257, 259–60; ASM Q 43, reports dated May 20, 1880, inspector to police chief (Milan), regarding Bernacchi and May 26, 1879, *Consigliere delegato* to royal prosecutor.

4. S.U., *Origini, vicende*, pp. 258, 260; *La Lotta di Classe*, September 30–October 1, 1893; ASM Q 50, police reports from May 11 to 21, 1893.

5. S.U., *Origini, vicende*, pp. 258, 261.

6. Ibid., pp. 255–56, 268.

7. ASM Q 43, report dated May 20, 1880, from inspector to police chief (Milan), re Bernacchi and the report of the *Consigliere delegato*, Milan, May 26, 1879, to royal prosecutor.

8. Hunecke, *Classe operaia*, p. 216.

9. Manon all-female strike, report by police dated May 5, and Zinelli and Centenari wage report, May 10, 1880, both in ASM Q 43.

10. *Fascio Operaio,* March 27–28, 1886; police report dated April 14, 1886, ASM Q 45. The branch continued to be disturbed, with a strike against Bilwiller in April 1892; see ASM Q 50.

11. ASM Q 43, declaration of workers to police, dated January 2, 1880, and inspector's report to police chief, January 26, 1880. Although it was too late to help them find other jobs, nine strikers arrested and accused of an illegal strike were acquitted almost a year later, when a court found their strike "just and reasonable." See also *La Plebe,* December 12, 1880, which reports the unhappiness of "Mosters" that his workers had joined the Textile Confederation as a cause of his move.

12. ASM Q 50, police reports dated from August 8 to September 5, 1890; *La Lombardia,* August 9, 1890; *La Lotta di Classe,* August 20–21, 1890. See also Renza Casero, "La Camera del lavoro di Milano dalle origini alla repressione del maggio 1898," in Marina Bonaccini and Renza Casero, *La Camera del lavoro di Milano dalle origini al 1904* (Milan: SugarCo, 1975), pp. 90–92.

13. ASM Q 51, 52, 53, and newspaper accounts. The dreary litany reveals some companies' workers striking time after time: Cova, 1887, 1895, 1898; Da Re, 1895, 1896; Strazza, 1894, 1896, 1897, 1898; Uglietti, 1890, 1895.

14. ASM Q 53; police strike reports dated August 21 and 23, 1898. The strikers returned to work without any concessions from their employers.

15. *Il Corriere della Sera,* December 20–21, 1895.

16. Annalucia Forti Messina, "Agitazioni e scioperi operai a Milano all' indomani dell' unità," *Nuova rivista storica* 52 (January–April 1968): 74–76.

17. Ibid., pp. 79–80. Leo Valiani, "Le Prime grandi agitazioni operaie a Milano e a Torino," *Movimento operaio,* October–November 1950, pp. 362–67, dates the Grondona agitation as March 2, 1860.

18. Eva Civolani, "Scioperi e agitazioni operaie dell' estate 1872 nei comparti manifatturieri di Milano e di Torino," *Movimento operaio e socialista* 27 (1977): 428, and Table 5, p. 451.

19. Brochure laying out factory regulations in ASM Q 46, "Sciopero Suffert, 1887."

20. Strike totals from Hunecke, *Classe operaia,* pp. 194–95. (The published strike statistics list only one strike, in 1882).

21. Ibid., p. 195.

22. Duccio Bigazzi, " 'Fierezza di mestiere' e organizzazione di classe: Gli operai meccanici milanesi (1880–1900)," *Società e storia* 1 (1978): 97–102.

23. Luciano Davite, "I Lavoratori meccanici e metallurgici in Lombardia dall' unità alla prima guerra mondiale," *Classe* 5 (1972): 383.

24. Record of Algiso Sormani, age 34, found on an undated and unattributed slip of paper with, however, the Grondona letterhead on its other side; in ASM Q 47, with police reports of the unemployment crisis in the machine shops.

25. Hunecke, *Classe operaia,* pp. 200–1; M. Gobbini, biography of Ernesto Breda in *Dizionario biografico degli italiani* (Rome: Istituto della Enciclopedia italiana, 1972), vol. 14; Società italiana Ernesto Breda per costruzioni meccaniche, Milano, *Per la millesima locomotiva,* November 1908, pp. 8–11, 18–19.

26. Breda quoted in Bigazzi, " 'Fierezza di mestiere'," pp. 93–94. On the Sesto San Giovanni plant, see Donald Howard Bell, *Sesto San Giovanni: Workers, Culture, and Politics in an Italian Town, 1880–1922* (New Brunswick, NJ: Rutgers University Press, 1986), pp. 14–16.

27. Bigazzi, " 'Fierezza di mestiere'," p. 96. Bigazzi, citing Direzione generale della statistica, "Notizie," mentions a mechanical work force of 14,300, but this is for the *circondario* of Milan, not the commune. Direzione generale della statistica, "Notizie," pp. 403–4, lists 11,394 metal and mechanical workers in the commune of Milan.

28. *Fascio operaio*, January 2–3, 1886; Giuseppe Berta, "Dalla manifattura al sistema di fabbrica; razionalizzazione e conflitti di lavoro," in *Storia d'Italia, Annali I: Dal Feudalismo al capitalismo* (Turin: Einaudi, 1978), pp. 1095–1106.

29. *Per la millesima locomotiva*, p. 9, quoted in Bigazzi, " 'Fierezza di mestiere'," pp. 91–92.

30. Bigazzi, " 'Fierezza di mestiere'," pp. 88–89; Davite, "I Lavoratori meccanici," p. 380; Carlo Carotti, "L'Introduzione dell' organizzazione 'scientifica' di lavoro in Italia e la prima lotta contro il cottimo," *Classe* 7 (1973): 281–82.

31. *Fascio Operaio*, September 9, 1883.

32. *Fascio Operaio*, February 6–7, 27–28 and March 13–14, 27–28, 1886.

33. S.U., *Origini, vicende*, pp. 3–10; Davite, "I Lavoratori meccanici," pp. 393–94.

34. S.U., *Origini, vicende*, p. XVII. See also Davite, "I Lavoratori meccanici," p. 394; Giuseppe Paletta, "Strategia rivendicativa di fabbrica e rapporto di delega nelle organizzazioni operaie milanesi (1900–1906)," in Alceo Riosa, ed., *Il Socialismo riformista a Milano agli inizi del secolo* (Milan: Franco Angeli, 1981), p. 145; Bigazzi, " 'Fierezza di mestiere'," notes (p. 90) that among 26 workers arrested in the 1891 strike whose place of birth is known, 16 were from Milan, 4 from Milan Providence, 4 from other parts of Lombardy, and 2 from other regions. This is 60 percent born in Milan, a higher proportion than among workers in the sector as a whole.

35. Davite, "I Lavoratori meccanici," pp. 398–99.

36. Hunecke, *Classe operaia*, p. 197.

37. ASM Q 48, report of inspector to police chief, March 5, 1889. This report, which ends with a tirade against subversives and the troubles they cause, was found in the carton for 1891, along with documents concerning the beginning of the mechanics' strike that August.

38. ASM Q 46, three reports from delegate to police chief, February 18 and 19, 1889; undated factory regulations (identified as 1872 version by another document) and "Nuovo Regolamento per gli operai," February 2, 1889; Hunecke, *Classe operaia*, p. 197; *La Lombardia*, February 20, 1889 mentions that the fact that workers were now responsible for two machines (a speedup) was another grievance.

39. ASM Q 46. Reports of delegate to police chief, April 2 and 3, 1889; article from *Il Corriere della Sera*, April 5–6, 1889, about the apology; and Hunecke, *Classe operaia*, p. 197.

40. Carotti, "L'Introduzione dell' organizzazione 'scientifica'," pp. 285–86.

41. Ibid., pp. 286–87.

42. Ibid., pp. 287–90. Anna Kuliscioff (1854–1925), a Russian-born aristocrat, medical doctor, and anarchist turned socialist, was one of the founders of the Socialist party, in which she continued to serve as theorist and publicist throughout most of her life.

43. ASM Q 49.

44. *Il Sole*, September 7–8, 1891, quoted in Hunecke, *Classe operaia*, p. 202.

45. Suffert 1895 strike in ASM Q 51, and *La Battaglia*, October 5, 1895, which accused other plants of filling Suffert's orders, thus enabling the company to hold out against its workers. There were two foundry strikes and a large layoff and shutdown at Grondona; also in ASM Q 51. *La Battaglia*, November 10, 1894, and January 25, 1896, reports on agitation at Stigler when workers were fined for not returning toolboxes to the factory toolroom at the end of the workday. Tools were no longer to be reserved for use by an individual worker to whose work style the tools became adapted. Suffert remained at loggerheads with his workers: 1896 strike in ASM Q 52, and *La Battaglia*, November 28, 1896; 1897 strike in S.U., *Origini, vicende*, p. 4.

46. S.U., *Origini, vicende*, pp. 4–5; Stigler strike in ASM Q 53, including police report dated March 23 1898, detailing the industrialists' committee, which included some former arbitration board members. See Guido Baglioni, *L'Ideologia della borghesia industriale nell' Italia liberale* (Turin: Einaudi, 1974), pp. 468–69. The Consorzio was founded on the initiative of Ernesto Breda. Giulio Prinetti (1851–1908), another engineering firm owner who played an active role in national politics as a member of Parliament and a minister, was its first chair. The Consorzio's ostensible purpose was to guide the business side of the Probiviri (a local labor arbitration board established according to norms set in 1893 by national legislation) and the accident insurance law (passed in 1898), but its founders also intended it as a countervailing force to workers' organizations. Baglioni discusses also (pp. 462–63) Milanese businessmen's opposition in the twentieth century to the extension of the model of Turin's activist Lega industriale to Lombardy and other parts of Italy.

47. S.U., *Origini, vicende*, pp. 9–10.

48. Bigazzi, " 'Fierezza di Mestiere'," p. 106. For the physical location of work groups in the plant, see the floor plan of the Breda factory in Milan in Società italiana Ernesto Breda, *Per la millesima locomotiva*, p. 101.

49. ASM Q 48: Pirelli's printed announcement to its workers, dated July 30, 1891; police reports dated July 30 and 31 and August 2, 3, and 21. See also Casero, "La Camera del lavoro di Milano," pp. 59–60; S.U., *Origini, vicende*, pp. 114–17. Pietro Gori (1865–1911), son of prosperous parents, became an anarchist while a student at the University of Pisa. He joined Filippo Turati's law office in Milan, but they eventually parted ways over politics. Gori continued to write and speak for the anarchist cause all his life, in Italy and in exile.

50. S.U., *Origini, vicende*, pp. 115–17. See also Carlo Dell' Avale's sarcastic open letter to Pirelli in *La Lotta di Classe*, April 23–24, 1898, mocking his paternalism and accusing him of giving crumbs to his workers while his stockholders prospered.

51. S.U., *Origini, vicende*, p. 301.

52. ASM Q 48, printed flyer.

53. Letter quoted in Carotti, "L'Introduzione dell'organizzazione 'scientifica'," p. 304.

54. *La Battaglia,* May 5 and 12, 1894. This strike was preceded in March by an organizing drive among women tailors. The news reports made no connection between the drive and the strike, but there likely was one. ASM Q 52, workers' dispute with foreman on April 20, 1896; police reports dated July 3 and 6, 1896, printed manifesto dated July 6, 1896, clipping from *Il Secolo* dated July 6–7, 1896.

55. S.U., *Origini, vicende,* pp. 301–3; *Avanti!* June 27, 1902; S.U., *Origini, vicende,* p. 303.

56. Hunecke, *Classe operaia,* p. 433, does not cite his source for his statement that "as early as 1882, the women tobacco workers attempted, but failed, to counterbalance an 'internal' [state-sponsored] mutual benefit society by an independent one." See also S.U., *Origini, vicende,* pp. 360–62.

57. ASM Q 48, newspaper clippings and police reports: "Disordini nella fabbrica di tabacchi."

58. S.U., *Origini, vicende,* p. 360.

59. *Avanti!* July 19, 20, 23–25, 27, 1901; S.U.; *Origini, vicende,* pp. 360–61.

60. S.U., *Origini, vicende,* p. xii.

61. Anna Maria Mozzoni, "Alle figlie del popolo" (Milan: C. Lazzari, 1885), reprinted in Anna Maria Mozzoni, *La Liberazione della donna,* ed. Franca Pieroni Bortolotti (Milan: Gabriele Mazzotta, 1975), p. 160. Mozzoni (1837–1920) was born in Milan, the well-educated daughter of an aristocratic Lombard family. A lifelong feminist, she worked first with Republicans, then with Socialists, and lastly with Socialist and bourgeois women's groups that were formed at the beginning of this century.

62. Anna Kuliscioff, *Il Monopolio dell' uomo,* 2nd (Milan: Ufficio della *Critica sociale,* France Pedone, Editore, 1894), p. 25; pamphlet quoted in Claire La Vigna, "The Marxist Ambivalence Toward Women: Between Socialism and Feminism in the Italian Socialist Party," in Marilyn J. Boxer and Jean H. Quataert, eds., *Socialist Women: European Socialist Feminists in the Nineteenth and Early Twentieth Centuries* (New York: Elsevier, 1978), p. 157.

63. Linda Malnati (1855–1921), a native of Milan, was one of the schoolteachers who were suspended after the Fatti di maggio while her possible role in the rebellion was being investigated. She continued to be active in Chamber of Labor and union activities throughout her life. Quotation from Malnati lecture in Anita Guarino, "Aspetti e problemi della questione femminile a Milano, attraverso la stampa operaia e socialista, dal 1890 al 1910;; (Ph.D. diss., Universita degli studi di Urbino, 1967), p. 53; Linda Malnati, "Le Tre Case," in Linda Malnati, *Scritti vari* (Milan: La Editrice libreria, 1922), p. 140.

Chapter 8

1. Gaetano Salvemini, "I Partiti dal 1859 ai nostri giorni," in *I Partiti politici milanesi nel secolo XIX* (Milan: Editori dell' "educazione politica," 1899); quotations from pp. 177, 179, 188.

2. Ibid., p. 182. Guiseppe Mussi (1836–1904) served also as deputy from Abbiategrasso and, later, from Milan. He was the democratic coalition mayor of Milan from 1899 to 1903.

3. Quoted in Franco Nasi, "1860–1899: Da Beretta a Vigoni. Quarant'anni di amministrazione comunale," *Città di Milano* 5 (1968): 43–44.

4. Fausto Fonzi, *Crispi e lo "Stato di Milano"* (Milan: Giuffrè, 1964), pp. 26–27.

5. Charity was distributed through the *opere pie,* private philanthropic institutions.

6. *La Plebe,* February 6–8, 1880. Enrico Bignami (1844–1921), in turn a Mazzinian, rationalist, and evolutionary socialist, started several study groups in Milan and was one of the founders of the Partito operaio. He spent the end of his life in Switzerland, writing for pacifist causes.

7. *La Plebe,* March 27, 1880.

8. *La Plebe,* April 4, 1880, for the call to congress; see *La Plebe,* May 15, 1880, for the editorial after the police injunction against the congress. Gastone Manacorda, *Il Movimento operaio italiano attraverso i suoi congressi: Dalle origini alla formazione del partito socialista (1853–1892)* (Rome: Edizioni rinascita, 1953), pp. 127–28, demonstrates, by a close reading of *La Plebe* in 1880, that Bignami actually planted the suggestion in his articles that the congress be forbidden. He also cites Osvaldo Gnocchi-Viani, *Ricordi di un internazionalista* (Milan, 1903), p. 129, for testimony that indeed Bignami and colleagues did try to provoke police action against themselves because they were not prepared for the congress and had received few responses to their call.

9. *La Plebe,* October 11, 1880 (emphasis in original).

10. *La Plebe,* December 5, 1880.

11. *La Plebe,* December 12, 1880; Manacorda, *Il Movimento operaio italiano,* pp. 133–35. The meeting was held in Switzerland because many Italian anarchists lived there in exile. The anarchist organization was renamed the Socialist Federation of Northern Italy.

12. *La Plebe,* July 4, 1880.

13. *La Plebe,* September 12 and November 21, 1880.

14. *La Plebe,* December 5 and 9, 1880.

15. *La Plebe,* February 7, 1881.

16. *La Plebe,* April 10 and June 30, 1881.

17. Manacorda, *Il Movimento operaio italiano,* p. 137.

18. *La Plebe,* 14 n.s., no 2, 1881 (no further dating). Manacorda, in *Il Movimento operaio italiano,* writes that "it may be that the author of the *Plebe* article was the same Osvaldo Gnocchi-Viani" (p. 139).

19. These lists and the minutes of the congresses are in the Archivio di Stato di Milano (ASM) Q 31. See also Franco Della Peruta, "Carte della Confederazione operaia lombarda nell' Archivio di Stato di Milano," *Movimento operaio* 2 (1950): 11–12.

20. *Congresso operaio lombardo tenutosi in Milano nei giorni 25 e 26 settembre 1881* (Como: Tipografia Bellasi e Buzzoro, 1881), p. 11.

21. Ibid., pp. 12, 14–15.

22. Manacorda, *Il Movimento operaio italiano,* p. 144 (emphasis in original).

23. *Congresso operaio lombardo*, pp. 35–38.

24. Ibid., pp. 54–55. The motion noted the importance of women's cooperation with the workers' movement; hence it was necessary to free women from the
unjust laws that sanctioned their inferiority and made them "objects of vice." It
called for women's sections in workers' societies, presumably to give them an
opportunity to discuss their own interests and gain experience in debate.

25. Manacorda, *Il Movimento operaio italiano*, pp. 145–48.

26. Quoted from *La Favilla*, May 21, 1882, in Manacorda, *Il Movimento operaio italiano*, p. 159.

27. *La Plebe*, June 15, 1882.

28. *La Plebe*, July 25, 1882 (reporting on meeting of July 16).

29. *La Plebe*, August 6, 1882.

30. *La Plebe*, August 11, 1882.

31. *La Plebe*, August 11 and 20, and September 10, 1882.

32. *La Plebe*, suppl., October 27, 1882.

33. *La Plebe*, October 28, 1882.

34. *La Plebe*, November 5, 1882.

35. Manacorda, *Il Movimento operaio italiano*, pp. 163–64.

36. Confederazione operaia lombarda, "Processo verbale del secondo congresso tenutosi in Como il 1 ottobre 1882 nella sala della Società generale di mutuo
soccorso ed istruzione degli operai," ASM Q 31.

37. *La Plebe*, November 12 and December 24, 1882.

38. *La Plebe*, February (no day given) 1883. Among the Partito operaio candidates were Emilio Kerbs (1854–?) and Ambrogio Galli (1841–?). Kerbs, born in
Schtetin, Pomerania, was trained as a lithographer. He became a Socialist in the
1870s, moved to Milan in 1879, and briefly joined the Democratic society, only to
be expelled. He was the Italian correspondent of the German *Sozialdemokrat*, contributed to the *Fascio operaio*, and worked with the Partito operaio. Galli, born in
Milan, a printer who headed the anarchist workers' circle, was one of the founders
of the P.O.I. but maintained his anarchist leanings into the 1890s, when he gradually moved to join the reformist socialists.

39. *Fascio Operaio*, July 1, 1883 (emphasis in original). Manacorda, *Il Movimento operaio italiano*, pp. 172–73, n. 52, attributes this ringing manifesto to
Plebe editor and socialist Osvaldo Gnocchi-Viani—not a manual worker—as based
on the testimony of Felice Anzi and Gnocchi-Viani's own inclusion of the article
in full in his own history of the P.O.I.: *Il Movimento operaio italiano, 1882–1885*
(Milan: Tipografia industriale di Stefani e Pizzi, 1885). Manacorda notes the importance of socialist intellectuals of petty bourgeois background in the founding of
the P.O.I.

40. This discussion of the congress is based on the "Verbale del 30 congresso
della Confederazione operaia lombarda, tenuto in Varese nei giorni 16 e 17 settembre, 1883," ASM Q 31. A firsthand report of the congress can be found in
Fascio Operaio, September 23, 1883; Manacorda, *Il Movimento operaio italiano*,
pp. 172–77; and Maria Grazia Meriggi, *Il Partito operaio italiano: Attività rivendicativa, formazione e cultura dei militanti in Lombardia (1880–1890)* (Milan: Franco
Angeli, 1985), pp. 23–24.

41. Manacorda, *Il Movimento operaio italiano*, pp. 174–75.

42. Ibid., p. 176.

43. The three-point "counterproject," published in the *Fascio Operaio* on December 22–23, 1883, included (1) freedom of "coalition" and strike, (2) all worker associations to acquire civil status through their simple existence, and (3) conflicts between workers and employers to be handled by the existing system of *probiviri*— leaving recourse to mediation up to the contending parties. Report on workers' meeting is in *Fascio Operaio*, February 2–3, 1884.

44. The manuscript list of societies present that can be found in ASM Q 31 includes 89 associations; *La Rivista Operaia* concurs. The *Fascio Operaio* account mentions 80. The *Rivista Operaia* reports 12 newspaper representatives, instead of the 8 on the manuscript listing. The following section is based on the conference report published in *La Rivista Operaia*, found in ASM Q 31; *Fascio Operaio*, February 9–10, 1884; Manacorda, *Il Movimento Operaio italiano*, pp. 177–83; and Meriggi, *Il Partito operaio italiano*.

45. *L'Operaio*, February 16, 1884, unsigned article attributed to Grando by Manacorda, *Il Movimento operaio italiano*, p. 183. See also Grando's letter dated Venice, September 1884, in ASM Q 31.

46. *Fascio Operaio*, February 9–10, 1884.

47. Letterio Briguglio, *Il Partito operaio italiano e gli anarchici* (Rome: Edizioni di storia e literatura, 1969), p. 11.

48. Andre Cocucci Deretta, "I Cappellai monzesi dell' avvento della grande industria meccanica alla costituzione della federazione nazionale," *Classe* 5 (1972): 204, citing the report on the strike in ASM Q 44; and the Minister of Agriculture, Industry, and Commerce, *Statistica degli scioperi avvenuti nell' industria e nell' agricolture durante gli anni dal 1884 al 1891* (Rome: Tipografia nazionale G. Bertero, 1897), p. 72.

49. Briguglio, *Il Partito operaio*, p. 12.

50. Manacorda, *Il Movimento operaio italiano*, pp. 197–98.

51. The following account is based on ibid., pp. 192–97, and *Fascio Operaio*, January 10, 1885.

52. Manacorda, *Il Movimento operaio italiano*, p. 200.

53. Ibid., pp. 197–98.

54. Briguglio, *Il Partito operaio*, pp. 33–42, 61,

55. "Resoconto del congresso," in Diana Perli, *I Congressi del Partito operaio italiano* (Padua: Tipografia Antoniana, 1972), p. 66.

56. Ibid., p. 70 ("Statuto [constitution] del Partito operaio italiano").

57. Quoted in Giancarlo Galli, *Il Movimento operaio milanese alla fine dell' ottocento* (Milan: IGIS, 1971).

58. Felice Anzi, *Il Movimento operaio italiano, 1882–1891: Episodi e apunti: Cronistoria autobiografica di un giornalaio-giornalista* (Milan: Edizioni dell' A.N.S., problemi del lavoro, 1933), p. xx. Anzi (1869–1958), born in Verona into a lower-middle-class family, ended his studies in 1884 and moved to Milan, where he worked for a press agency. He attended Consolato operaio classes and joined the Lega de figli del lavoro. He wrote for years for socialist newspapers but abandoned politics early in the Fascist period, although he later published several histories of the Partito operaio.

59. Meriggi, *Il Partito operaio*, pp. 189–94.

60. Ibid., pp. 154–57; Deretta, "I Capellai monzesi," pp. 161. 182–83, 205–8.

61. Manacorda, *Il Movimento operaio italiano*, p. 205; Perli, *I Congressi*, p. 80, taking her language from a different newspaper account, states "will participate in public struggle as a class, apart from any bourgeois party." The newspaper accounts used by both Manacorda and Perli are the *Fascio Operaio*, December 12–13, 1885, and *La Favilla*, December 8, 1885.

62. Manacorda, *Il Movimento operaio italiano*, p. 207.

63. Perli, *I Congressi*, p. 81.

64. Ibid., p. 82.

65. Ibid., p. 83.

66. Manacorda, *Il Movimento operaio italiano*, p. 208.

67. ASM Q 106, report dated January 25, 1886. (The document mentions ninety other reports in 1885, but these were apparently not deposited in the Archivio di Stato.)

68. Announcement in *Fascio Operaio*, January 30–31, 1886, invited "sisters of labor" to join them.

69. *La Lombardia*, March 28, 1886.

70. Report on the private Festa del lavoro from *La Lombardia*, March 19, 1886.

71. *Fascio Operaio*, March 27–28, 1886.

72. *Fascio Operaio*, April 3–4, 1886.

73. *La Lombardia*, April 2, 1886.

74. ASM Q 45, folder on April 1, 1886, "Manifesto suvversivo."

75. The following account is from *La Lombardia*, April 2, 1886. See also Jonathan Morris, "The Context of Shopkeeper Mobilization in Milan, 1885–1905" (unpublished paper, Cambridge University, 1989), for an analysis that emphasizes the entry in this period of shopkeepers as a group into the political arena. Morris argues that competing interests of the internal and external *circondari*, not class divisions, were at issue. The Democratic Radicals—through articles in *Il Secolo*—attempted to build united worker and shopkeeper opposition to Negri's policy in the external *circondario*.

76. *La Lombardia*, April 3, 1886.

77. Nasi, "1860–1899," p. 71.

78. Poster dated April 3, 1886, found in ASM Q 45.

79. *Fascio Operaio*, April 10–11, 1886.

80. Arrest lists in ASM Q 45.

81. *La Lombardia*, April 5, 1886.

82. Signers included Alfredo Casati for the Partito operaio. One might ask how the several middle-class men, professional actors, and singers who were arrested came to be involved. Newspaper accounts remark that the police arrested bystanders who refused to move, as well as active demonstrators, which may account for these arrests.

83. *Fascio Operaio*, April 24–25, 1886.

84. Guido Cervo, "Le Origini della Federazione socialista milanese," in Alceo Riosa, ed., *Il Socialismo riformista a Milano agli inizi del secolo* (Milan: Franco Angeli, 1981), p. 25.

85. Quoted in Galli, *Il Movimento operaio milanese*, p. 42.

86. *Fascio Operaio*, special issue, May 19, 1886 (emphasis in original).

87. *Fascio Operaio*, May 29–30, 1886.

88. *Fascio Operaio*, June 5–6, 6–7, 12–13, and 19–20, 1886. The only secret contribution to the Partito operaio was from Prospero Moise Loria, who later endowed the Società Umanitaria.

Chapter 9

1. *La Tipografia*, December 10, 1889, quoted in Renza Casero, "La Camera del lavoro di Milano dalle origini alla repressione del maggio 1898," in Marina Bonaccini and Renza Casero, eds., *La Camera del lavoro di Milano dalle origini al 1904* (Milan: SugarCo, 1975), p. 22.

2. Filippo Turati, "Alleanze impossibili e candidature di classe," *Il Cerino* (Cremona), election special, May 12, 1886, quoted in Luigi Cortesi, "La Giovinezza di Filippo Turati," *Rivista storica del socialismo* 1 (1958): 26–27.

3. Quoted in Gastone Manacorda, *Il Movimento operaio italiano attraverso i suoi congressi: Dalle origini alla formazione del Partito socialista (1853–1892)* (Rome: Edizioni Rinascita, 1953), pp. 223, 224.

4. According to Anzi, Mozzoni was the author of the *dichiarazione d'onore*, which denied Cavallotti's charges against the Partito operaio. See Felice Anzi, *Il Partito operaio italiano, 1882–1891: Episodi e appunti cronistoria. Autobiografia di un giornalaio-giornalista* (Milan: Edizioni dell' A.N.S. problemi del lavoro, 1933), pp. 47–48. See also Felice Anzi, *Battaglie d'altri tempi: 1882–1892* (Milan: Libreria editrice Avanti! 1917), p. 125, in which Anzi writes simply that Mozzoni "promoted" the declaration. See also Manacorda, *Il Movimento operaio italiano*, p. 223; Guido Cervo, "Le Origini della Federazione socialista milanese," in Alceo Riosa, ed, *Il Socialismo riformista a Milano agli inizi del secolo* (Milan: Franco Angeli, 1981), p. 26.

5. Anzi, *Il Partito operaio*, pp. 50–54. Anzi claims that there was a secret meeting of regional leaders on a boat on Lake Como.

6. *Fascio Operaio*, October 16–17, 1886 (emphasis in original).

7. Manacorda, *Il Movimento operaio italiano*, p. 240; Volker Hunecke, *Classe operaia e rivoluzione industriale e Milano, 1859–1892* (Bologna: Il Mulino, 1982), p. 354.

8. *Fascio Operaio*, February 5–6, 1887.

9. *Fascio Operaio*, February 1–2, 1887; Cervo, "Le Origini della Federazione socialista milanese," p. 29. Note that before the election of 1886, at the Mantua congress of the Revolutionary Socialist party (April 25, 1886), the *operaisti* had rejected Costa's tentative proposal for unification or coalition. At that time Augusto Dante announced that although the Partito operaio and the Romagnol Socialist party had the same goals, continuing diverse methods and a division of labor (the P.O.I. to organize the masses, the socialists to propagate the ideals that could inspire them) were necessary. Manacorda, *Il Movimento operaio italiano*, pp. 210–20 (Dante's speech quoted on p. 215).

10. Cervo, "Le Origini della Federazione socialista milanese," p. 32.

11. Minutes of the congress from *Fascio Operaio*, October 1–2, 1887, printed

in full in Diana Perli, *I Congressi del Partito operaio italiano* (Padua: Tipografia Antoniana, 1972), pp. 87–88. See also Manacorda, *Il Movimento operaio italiano,* pp. 239–45.

12. Perli, *I Congressi del Partito operaio,* pp. 90–91.

13. Manacorda, *Il Movimento operaio italiano,* pp. 245–46. See also Cervo, "Le Origini della Federazione socialista milanese," p. 33.

14. "Quarto Congresso, Bologna 8, 9, 10 settembre, 1888: Resoconto," in Perli, *I Congressi del Partito operaio,* pp. 112–13, 114–15.

15. Perli, *I Congressi del Partito operaio,* pp. 26–27.

16. Manacorda, *Il Movimento operaio italiano,* p. 249; letter quoted by Letterio Briguglio, *Il Partito operaio Italiano e gli anarchici* (Rome: Edizioni di Storia e Litteratura, 1969), pp. 99–100.

17. Cervo, "Le Origini della Federazione socialista milanese," pp. 37–38, 41.

18. Ibid., pp. 42–44,

19. Ibid., pp. 47–48.

20. *L'Italia del Popolo,* June 19–20 and 22–23, 1890, cited by Cervo, "Le Origini della Federazione socialista milanese," p. 49.

21. Manacorda, *Il Movimento operaio italiano,* pp. 249–68.

22. Ibid., pp. 269–72, quotation on p. 270.

23. Ibid., pp. 275–76.

24. Ibid., 278.

25. The full text of the *Fascio Operaio* (November 9, 1890) account of the congress is published in Perli, *I Congressi del Partito operaio,* pp. 130–37.

26. *Fascio Operaio,* November 9, 1890; *L'Italia del Popolo,* November 2–3, 1890.

27. Manacorda, *Il Movimento operaio italiano,* p. 282; Perli, *I Congressi del Partito operaio,* p. 34; Cervo, "Le origini della Federazione socialista milanese," p. 59; Maria Grazia Meriggi, *Il Partito operaio italiano: Attività rivendicativa, formazione e cultura dei millitanti in Lombardia (1880–1890)* (Milan: Franco Angeli, 1985), p. 56.

28. Manacorda, *Il Movimento operaio italiano,* pp. 283–89. Moderate anarchists at the congress favored collaboration with the Romagnol revolutionary socialists (on their own terms, however, which amounted to full capitulation of the latter on electoral questions and refusal of the party's elected deputies to serve in Parliament). They were no more receptive to the newly formed group of Republican collectivists that had broken away from the Mazzinians. Extremist anarchists were committed to provocation and terrorism, totally opposed to any relationship with other groups or a formal organization of anarchists themselves. A third group, which prevailed, was led by Errico Malatesta and Saverio Merlino. Their program supported—with rhetorical flourish—the poor against the rich, the oppressed against their oppressors. They eschewed any criticism of capitalism; hence there was no mention of other socialists' principle of class struggle. The anarchist majority was collectivist, however, for Malatesta prevailed in his argument that individual revolutionary efforts were insufficient.

29. Ibid., p. 292.

30. Turati in *Critica sociale* (1891), p. 5, quoted in Manacorda, *Il Movimento operaio italiano,* p. 293.

31. Full analysis in Manacorda, *Il Movimento operaio italiano,* pp. 287–302.

32. *L'Italia del Popolo,* March 9–10 and 17–18, 1891; see also Perli, *I Congressi del Partito operaio,* p. 35.

33. *L'Italia del Popolo,* July 4–5, 1891.

34. Cervo, "Le Origini della Federazione socialista milanese," pp. 65–66.

35. Ibid., pp. 66–67.

36. Ibid., p. 71.

37. Manacorda, *Il Movimento operaio italiano,* pp. 303–4.

38. Camillo Prampolini (1857–1930) was born in Reggio Emilia to a bourgeois liberal patriot family, studied jurisprudence and economics, and entered workers' politics as he moved toward Socialism. He later served as a reformist Socialist member of Parliament.

39. Cervo, "Le Origini della Federazione socialista milanese," p. 73.

40. Ibid., pp. 73–74.

41. Ibid., pp. 76–77.

42. Ibid., p. 77; Manacorda, *Il Movimento operaio italiano,* p. 305. The following paragraphs are based also on the accounts in Cervo and Manacorda.

43. *L'Italia del Popolo,* May 1, 2, and 3, 1892.

44. Adalberto Nascimbene, *Il Movimento operaio in Italia: La Questione sociale in Milano dal 1890 al 1900* (Milano: Cisalpino–Goliardica, 1972), pp. 156–57; Cervo, "Le Origini della Federazione socialista milanese," pp. 86–87.

45. Cervo, "Le Origini della Federazione socialista milanese," pp. 87–89.

46. Motion in ibid., p. 94, quoting *La Lombardia* and Labriola letter, p. 95.

47. Manacorda, *Il Movimento operaio italiano,* p. 312.

48. Quoted in Cervo, "Le Origini della Federazione socialista milanese," p. 104.

49. Ibid.

50. Quoted in Letterio Briguglio, *Congressi socialisti e tradizione operaista (1892–1904),* 2nd ed. (Padua: Tipografia Antoniana, 1972), p. 28.

51. Quoted in Cervo, "Le Origini della Federazione socialista milanese," p. 106.

52. Manacorda, *Il Movimento operaio italiano,* p. 316.

53. Cervo, "Le Origini della Federazione socialista milanese," pp. 116–17; Manacorda, *Il Movimento operaio italiano,* pp. 317–18.

54. Cervo, "Le Origini della Federazione socialista milanese," p. 109.

55. Manacorda, *Il Movimento operaio italiano,* p. 322; Cervo, "Le Origini della Federazione socialista milanese," pp. 111–12.

56. Cervo, "Le Origini della Federazione socialista milanese," pp. 116–17.

57. Roberto Michels, *Il Proletariato e la borghesia nel movimento socialista italiano: Saggio di scienza sociografico–politica* (Turin: Fratelli Bocca, 1908), pp. 83 and 105, among others. Michels also shows that no more than half of proletarian votes in Italy went to the Socialists (p. 248).

58. Maurizio Punzo, "Il Socialismo milanese (1892–1899)," in *Anna Kuliscoff e l'età del riformismo: Atti del convegno di Milano della Fondazione Giacomo Brodolini* (Rome: Mondo operaio-Avanti!, 1978), p. 434.

59. *Lo Lotta di Classe,* September 24–25, 1892.

60. Manacorda, *Il Movimento operaio italiano,* pp. 326–27.

61. *La Lotta di Classe*, September 3–4, 1892.

62. *La Lotta di Classe* September 3–4 and November 12–13, 1892.

63. *La Lotta di Classe*, June 3–4, 1893.

64. *La Lotta di Classe*, June 24–25, 1893.

65. *La Lotta di Classe*, July 29–30, 1893.

66. *La Lotta di Classe*, August, 26–27, 1893.

67. *La Lotta di Classe*, September 2–3, 1893.

68. *La Lotta di Classe*, September 9–10, 1893.

69. Motion from official minutes of the congress reprinted in Luigi Cortesi, ed., *Il Socialismo italiano tra riforme e rivoluzione: Dibattiti congressuali del PSI, 1892–1921* (Bari: Laterza, 1969), pp. 38–39; Briguglio, *Congressi socialisti*, p. 46.

70. *La Lotta di Classe*, September 16–17, 1893; Briguglio, *Congressi socialisti*, p. 50.

71. Briguglio, *Congressi socialisti*, p. 51.

72. The words are those of the congress report in *La Lotta di Classe*, September 16–17, 1893. Despite their precision, Briguglio, in *Congressi socialisti*, p. 52, concludes: "One can speak here of necessary coordination of resistance leagues and regional and provincial federations, but no more." Much more important "is that in the economic sphere also the concept of class struggle prevailed."

73. *La Lotta di Classe*, September 16–17, 1893.

74. *La Lotta di Classe*, September 16–17, 1893. Roberto Michels, *Storia critica del movimento socialista italiano* (Florence: "La Voce," 1921), pp. 126, 127.

75. This account is based on *La Lombardia*, October 1, 1893, and *manifesti* found in ASM (Archivio dello Stato, Milano) Q 51.

76. Franco Catalano, "Vita politica e questioni sociale (1859–1900)," pt. 2 of *Storia di Milano*, vol. 15 (Milan: Treccani degli Alfieri, 1961), p. 284.

77. Direzione generale della pubblica sicurezza, Rome, December 6, 1883, ASM Q 51.

78. *La Lotta di Classe*, January 6–7, 1894; telegram from *delegato* to police chief, January 4, 1894, ASM Q 51.

79. Call to meeting in *La Lotta di Classe*, January 13–14, 1894; "Agitazione socialista pei fatti di Sicilia," unlabeled newspaper clipping in ASM Q 51; *La Lotta di Classe*, January 20–21, 1894.

80. *Critica sociale*, January 16, 1894; and Luigi Cortesi, "Il Partito socialista e il movimento dei Fasci," *Movimento operaio*, November–December 1954, cited in Lorenzo Strik Lievers, "Turati, la politica delle alleanze e una celebre lettera di Engels," *Nuova rivista storica* 57 (January–April 1973): 146–47.

81. Strik Lievers, "Turati, la politica delle alleanze," pp. 147–48.

82. *La Lotta di Classe*, April 14–15, 1894.

83. *La Battaglia*, May 19, 1894 (this newspaper, a Milanese Socialist weekly, had begun publication in early 1894); Punzo, "Il Socialismo milanese," p. 439; Strik Lievers, "Turati, la politica delle alleanze," pp. 151–52; Turati in *Critica sociale*, 1894, quoted in Catalano, "Vita politica," p. 291.

84. *La Battaglia* 1 (30) 2nd ed., n.d.

85. "Per la difesa della liberta," *Critica sociale* 1 (November 1894), quoted in Strik Lievers, "Turati, la politica delle alleanze," pp. 152–53.

Chapter 10

1. Giuliano Procacci, "La Classe operaia italiana agli inizi del secolo XX," in Giuliano Procacci, *La Lotta di classe in Italia agli inizi del secolo XX* (Rome: Editori riuniti, 1970), pp. 4–5, 12.

2. Leo Valiani, "Il Partito socialista italiano dal 1900 al 1918," in *Il Movimento operaio e socialista: Bilancio storiografico e problemi storici* (Milan: Gallo, 1965), p. 201, quoted in Alfredo Canavero, *Milano e la crisi di fine secolo (1896–1900)* (Milan: SugarCo, 1978), p. 138. Roberto Michels, *Storia critica del movimento socialista italiano: Dagli inizi fino al 1911* (Florence: "La Voce," 1921), p. 286, reports that only 15 percent of respondents to the survey were landless peasants and that 6 percent were peasants with tenancy contracts. See also Roberto Michels, *Il Proletariato e la borghesia nel movimento socialista italiano: Saggio di scienza sociografica–politica* (Turin: Fratelli Bocca, 1908), p. 248. See Aristide Zolberg, "How Many Exceptionalisms?" In Ira Katznelson and Aristide Zolberg, eds., *Working-Class Formation: Nineteenth-Century Patterns in Western Europe and the United States* (Princeton, NJ: Princeton University Press, 1986), pass., for comparisons of the proportion of workers voting for Socialist parties in Germany, England, France, and the United States.

3. Renza Casero, "La Camera del lavoro di Milano dalle origini alla repressione del maggio 1898," in Marina Bonaccini and Renza Casero, *La Camera del lavoro di Milano dalle origini al 1904* (Milan: SugarCo, 1975), pp. 117–18; *L'Italia del Popolo,* October 25–26, 1894, quoted in Adalberto Nascimbene, *Il Movimento operaio in Italia: La Questione sociale a Milano dal 1890 al 1900* (Milan: Cisalpino–Goliardica, 1972), p. 182.

4. *La Battaglia,* November 2, 1894; Fausto Fonzi, *Crispi e lo "Stato di Milano"* (Milan: Dott. A. Giuffrè, 1965), p. 230.

5. Fonzi, *Crispi e lo "Stato di Milano,"* pp. 232–35; Maurizio Punzo, "Il Socialismo milanese (1892–1899)," in *Anna Kuliscioff e l'età del riformismo: Atti del convegno di Milano della Fondazione Giacomo Brodolini* (Rome: Mondo operaio-Avanti, 1978), pp. 441–42.

6. Arturo Labriola, "Le Future elezioni e la tattica del Partito socialista," *Critica sociale,* January 1 1895; and Leonida Bissolati, "Le Future battaglie elettorali e il nostro partito," *La Lotta di Classe,* January 5–6, 1895, both quoted in Franco Catalano, "Vita politica e questioni sociali (1859–1900)," pt. 2 of *Storia di Milano,* vol. 15 (Milan: Treccani degli Alfieri, 1961), p. 291. See also Fonzi, *Crispi e "lo Stato di Milano,"* pp. 240–41. Bissolati (1857–1920), a native of Cremona, earned a law degree but soon entered reformist democratic politics, from which he eventually converted to Socialism. He continued to emphasize democratic rather than Marxist principles in his organizational and journalistic work.

7. This argument is based on Punzo, "Il Socialismo milanese," p. 443. See also Fonzi, *Crispi e "lo Stato di Milano,"* pp. 240–41, 247. Fonzi puts much more emphasis on the personality and beliefs (in democratic decentralization) of Filippo Turati in this matter than I would accept. After all, the *operaisti,* despite their principled and determined opposition to the Democratic Radicals (this hostility was not one way, of course), similarly preached decentralized democracy but rejected alliances. This *ad hoc* theorizing fits Gabriele Turi's description of the P.S.I.'s

directive group's "aversion of theoretical elaborations, together with its political disorientation" that he attributes to the "eclectic nature of the Socialist International literature available to them in this period." "Aspetti dell' ideologia del PSI (1890–1910)," *Studi storici* 21 (January–March 1980): 71. It could also be attributed to the circumstances and structural conditions that they experienced in the period, especially arbitrary repression. Contrast the German Social Democrats, who, with a different pattern of repression, developed and elaborated Marxist theory, but without agreement among themselves, just like the Italians.

8. Michels, *Storia critica*, p. 173; Letterio Briguglio, *Congressi socialisti e tradizione operaista (1892–1921)*, 2nd ed. (Padua: Tipografia Antoniana, 1972), pp. 51–56. Luigi Cortesi, *Il Socialismo italiano tra riforme e rivoluzione: Dibattiti congressuali del PSI, 1892–1921* (Bari: Laterza, 1969), pp. 44–45.

9. Quoted in Cortesi, *Il Socialismo italiano*, pp. 382–83.

10. Open letter in *L'Italia del Popolo*, April 10–11, 1895. Briguglio, *Congressi socialisti*, p. 62.

11. Briguglio *Congressi socialisti*, pp. 65–66. Gnocchi-Viani article in *Critica sociale*, January 16, 1895.

12. Cortesi, *Il Socialismo italiano*, pp. 50–53.

13. *La Battaglia*, February 9, 1895. This newspaper, which supported the majority motion on electoral tactics at Parma, made no mention of the democratic or republican candidates in the election. It did not repeat, however, its sarcastic comment from the previous (June 1894) election that their list was a *minestrone democratico–massonico–repubblicano*.

14. According to D. Farini, *Diario di fine secolo*, ed. E. Morelli (Rome 1961), vol. 1, p. 738, Crispi labeled the Milanese opposition in 1894 the "proclamation of the state of Milan," quoted in Fonzi, *Crispi e "lo Stato di Milano,"* p. vii; Fonzi's discussion of 1895 and the *Stato di Milano*, esp. pp. 352 and 359.

15. Fonzi, *Crispi e "lo Stato di Milano,"* p. 365.

16. Punzo, "Il Socialismo milanese," pp. 444–45; Fonzi, Crispi e "lo Stato di Milano," p. 383.

17. "Alla conquista del commune. La nuova orientazione dei partiti nelle elezioni amministrative di Milano," signed "La Critica sociale," *Critica sociale*, February 16, 1895; "Per l'organizzazione elettorale socialista," *La Lotta di Classe*, February 23–24, 1895, both cited by Punzo, "Il Socialismo milanese," p. 445.

18. *La Battaglia*, April 13 and June 22, 1895; Punzo, "Il Socialismo milanese," pp. 446–47, points out that *La Lotta di Classe* reported the Turati position as though it had prevailed, as most elections would be decided in runoffs. Giuseppe Mammarella, *Riformisti e rivoluzionari nel Partito socialista italiano, 1900–1912* (Florence: Marsilio, 1968), p. 30, writes that Socialist-sponsored protest candidacies outside Sicily of Fasci leaders in the 1895 election (in addition to Barbato, Garibaldi Bosco and Giuseppe De Felice also ran) were indicative of the Socialist party's "going national."

19. "Una riforma che i socialisti potrebbero introdurre nel commune," *La Battaglia*, November 9, 1895; "L'Amministrazione borghese dal 1885 al 1894," *La Battaglia*, November 12, 1895.

20. Punzo, "Il Socialismo milanese," pp. 452–53; The city council met on December 16, 17, and 18, 1895; debates reported in *Atti del Municipio di Milano:*

Annata 1895–1896; La Battaglia, November 23, 1895. Jonathan Morris discusses the relations of shopkeepers, industralists, and workers in "The Context of Shopkeeper Mobilisation in Milan, 1885–1886," (unpublished paper, 1989), and in "The Political Economy of Shopkeeping in Milan, 1885–1905" (Ph.D. diss., Cambridge University, 1989).

21. *Atti del Municipio di Milano: Annata 1895–96,* December 17, 1895, session, pp. 87–95; December 18, 1895, session, p. 122; December 19, 1895, session, pp. 160–65.

22. *La Battaglia* October 3 and 31, December 25, 1896; *Atti del Municipio, 1896–97,* sessions of December 21 and 23 1896; Punzo, "Il Socialismo milanese," pp. 453–54.

23. *Atti del Municipio, 1896–97,* session of January 5, 1897, pp. 145–49; *La Battaglia,* January 16, 1897.

24. *La Battaglia,* January 23 and 30, 1897; Canavero, *Milano e la crisi,* pp. 84–85.

25. *La Battaglia,* February 20, 1897; Canavero, *Milano e la crisi,* pp. 85–91.

26. *La Battaglia,* May 23, 1896.

27. *La Battaglia,* October 3 and 10 and November 21 and 28, 1896.

28. Catalano, "Vita politica," pp. 293–94.

29. Fonzi, *Crispi e "lo Stato di Milano,"* p. 514.

30. *La Battaglia,* February 29, 1896.

31. Fonzi, *Crispi e "lo Stato di Milano,"* pp. 517–21; Catalano, "Vita politica," p. 296; *Il Corriere della Sera,* March 4–5, 1896; *La Battaglia,* March 7, 1896.

32. Filippo Turati, "La Circolare Rudini e il nostro partito," *Critica sociale,* April 1, 1896; letter to Cavallotti dated April 4, 1896, in Liliana dalle Nogare and Stefano Merli, eds., *L'Italia radicale: Carteggi di Felice Cavallotti, 1867–1898* (Milan: 1959), p. 360; both cited in Punzo, "Il Socialismo milanese," p. 448.

33. *Il Corriere della Sera,* March 5–6 and 6–7, 1896.

34. Antonio di Rudini (1839–1908) supported Garibaldi in 1859–60, became mayor of Palermo at age 25, prefect of Naples at age 29, and he had already participated in several cabinets.

35. Filippo Turati, "Il Domani," *Critica sociale,* March 16, 1896, and "Piccole polemiche," *Critica sociale,* April 16, 1896; Ivanoe Bonomi, "Il Domani e la democrazia," *Critica sociale,* April 1, 1896, all quoted in Fonzi, *Crispi e "lo Stato di Milano,"* pp. 522–23.

36. "La Parola e a Nicola Barbato," *La Battaglia,* March 21, 1896.

37. *La Battaglia,* April 25, 1896.

38. *La Battaglia,* May 1, 1896. Emilio Caldara, a young Socialist lawyer (born in 1868 to a family of modest means in Soresino), supported Turati's position as well. See Maurizio Punzo, *La Giunta Caldara: L'Amministrazione comunale di Milano negli anni 1914–1920* (Milan: Cassa di Risparmio delle Provincie Lombarde, and Bari: Laterza 1986), p. viii.

39. *La Battaglia,* May 1 and 16, 1896.

40. See Morris, "The Political Economy of Shopkeeping in Milan," who concludes that the interests of shopkeepers and of reformist socialists were, in the end, incompatible.

41. "La Prossima battaglia elettorale nel V Collegio di Milano," in *La Battaglia*, May 16 and 23, 1896; Punzo, "Il Socialismo milanese," p. 450; Canavero, *Milano e la crisi*, pp. 48–51.

42. *La Battaglia*, May 30, 1896. In the next number, published on June 6, 1896, *La Battaglia* laid out the 6 concrete political reforms and 15 economic, health, and educational reforms that comprised the national-level minimum program.

43. *La Battaglia*, June 20, 1896.

44. Journalistic *tour d'horizon* in *La Battaglia*.

45. *La Battaglia*, June 27, 1896; Catalano, "Vita politica," p. 300, quotes Turati's idealistic statement: "Some accuse me of being an aristocrat, and I *wish* to be such. I think socialism is essentially aristocratic, because it aspires to the good, the true, the just, the ideal" (emphasis in original).

46. Cortesi, *Il Socialismo italiano*, p. 55; Briguglio, *Congressi socialisti*, p. 70.

47. *La Battaglia*, September 5, 1896.

48. Filippo Turati, "Democrazia in vacanza," *Critica sociale*, September 1, 1896, quoted in Canavero, *Milano e la crisi*, p. 59.

49. "In commemorazione del XX settembre, 1870," *La Battaglia*, September 22, 1896.

50. *La Battaglia*, December 12, 1896.

51. Don Davide Albertario, "Le Persecuzioni contro i socialisti," *Osservatore Cattolico*, January 11–12, 1897; Un Deputato [Sidney Sonnino], "Torniamo allo Statuto!" *Nuova Antologia*, January 1 1897, p. 11, both quoted in Canavero, *Milano e la crisi*, pp. 71–72; cf. Umberto Levra, *Il Colpo di Stato della borghesia: La Crisi politica di fino secolo in Italia, 1896–1900* (Milan: Feltrinelli, 1975), p. 21.

52. *La Battaglia*, February 6, 1897. The school lunch debate, the Anzi letter, and Malnati's speech can be seen as part of a debate about workers' culture and access to education in which the *operaisti* and bourgeois intellectuals participated, as Carl Levy points out in his concluding essay in *Socialism and the Intelligentsia, 1880–1914* (London: Routledge & Kegan Paul, 1987), p. 280.

53. Quoted in Canavero, *Milano e la crisi*, p. 59.

54. Filippo Turati, "Aspettando la battaglia," *Critica sociale*, February 16, 1897, and "La Prossima lotta: Per una piattaforma elettorale," *La Lotta di Classe*, February 6–7, 1897, quoted in Canavero, *Milano e la crisi*, pp. 104–5; *La Battaglia*, February 20 and March 7, 1897.

55. *La Battaglia*, March 13 and 14, 1897.

56. Levra, *Il Colpo di Stato*, p. 17.

57. August 1, 1898, quoted in Canavero, *Milano e la crisi*, p. 136; resolution of the sections of the Chamber of Labor, meeting on August 7, 1897, quoted in Casero, *La Camera del lavoro*, pp. 142–43.

58. Filippo, Turati, "La Via maestra di socialismo," quoted in Briguglio, *Congressi socialisti*, p. 72. Briguglio also quotes Gnocchi-Viani's letter to *La Lotta di Classe*, July 24–25, 1897, which concluded that the Socialists had come to see politics as "an all-absorbing force, shaping all economic arrangements according to a theoretical and ideal plan, preordained by doctrine." He called instead for a party with a dual base in politics and economics.

59. Cortesi, *Il Socialismo italiano*, pp. 86–87; Briguglio, *Congressi socialisti*, p. 75, notes with irony Kuliscioff's advocacy of protective labor legislation, opposed for years by the Partito operaio and other workers' organizations.

60. Cortesi, *Il Socialismo italiano*, pp. 87–88.

61. Ibid., pp. 89–92, 82.

62. Decision of the Federazione socialista milanese on August 24, 1897, cited in Raffaele Colapietra, *Il Novant'otto: La Crisi politica di fine secolo (1896–1900)* (Milan: Edizioni Avanti!, 1959), p. 211. See Guglielmo Ferrero, "Tormenti e tormentati," *Il Secolo*, October 8–9, 1897, and "I Sassi di Roma," *Il Secolo*, October 15–16, 1897, quoted in Canavero, *Milano e la crisi*, pp. 138 and 139 for a discussion of middle-class protest against higher taxes.

63. Luigi Pelloux (1839–1924), born in Savoy, became a career army officer. He was elected deputy from Livorno in 1881 and served as minister of war in several cabinets, including Rudini's.

64. *La Lombardia*, January 10 and 19, 1898; ASM (Archivio dello Stato, Milano) Q 53, dossier on "Rincaro del Pane."

65. ASM Q 53, dossier on "Rincaro del Pane"; *Il Secolo*, January 27–28, 1898, quoted in Casero, "La Camera del lavoro," p. 147.

66. Quoted in Canavero, *Milano e la crisi*, 155–56.

67. ASM Q 53, dossier on "Fatti di maggio"; Canavero, *Milano e la crisi*, pp. 163–65.

68. This description of the Fatti is based primarily on Canavero, *Milano e la crisi*, pp. 166–76, and newspaper accounts.

69. Both some contemporaries and historians have argued that it was a lockout. Lucio Villari, "I Fatti di Milano del 1898: La Testimonianza di Eugenio Torelli Viollier," *Studi storici* 8 (1967): 542; Levra, *Il Colpo di Stato*, pp. 107–8 believes it was a deliberate act of the Milanese *"consorteria moderata . . . to push workers into the streets to demonstrate and provide a pretext for a draconian repression."*

70. Bava Beccaris, "Relazione dell'autorità militare," in ACS (Archivio centrale di stato), Ministero del' interno, Direzione generale di pubblica sicurezza, ufficiale riservata, 1879–1912, b. 4, f. 10, s.f. 1, cited in Canavero, *Milano e la crisi*, p. 171.

71. Turati letter dated March 18, 1899, from prison to Anna Kuliscioff, in Filippo Turati and Anna Kuliscioff, *Carteggio: I. Maggio, 1898–giugno, 1899* (Turin: Einaudi, 1949), pp. 347–48, quoted in Levra, *Il Colpo di Stato*, p. 110 (emphasis in original).

72. Casero, "La Camera del lavoro," pp. 153–55; Canavero, *Milano e la crisi*, pp. 179, 185.

73. *Atti del Municipio di Milano: Annata 1897–1898* (Milan: 1899), p. 814.

74. Quoted in Levra, *Il Colpo di Stato*, p. 104.

75. Torelli letter in Villari, "I Fatti di Milano del 1898," p. 546. Gaetano Salvemini, "I Partiti politici milanesi prima e dopo il 1890," originally printed in *Critica sociale*, 1899; excerpt in Luciano Cafagna, *Il Nord nella storia d'Italia* (Bari: Laterza, 1962), p. 324, objected that Torelli's resignation was a trick to preserve his liberal reputation while permitting *Il Corriere della Sera* to express reactionary views.

76. Nicola Badaloni in *Atti del Parlamento italiano: Camera dei deputati*, session 1897–98, vol. 6 (Rome, 1898), pp. 6298, 6303.

77. Napoleone Colajanni, *L'Italia nel 1898 (tumulti e reazione)* (Milan: Società editrice lombarda, 1898), p. 225.

78. A list of the names of the 80 identified dead and their places of birth appears in *I Disordini di Milano e le sentenze del tribunale del Milano: Dati riassuntivi* (Milan: Tipografia ambrosiana, 1898). Military and police casualties were 2 dead and about 50 wounded. Don Davide Albertario, the priest who edited the firmly anti-state, social Catholic *Osservatore Cattolico* was arrested on May 24, and repressive measures were taken against Catholic associations in Lombardy.

79. *La Lombardia*, September 2, 1898.

80. Tables displaying complete participation rates by occupation and sex; age, sex, and occupational distribution; and nativity by occupation can be found in Louise A. Tilly, "I Fatti di Maggio: The Working Class of Milan and the Rebellion of 1898," pp. 124–158 in Robert Bezucha, ed., *Modern European Social History* (Lexington, MA.: Heath, 1972).

81. Colajanni, *L'Italia nel 1898*, p. 298.

82. *I Disordini di Milano.*

83. See Giovanni Montemartini, *L'Industria delle calzature in Milano* (Milan: Editore l'Ufficio del lavoro [Societè Umanitaria], 1904), pass.; and Società Umanitaria, *Origini, vicende e conquiste delle organizzazioni operaie aderenti alla Camera del lavoro in Milano* (Milan, 1909).

84. In this regard, Merli's claim that in 1898, Socialist and labor leaders and the worker base separated, and the workers; spontaneous uprising was doomed to failure because of their fearful leaders' abdication goes far beyond the evidence. Stefano Merli, *Proletariato di fabbrica e capitalismo industriale* (Florence: La Nuova Italia, 1972), vol. 1.

Chapter 11

1. A. Manzoni, *I Promessi sposi: Storia milanese del secolo XVII* (Lugano Mendrisio: Tip. F.lli fu F. Traversa, 1898), pp. 302–3 (an account of the Fatti di maggio published under a false title), cited in Rafaele Colapietra, *Il Novantotto: La Crisi politica di fine secolo (1896–1900)* (Milan: Edizioni Avanti!, 1959, pp. 218–20. Of the 56 "convicted Socialists" on Colapietra's list, 27 did not appear in my list based on newspaper reports in *Il Corriere della Sera* and *La Lombardia* of all those tried. (Two others were listed with different first names, but because their sentences were the same as those on my list, I counted them as the same person. One other had a different first name and sentence so was not assumed to be the same person.)

2. *Atti del Municipio di Milano, annata 1897–98*, session of June 3, 1898, p. 815.

3. Alfredo Canavero, *Milano e la crisi di fine secolo (1896–1900)* (Milan: SugarCo, 1976), pp. 284–85.

4. Umberto Levra, *Il Colpo di Stato della borghesia: La Crisi politica di fine secolo in Italia, 1896–1900* (Milan: Feltrinelli, 1975), pp. 350–51; Canavero, *Mil-*

ano e la crisi, pp. 291–92. Enrico Ferri (1856–1929) was born in Mantua province into a family of means and became a lawyer and a positivist criminologist. A highly effective speaker, in the mid-1880s he supported peasants tried for striking and was elected to Parliament first as a Radical and later as a Socialist. He worked briefly, between 1902 and 1904, with Arturo Labriola in the Milanese revolutionary syndicalist group that published *Avanguardia Socialista,* edited *Avanti!* (also briefly), and, with the passage of time, moved closer first to the reformist Socialists and later to the Fascists.

5. Canavero, *Milano e la crisi,* pp. 286–87; Levra, *Il Colpo di Stato,* pp. 350–52.

6. G. Ferrero, "I Partiti," *Il Secolo,* December 23–24, 1898, cited in Canavero, *Milano e la crisi,* p. 310.

7. Canavero, *Milano e la crisi,* pp. 324–29.

8. Ibid., p. 337.

9. *La Lotta,* June 17–18, 1899 (this edition of the newspaper was seized by the censor), quoted in Canavero, *Milano e la crisi,* pp. 363, 376.

10. *La Lotta,* May 27–28, 1899, quoted in Levra, *Il Colpo di Stato,* p. 350.

11. Levra, *Il Colpo di Stato,* pp. 355–56, quoting *** [Salvemini], "Perchè defendiamo lo Statuto," *Avanti!,* February 21, 1899.

12. *Critica sociale,* December 1, 1899, quoted in Canavero, *Milano e la crisi,* p. 383.

13. "La Dichiarazione dei socialisti," *Il Tempo,* November 11, 1899, quoted in Canavero, *Milano e la crisi,* p. 383. Caldara was elected to the city council in the December 1899 election. Maurizio Punzo, *Socialisti e radicali a Milano: Cinque anni di amministrazione democratica (1899–1904)* (Milan: Franco Angeli 1979), pp. 63–64.

14. Filippo Turati and Anna Kuliscioff, "Dichiarazioni necessarie: Rivoluzionari od opportunisti?" *Critica sociale,* 1 January 1900, quoted in Levra, *Il Colpo Di Stato,* pp. 354–55.

15. Levra, *Il Colpo di Stato,* p. 230. See also Luigi Cortesi, *Il Socialismo italiano tra riforme e rivoluzione: Dibattiti congressuali del PSI, 1892–1921* (Bari: Editori Laterza, 1969), pp. 94–95; Giuseppe Mammarella, *Riformisti e rivoluzionari nel Partito socialista italiano, 1900–1912* (Florence: Marsilio, 1968), pp. 67–81.

16. Cortesi, *Il Socialismo italiano,* pp. 96–97; Mammarella, *Riformisti e rivoluzionari,* pp. 83–89.

17. Alexander De Grand, *The Italian Left in the Twentieth Century: A History of the Socialist and Communist Parties* (Bloomington: Indiana University Press, 1989), pp. 23–26; see also James Edward Miller, *From Elite to Mass Politics: Italian Socialism in the Giolittian Era, 1900–1914* (Kent, OH: The Kent State University Press, 1990); Alceo Riosa, *Il Partito socialista italiano dal 1892 al 1918* (Rocca San Casciano: Cappelli, 1969) and *Il Sindacalismo rivoluzionario in Italia e la lotta politica nel Partito socialista dell' età giolittiana* (Bari: De Donato, 1976); Brunello Vigezzi, *Il PSI, le riforme e la Rivoluzione: Filippo Turati e Anna Kuliscioff dei Fatti del 1898 alla Prima guerra mondiale* (Florence: Sansoni, 1981).

18. Giuseppe Paletta, "Dinamiche occupazionali e sindacali nell' industria a Milano tra i censimenti del 1901 e del 1911," *Economia e Lavoro* 16 (1982): 91–

104, argues that Milan's unions failed to serve the interests of the changing mix of workers in that city in the first decade of this century.

19. Aristide R. Zolberg, "How Many Exceptionalisms?" in Ira Katznelson and Aristide R. Zolberg, eds., *Working-Class Formation: Nineteenth-Century Patterns in Western Europe and the United States* (Princeton, NJ: Princeton University Press, 1986).

Index

337